GEOGRAPHIES OF RESISTANCE

Radical geographers have always been interested in the social bases of injustice. They have produced finely tuned and insistent critical theories of the geographies of power. Less attention, however, has been paid to notions of resistance. *Geographies of Resistance* not only introduces new understandings of resistance, it also presents radical reinterpretations of the relationships between political identites, political spaces and radical politics.

Too often, political activism has been seen as the direct outcome of opposition to the things that the powerful do. Too often, the terrain of political struggle has been thought of as being constituted by practices of the dominant. Geographies of resistance challenge any sense that political struggles can be understood solely in terms of the practices and institutions of those in power. In this collection, the contributors show that political identities and political actions take place on grounds other than those defined only through the effects of the powerful.

By bringing a geographical imagination to bear on understandings of resistance, the contributors gathered here are able to open up other spaces of political practice, to explore how radical political identities are formed and acted out, and to show that resistance involves not only heroic struggles or grand gestures of opposition, but often everyday battles for survival and commonplace struggles for empowerment. Moreover, thinking through geographies of resistance shows precisely that resistance seeks to occupy new spaces, to create new geographies, to make its own place on the map.

Bringing together some of the most radical voices, both new and established, *Geographies of Resistance* questions received notions of resistance. Presenting a wide range of detailed case studies, the authors offer clearly focused and path breaking theoretical positions and a wide range of international empirical examples, which will surprise and transform common understandings of the geographies of resistance.

Contributors: Michael Brown, Michael Dear, Shlomo Hasson, Jane M. Jacobs, Michael Keith, Lawrence Knopp, Lisa Law, Donald S. Moore, Steve Pile, Gillian Rose, Paul Routledge, David Slater, Nigel Thrift, Michael Watts.

Steve Pile is Lecturer in the Faculty of Social Sciences, Open University; **Michael Keith** is Senior Lecturer in Sociology, Goldsmiths College, University of London.

GEOGRAPHIES OF RESISTANCE

Edited by
STEVE PILE and
MICHAEL KEITH

LONDON AND NEW YORK

First published 1997
by Routledge
11 New Fetter Lane, London EC4P 4EE

Simultaneously published in the USA and Canada
by Routledge
29 West 35th Street, New York, NY 10001

Typeset in Garamond by
Solidus (Bristol) Limited
Printed and bound in Great Britain by
T.J. International Ltd, Padstow, Cornwall

British Library Cataloguing in Publication Data
A catalogue record for this book is available from the British Library

Library of Congress Cataloging in Publication Data
A catalogue record for this book has been requested

ISBN 0-415-15496-0 (hbk)
ISBN 0-415-15497-9 (pbk)

CONTENTS

List of figures and tables vii
List of contributors ix
Preface xi

1 INTRODUCTION
Opposition, political identities and spaces of resistance 1
Steve Pile

2 BLACK GOLD, WHITE HEAT
State violence, local resistance and the national question in Nigeria 33
Michael Watts

3 A SPATIALITY OF RESISTANCES
Theory and practice in Nepal's Revolution of 1990 68
Paul Routledge

4 REMAPPING RESISTANCE
'Ground for struggle' and the politics of place 87
Donald S. Moore

5 DANCING ON THE BAR
Sex, money and the uneasy politics of third space 107
Lisa Law

6 THE STILL POINT
Resistance, expressive embodiment and dance 124
Nigel Thrift

7 RADICAL POLITICS OUT OF PLACE?
The curious case of ACT UP Vancouver 152
Michael Brown

8 RINGS, CIRCLES AND PERVERTED JUSTICE
Gay judges and moral panic in contemporary Scotland 168
Lawrence Knopp

9 PERFORMING INOPERATIVE COMMUNITY
The space and the resistance of some community arts projects 184
Gillian Rose

10 RESISTING RECONCILIATION
The secret geographies of (post)colonial Australia 203
Jane M. Jacobs

11 IDENTITY, AUTHENTICITY AND MEMORY IN PLACE-TIME 219
Michael Dear

12 LOCAL CULTURES AND URBAN PROTESTS 236
Shlomo Hasson

13 SPATIAL POLITICS/SOCIAL MOVEMENTS
Questions of (b)orders and resistance in global times 258
David Slater

14 CONCLUSION
A changing space and a time for change 277
Michael Keith

Bibliography 287
Index 309

FIGURES AND TABLES

FIGURES

2.1 The geography of oil, Ogoni and Maitatsine 35
2.2 Nigerian annual oil production and prices (Bonny Light), 1966–1996 39
2.3 Location of Ogoniland and Shell oilfields 51
2.4 Ethnic geography of Eastern Nigeria (Rivers, Delta, Imo, Abia, Cross-River and Akwa-Ibom States) 51
7.1 ACT UP Vancouver protests 157
8.1 Gay threat to justice 169
8.2 I *did* have gay fling 170

TABLES

2.1 Oil companies holding Nigerian concessions, 1969 and 1993 39
2.2 Reasons for the neglect of the oil producing areas in Rivers State 53
2.3 Spillage of crude oil in Nigeria, 1972–1979 54
8.1 Chronology of events 171

CONTRIBUTORS

Michael Brown is a lecturer in geography at the University of Canterbury in Christchurch, New Zealand. His research interests centre on urban political and cultural geographies.

Michael Dear is Director of the Southern California Studies Center and Professor of Geography and Urban Planning at the University of Southern California. He is founding editor of the journal *Society and Space*.

Shlomo Hasson is Professor of Geography at the Hebrew University of Jerusalem. He is the author of *Urban Social Movements in Jerusalem* (1993) and co-author of *Neighbourhood Organizations and the Welfare State* (1994). He is currently writing on issues of identity and territory.

Jane M. Jacobs is a lecturer in the Department of Geography and Environmental Studies at the University of Melbourne. She is the author of *Edge of Empire: Postcolonialism and the City* (1996). Her current research is focusing on 'postcolonial racism' in contemporary Australia.

Michael Keith works in the Centre for Urban and Community Research at Goldsmiths College, University of London. He is currently researching the relationship between narratives of the city and forms of racialised politics and identity that characterise contemporary inner-city London. He is author of *Lore and Disorder: Policing Multi-racist Britain* (1993) and the co-editor of *Hollow Promises: Rhetoric and Reality in the Inner City* (1991); *Racism, the City and the State* (1992) and *Place and the Politics of Identity* (1993).

Lawrence Knopp is Associate Professor and Head of the Department of Geography and Director of the Center for Community and Regional Research at the University of Minnesota–Duluth. His research interests are broad but have tended to coalesce around questions of social power as they relate to the spatiality of sexuality, gender and class. Currently he is working on a book examining gay male cultural identities and politics in the US, UK and Australia.

Lisa Law recently completed her doctorate in human geography at the Australian National University. Her doctorate focuses on how the increasingly global discourses of HIV/AIDS prevention construct the 'community' of prostitutes in the Philippines, as well as their relationships with foreign men.

Donald S. Moore is currently a postdoctoral fellow at the Institute of International Studies, University of California, Berkeley, USA. His ethnographic research has focused on the simultaneity of symbolic and material struggles over landscape and on the cultural politics of place and identity in Zimbabwe's Eastern Highlands.

Steve Pile is a lecturer in the Faculty of Social Sciences at the Open University. He has published work relating to identity, politics and geography. He has just co-edited *Mapping the Subject* (1995) with Nigel Thrift, and is currently recovering from *The Body and the City: Psychoanalysis, Space and Subjectivity* (1996).

Gillian Rose teaches feminist and cultural geographies at the University of Edinburgh. Her research focuses on the politics of the production of geographical knowledges. She is the author of *Feminism and Geography* (1993) and has published extensively on feminist cultural geographies.

Paul Routledge is a lecturer in geography at the University of Glasgow. His research interests include political/cultural geography and South Asia, with a particular focus on contemporary forms of popular resistance. He is author of *Terrains of Resistance* (1993).

David Slater is Professor of Social and Political Geography at Loughborough University. He is author of *Territory and State Power in Latin America* (1989) and editor of two special issues on 'Social movements and political change in Latin America' published in *Latin American Perspectives* (1994). His current research interests include geopolitics and North–South relations, and the spatialities of power in the context of globalisation.

Nigel Thrift is Professor of Geography at the University of Bristol. His interest in embodied practices and the politics of subjectivity goes back to the 1970s. His books include *Times, Spaces and Places* (1979), *Writing the Rural* (1994) and *Spatial Formations* (1996).

Michael Watts was raised in the southwest of England and is a graduate of University College, London. He is currently Director of the Institute of International Studies at the University of California, Berkeley, and is working on questions of political Islam and globalisation, and more frivolously a short history of post-war capitalism seen through the chicken (that's right).

PREFACE

In January 1992, we organised a conference session at the annual conference of the late Institute of British Geographers, entitled 'Communities of Resistance'. The session was engaging and lively, and we tried to capture the debates in an edited collection based on, but not limited to, papers given at the conference. In the course of producing this collection, we realised that the original title 'Communities of Resistance' did not describe the contents of the book and we chose instead, *Place and the Politics of Identity* (Keith and Pile, 1993a). Though we were delighted with the book, we were also aware that an underlying interest in problematising the relationships between 'resistance' and 'geography' had not been directly addressed. We decided to try again – and this book is the result. Though this collection has been some time in the making, issues of 'resistance' have only recently become foregrounded in geographers' discussions of power relationships, political identities and spaces, and radical politics.

This collection moves from a period when few geographers used or thought about issues of resistance – instead analysing and criticising structural relationships of power – to one where everyone seems to be talking about resistance and domination. The term resistance may become increasingly significant but only, we would argue, if two conditions can be met. First, resistance needs to be considered in its own terms, and not as simply the underside of domination or of all social relationships – the inevitable outcome of any social relation of power. Second, it must be recognised that a spatial understanding of resistance necessitates a radical reinterpretation and revaluation of the concept: by thinking resistance spatially, it becomes both about the different spaces of resistance and also about the ways in which resistance is mobilised through specific spaces and times. For us, the term resistance draws attention not only to the myriad spaces of political struggles, but also to the politics of everyday spaces, through which political identities constantly flow and fix. These struggles do not have to be glamorous or heroic, about fighting back and opposition, but may subsist in enduring, in refusing to be wiped off the map of history.

If there is a fellow traveller in our journey through geographies of resistance around the world, then maybe – just conceivably – it could be Mickey Mouse. Is this Mickey Mouse a revolutionary, or a plastic child's toy, or Walt Disney's money-spinner, or maybe the wizard's apprentice who tried to use the Master's powers but only got himself into more trouble than he could handle? Whatever, our Mickey Mouse graces the cover of this book, standing with his back to us, waving. Let us elaborate, but only briefly, firstly with a few images:

- Stanley Kubrick ends his stunning Vietnam movie, *Full Metal Jacket*, with the brutalised young Americans trudging wearily through a battle-scarred semi-urban landscape – singing, like good little Mouseketeers, the Mickey Mouse March (1955): 'Who is the leader of the club that's made for you and me? M-I-C-K-E-Y M-O-U-S-E'.
- To say something is 'Mickey Mouse' is to say it is not up to much, not very sophisticated, not modern, inferior, temporary, small, childish.
- Walter Benjamin, a cultural critic associated with the Frankfurt School of critical theory, apparently used to keep a Mickey Mouse on his desk. He reasoned that something that could take such a hold on the popular imagination must have something going for it.

In the surreal juxtaposition of battle-weary American boys singing to their childhood hero, Kubrick seeks to sicken his audience over the loss of innocence, with the waste of youth, of life, of blood, of bodies in pieces. Rest in pieces. The audience who is meant to identify with this scene is, of course, white America. Indeed, Kubrick uses all the familiar tropes of the Vietnam war to guarantee this identification – nothing is strange about this sur-real scene. In it, Mickey Mouse simultaneously stands for America, for youth and for boyhood. Mickey Mouse is a point of resistance – to war and its destructions – but simultaneously the re-assertion of an American account of culture and of the innocence of youth. Not all audiences will identify with this message: it is difficult to see Vietnamese veterans appreciating the scene, for example. Further, English audiences might well recognise that the scenes of conflict were filmed in London's docklands – itself the scene of another urban conflict, only this time between docklands' communities and financial capital (see Keith and Pile, 1993b). Perhaps the most surreal journey in the film is not between a Mickey Mouse youth and a vicious futile war, but between a conflict in the margins of the so-called Third World and one in the margins in the heart of the West. Mickey has travelled the world, twice over: as a vicious cultural imperialist, wiping out the stories of others, partly using the unsophisticated weaponry of the nostalgic fairy tale of shared youthful innocence (lost) – in the words of the Mickey Mouse March: 'all around the world we're marching'.

Dear Mickey Mouse, the maker of much money for poor old Walt Disney, is an ambiguous figure in other ways too. From the title of the picture, *The End of History*, and the series title, *Capital*, it can be surmised that the juxtaposition of Mickey and the building is meant to mean something ... but what? The building itself, judging from the road markings and the inscriptions is a commercial institution somewhere in the heart of the City of London, possibly an exchange like the Baltic Exchange – which incidentally the Irish Republican Army blew up some years ago and which, somewhat paradoxically, is the proposed site for Norman Foster's latest tall building (a development which has already attracted some resistance: the fantasised building has been described as an 'eroticised gherkin'!). The Mouse has travelled to London: now, Mickey could be waving 'hello' or 'good-bye' to this powerful institution. Could it be that Mickey Mouse killed history, and he has now arrived from across the Atlantic to claim the spoils of his victory – a victory over the representation of history ... and geography? If Mickey isn't the victor, then maybe he also got the boot from history. Maybe our little plastic friend didn't survive the journey across the Atlantic, and he's on his way out, maybe in search of other places to conquer ... or maybe to take a vacation and put his feet up to watch the latest Disney kiddie flick. Which is stronger, we might wonder, the concrete of the City of London or the plastic of Disney culture? Moreover, the geographies of these characters – the building, Mickey – do not just provide the places on which history takes place, they both interrupt the telling of history as one story about the powerful and also speak of the ways in which geography makes history happen. Far from the end of history, Mickey's wave begs questions about the beginnings of geography – who will make more difference to the world?

In some other way, the spatial juxtaposition of Mickey Mouse outside the closed doors of a London institution parodies both the exclusions of the City of London and the seeming innocence of an American culture aspiring to global dominance – there are, then, two histories of money, two spaces of power, and two geographies of resistance – each standing face to face, but neither really seeing the other. If the building represents its own commercial links to the past through architectural references to Roman and Greek civilisations and mercantile capitalism, then Mickey Mouse – however plastic – has his own (wayward) history: the naughty Mouse made good; the popular hero making out against the odds; or, the first of a devil's brood of cartoon characters that symbolise, dramatise and enact middle American values. If the building represents the security of money circulated through the timeless and exclusive exchanges within its doors, then Mickey is the shallow and plastic circulation of money in pursuit of trivial pleasures, which ultimately has the capacity to create worlds in its own image. Serious money resists frivolous waste, while gambling itself on the future of an unreal market; plastic money resists investment in the serious, while investing in a frivolous present ... for the kids? Mickey and the building are assiduous world travellers in their own ways, connected and interconnected across the surface of the world, in the dreams of people, in specific ways: hopes, fears, dangers are all embodied in the fabric of the stand off between the concrete and the plastic, the grand and the small.

If there is something serious here, then it may well be that resistance could be found in both spaces, and in neither. Behind the doors, maybe shareholders challenge the actions of a multinational corporation; maybe they sack the chairman (they're usually men). In the open, maybe people identify with Mickey Mouse's exclusion from the places where real money is made, and Mickey really does become the leader of our revolutionary gang. More likely, neither is true: money is made – in the building, by Mickey – and most people are excluded most of the time from this. In this respect, both Mickey and the building occupy the same space, though differently and possibly uncomfortably (who is your flexible friend?). *The End of History* symbolises an awkward reading of geographies of resistance, where the spatialities of resistance and of domination cannot be so easily decided, but where resistance might subsist in spaces of its own, in its engagements in struggle, in the struggle to change power relations, in the struggle to take up a place in history, in the production of *geographies of resistance*.

Steve Pile and Michael Keith,
London, September 1996

1

INTRODUCTION

opposition, political identities and spaces of resistance

Steve Pile

> A place on the map is also a place in history.
>
> (Rich, 1984: 212)[1]

> Costumes and witchcraft are precisely what people need to walk happily on the uncertain edge of blurred boundaries.
>
> (Castells, 1983: 162)

INTRODUCTION: GEOGRAPHY AND RESISTANCE

Perhaps the first images that spring to mind when we hear the word 'resistance' are of French men and women of the resistance fighting the Nazi occupation, 1939–1945, secretly carrying weapons on bicycles, or listening intently to coded messages on imperfect wirelesses, or lying in wait to blow up armoured trains carrying German troops and weaponry to the front. Then again, these heroics remind us of the barbarity with which the German forces of occupation attempted to suppress the resistance – by massacring villagers, by sending people to the concentration camps, by using vicious forms of torture on prisoners and by intimidating people into informing on friends, relatives, lovers. Or maybe other kinds of resistance are more prominent in our minds, perhaps we think of unemployed people marching to demonstrate their plight, or of the riots that take place in inner cities, or of people tying peace symbols to the barbed wire of military bases, or of a lone man standing in front of a tank as it rolls onwards to Tiananmen Square. Clearly, in these cases, 'resistance' stands in implacable opposition to 'power': so, 'power' is held by an élite, who use oppressive, injurious and contemptible means to secure their control;[2] meanwhile, 'resistance' is the people fighting back in defence of freedom, democracy and humanity. Since resistance opposes power, it hardly seems worth mentioning that acts of resistance take place through specific geographies: in the spaces under the noses of the oppressor, on the streets, outside military bases, and so on; or, further, around specific geographical entities such as the nation, or 'our land', or world peace, or the rainforests; or, over other kinds of geographies, such as riots in urban places or revolts by peasants in the countryside or jamming government web sites in cyberspace. There are, somewhat plainly, *geographies of resistance*.

In a very crude sense, then, resistance could be mapped – partly because it seems to have visible expressions (explosions, marches, riots, graffiti, and so on) and partly because it always *takes place*. Unfortunately, an exercise in mapping resistance would only capture particular forms of resistance, mobilised through specific forms of geography. Beyond this, the sense that resistance might happen *under* authority's nose or *outside* tightly controlled places implies that resistance might have its own distinct spatialities. This is to say that when geographies of resistance are examined then new questions arise not only about the ways in which resistance is to be understood and about the geographical expressions of identifiable acts of resistance, but also about the ways in which geography makes possible or impossible certain forms of resistance and about the ways in which resistance makes other spaces – other geographies – possible or impossible. This book is attendant, then, to the ways in which resistance uses extant geographies and makes new geographies and to the geographies that make resistance. This in itself unsettles discussions of resistance that see it as the inevitable outcome of domination, since power – whether conceived of as oppression or authority or capacity or even resistance – spread through geography can soon become uneven, fragmentary and inconsistent (see Allen, forthcoming). So, for example, Watts shows (this volume) that uneven development and regional rivalries in Nigeria provoked different and incommensurable responses amongst different groups of people. So, it is not clear where 'power' is, or even what it is, since it is working through many spatialised interrelationships – in this case, an oil-centred capitalism, a nation-centred polity arising out of oil and post-colonialism, but also through the Ogoni people's mobilisation, Muslim ideals and military orders. So, while there are different forms of control that work through distinct geographies, geographies of resistance do not necessarily (or even ever) mirror geographies of domination, as an upside-down or back-to-front or face-down map of the world.

There is – it is argued here – a more troubling effect of thinking through the geographies of resistance, that resistance is 'uncoupled' from domination. This is not to say that domination and resistance have nothing to do with one another, but that there are distinct spatialised modalities of control, and that resistance might have its own spatialities – and that this becomes much clearer once domination and resistance are dislocated and understood geographically. The introduction of 'geography' into discussions of 'resistance' forces reconsideration of the presumption that domination and resistance are locked in some perpetual death dance of control: a dance where domination and resistance hold each other's hands, each struggling to master the steps of the dance, each anticipating and mirroring the moves of the other, but neither able to let go – for dancers are nothing without the dance. Instead, *Geographies of Resistance* shows that people are positioned differently in unequal and multiple power relationships, that more and less powerful people are active in the constitution of unfolding relationships of authority, meaning and identity, that these activities are contingent, ambiguous and awkwardly situated, but that resistance seeks to occupy, deploy and create alternative spatialities from those defined though oppression and exploitation. From this perspective, assumptions about the domination/resistance couplet become questionable.

In a sense, this is a plea for recognising that the spatial technologies of domination – such as military occupation or, alternatively, urban planning – need to continually resolve specific spatial problems, such as distance and closeness, inclusion and exclusion, surveillance and position, movement and immobility, communication and knowledge, and so on. This is to say that authority produces space through, for example, cutting it up, differentiating between parcels of space, the use and abuse of borders and markers, the production of scales (from the body, through the region and the nation, to the globe), the control of movement within and across different kinds of boundaries and so on (see Slater, this volume, and Thrift, this volume). Nevertheless, these spatial practices of oppression do not mean that resistance is forever confined to the authorised spaces of domination. Indeed, one of authority's most insidious effects may well be to confine definitions of resistance to only those that appear to oppose it directly, in the open, where it can be made and seen to fail.

Geographies of Resistance begins with a sense that it is no longer enough to begin stories of resistance with stories of so-called power. From this perspective, resistance becomes a mode through which the symptoms of different power relations are diagnosed and ways are sought to get round them, or live through them, or to change them. For example, Law, this volume, shows how women workers in the sex industry diagnose power relations and seek strategies to manipulate, endure and benefit from them. So, thinking through geographies of resistance involves breaking assumptions as to what constitutes resistance. Now, this does not necessarily mean that resistance becomes 'anything' or 'everywhere', but precisely that resistance is understood where it takes place, and not through abstract theories which outline the insidious mechanisms, strategies and technologies of domination. Thus, it is no longer sufficient to assume that resistance arises from innate political subjectivities which are opposed to, or marginalised by, oppressive practices; whereby those who benefit from relations of domination act to reproduce them, while the oppressed have a natural interest in over-turning the situation. Instead, resistant political subjectivities are constituted through positions taken up not only in relation to authority – which may well leave people in awkward, ambivalent, down-right contradictory and dangerous places – but also through experiences which are not so quickly labelled 'power', such as desire and anger, capacity and ability, happiness and fear, dreaming and forgetting. Nor can resistance be so easily located as existing only in certain spatial practices or places (such as, for example, mobility or the permeability of boundaries or the local) and not others (such as, for example, stasis or the construction of clear boundaries or the global). Each of the authors in *Geographies of Resistance* takes issue with any presumption that resistance is self-evident, that geography is an inert, fixed, isotropic back-drop to the real stuff of politics and history, and that the relationship between geographies of domination and geographies of resistance is as simple as that between a hammer and a nail.

If this is the basic agenda of the book (and some authors may not agree that it is ... just to keep the issues in the air for a bit longer!), then it also useful to map out some key points of debate in discussions of radical politics. Geographers have, for a long time, engaged in production of radical knowledges – from Marxism, through feminism, to post-colonialism and queer theory (to pass only a few places on the map). They have been attendant to the geographies of domination and exploitation and the possibilities of political struggle. What follows is not a review of all radical geographers, nor of all radical geographies, but instead a developing argument about alternative ways of conceptualising geographies of resistance. The story begins, in the section entitled 'spaces of power and opposition', with the ways in which radical geographies have been produced through analyses of power, where what is radical is presumed to be the uncovering of the constitution of power relations and the production of space. Resistance, from this perspective, is about mass mobilisation in defence of common interests, where resistance is basically determined by the action: the strike, the march, the formation of community organisations, and so on. For geographers, political mobilisation is commonly seen as embedded in either geographically circumscribed communities (as in the politics of turf) or in spatialised communities (as in the geography of class). Nevertheless, notions of community and their geographies cannot be assumed, nor can their operativeness in resistance – see Rose's analysis, this volume.

The intentions and meanings of political acts are moreover not straightforward, especially since seemingly subversive actions can be constituted out of, and actively maintain, brutal forms of oppression, while sickeningly sadistic regimes can be grounded in emancipatory values. So, in the second subsection of this introduction, Fanon's discussion of the changing meanings of the veil during the anti-colonial struggle in Algeria is described in order to show how political identities are constituted through political struggles, and how both subjectivities and struggles are constituted spatially. Frantz Fanon's analysis has been both praised and condemned, but he nevertheless demonstrates that the spatial technologies of oppression and exploitation can be resisted on grounds which are not exclusively produced through the practices of domination. Further, this analysis shows how political identities are constituted through power relationships and that, as much as resistance seeks to undermine or throw off control over exterior spaces, the interior world is also colonised by hegemonic norms and values. From this perspective, it is possible to recognise that resistance can involve resistance to any kind of change, to progressive and radical politics and to social transformation.

This introduction traces the argument sketched above. First, by looking at a spatiality of resistance understood in terms of those defined by structural power relations; second, by thinking about the ways in which spaces of resistance are distinct from the spaces of domination; finally, by thinking about the distinct spatialities of resistance and by suggesting that resistance may involve spatialities that lie beyond 'power' – and it is on these alternative grounds that alternative places on the map are being produced.

SPACES OF POWER AND OPPOSITION

Many radical geographers have sought to contribute to (an understanding of) resistance through their politically informed and committed analyses of the power relations involved in the production of space – combined with often angry and anguished calls to recognise the brutality and cynicism of those in power.[3] Exemplary amongst recent writings is David Harvey's plea for radical politics to address the historical and geographical situatedness of exploitation (Harvey, 1993a). Harvey's work is an excellent place to begin a discussion of geographies of resistance not only because he asks that radical politics should judge the importance of different kinds of oppression but also because he argues that resistance can only be effective when organised in opposition to only one structure of power relations. Resistance, it seems, comes from being able to recognise the real enemy amongst a frightening array of enemies. In this war, resistance is about taking up one position of opposition – for Harvey, that is a class position in opposition to the hidden injuries of capitalist social relations.

Oppositional positions

On the day after Labor Day in 1991, the Imperial Foods chicken processing plant in Hamlet, North Carolina, caught fire. The exits from the factory were locked and, as a result, 25 of the 200 workforce were killed and a further 56 seriously injured (see Harvey, 1993a: 41). For Harvey, this horrific industrial accident revealed much about the harsh labour conditions experienced by workers in the American south. He argues that industrial interests used, and reproduced, inequalities between the countryside and the city by preying on rural poverty in order to provide inexpensive fast food for richer urban workers. Capitalist interests exploited the fact that people in rural areas have little choice of employment, which meant that the labour force was trapped into low wages, accepting poor conditions and was unable or unwilling to fight back. At every turn, the logic of capital investment involved the cold-blooded necessity of cutting costs – the ultimate price, in this case, was paid in workers' lives. From this perspective, this industrial disaster was no stroke of bad luck, but an accident waiting to happen.

What distresses Harvey is that a century of labour activism in North America – which had gained some ground in the defence of its own interests through health and safety legislation – had been openly disregarded by the

North Carolina state. Instead, the state not only had 'the habit of openly touting low wages, a friendly business climate, and "right-to-work" legislation which keeps the unions at bay', but had also failed to enforce health and safety legislation: so, the plant had not been inspected once in the 11 years that it had been running (pages 42–43). Astonishingly and tragically, according to Harvey, labour conditions in late twentieth-century North America do not differ substantially from Marx's description of work in late nineteenth-century Europe. Though Harvey does not quote him, Marx's words are evocative.

> As a capitalist, he is only capital personified. His soul is the soul of capital. But capital has one sole driving force, the drive to valorize itself, to create surplus-value . . . Capital is dead labour which, vampire-like, lives only by sucking living labour, and lives the more, the more labour it sucks. The time during which the worker works is the time during which the capitalist consumes the labour power he has bought from him.
>
> (Marx, 1867: 342)

The equation balances the profits of vampyric capital with the worker's impoverished, blood-sucked life, and Harvey sorrowfully points out that workers in the Imperial Foods plant were also vulnerable to such exploitation. Only, in this sickening accident-which-was-not-an-accident, capital consumed workers by fire. The reasonable question that Harvey asks concerns what has changed. He begins to look for an answer in the different responses to the Imperial Foods accident and a similar fire which occurred at the Triangle Shirtwaist Company, killing 146 workers, in New York in 1911. The fire in New York provoked one of the 'classic' acts of labour resistance: 100,000 people marched down one of New York's most famous streets, Broadway, to protest about working conditions and to demand better protection in the workplace. However, the fire in Hamlet provoked no such protest: no march, no demands, so no greater protection. In both situations, a soulless capitalism defines the conditions under which workers live out their lives, and attempts to take greater and greater control over those lives – unless the workers resist.

Capital strives to valorize itself through sucking the blood of the most vulnerable workforces, wherever these are, though they will be in specific places – whether in an impoverished countryside or destitute/d inner city.[4] Workforces remain vulnerable where they cannot organise and agitate in defence of their own interest and where alliance cannot be formed with groups sharing those common interests. The question, for Harvey, is what is preventing worker resistance to capitalist blood-sucking. Partly the reason lies in capital's avoidance of organised labour and tighter regulation in cities but also, significantly, the lack of a political response was due to the relative marginalisation of rural issues by political agendas dominated by urban events, such as the beating of Rodney King (see Gooding-Williams, 1993). While this political economy clearly (re)produces a geography of rural labour vulnerability, Harvey is most concerned that the failure to mobilise in defence of labour interests exemplifies the marginalisation of class-focused issues in contemporary radical politics. He argues passionately that – while 18 of the 25 who died were women, and 12 were African-Americans – the position that the dead shared, irrespective of other differences, was that they were workers.

> The commonality that cuts across race and gender lines in this instance is quite obviously that of class, and it is hard not to see the immediate implication that a simple, traditional form of class politics could have protected the interests of women and minorities as well as those of white males. And this in turn raises important questions of exactly what kind of politics, what definition of social justice and of ethical and moral responsibility, is adequate to the protection of such exploited populations, irrespective of their race and gender.
>
> (Harvey, 1993a: 44)

In this situation, Harvey urges that old-style, traditional forms of labour resistance would have saved lives, and even improved them. Resistance, then, would have sought out the vampire and nailed its heart with a stake: first, through exposing to sunlight that the most significant power relation was class, rather than race and/or gender; and, then, by using the means of labour struggle – lightning and/or long-term strikes, pickets outside factory gates, mass marches, trades unionism, the politicisation of workers and public alike: famously, 'Agitate! Educate! Organise!'. In this analysis, mobilisation around the blood-soaked experiences of racial and gendered oppression would not have proved effective in resisting workplace exploitation: indeed, these experiences might have to be marginalised in order to form an effective resistance to class injustices. For,

> when a relatively coherent class force encounters a fragmented opposition which cannot even conceive of its interests in class terms, the result is hardly in doubt.
>
> (Harvey, 1993a: 46)

The result: labour loses (lives). What troubles Harvey is that class (op)positions have become separated off from one another by privileging other lines of power, such as race, gender, sexuality. Moreover, because the working class can no longer conceive of itself as a working class, other political identities – formed around the rallying cries of the civil rights, women's, peace and environmental movements (aka the 'new social movements'[5]) – have had the unfortunate, if unintended, consequence of obscuring the injustices of class relations. Instead, in order to enable 'progressive' politics that is capable of redressing (and seeking redress for) injustice, Harvey proposes that four tasks should be undertaken (pages 63–64[6]): first, social justice must be defined from the perspective of the oppressed; second, a hierarchy of oppressions has to be defined, with the full recognition that this will necessitate the marginalisation of certain issues;[7] third, political actions need to be understood and undertaken in terms of their situatedness and position in dynamic power relations; and, finally, an epistemology capable of telling the difference between different differences has to be developed.

For Harvey, this 'epistemology' would be the Professor Abraham Van Helsing of vampire killers – wary, experienced, resourceful and successful. Conceivably, in this plot, Van Helsing is played by Karl Marx, while Bob Dole plays – with a hint of dementia – Dracula (a real charmer, but ultimately doomed) and the Republican Party play a supporting role: the vampire brood – from dusk till dawn. Meanwhile the angry, but bemused and victimised, villagers are played by a rainbow coalition of feminists, civil rights activists and environmentalists. In this story, the spaces of power are well defined, like Dracula's castle standing ominously above the village, cast in a doom-laden sky, the skyscrapers of capitalist exploitation dominate the landscape, like monuments to the terrible death toll inflicted on a quiescent and unwitting population. Having unmasked the enemy, and having understood its blood-sucking ways, then maybe the workers will be able to use its weak weapons – what could be weaker than garlic, or a wooden stake, or sunlight? – to strike death into the soulless heart of capitalism.

However, the villagers in this story do not seem to agree that Dracula is necessarily the only or the most dangerous of the terrors that might befall them. Harvey has picked his exemplary story of vampire blood lust well: without question, he is right – there is class exploitation and failure to organise around issues of class will be disastrous. On the other hand, these may not be the only grounds on which it is best to oppose the rape of women, white supremacist activities, gay bashing, the threat of nuclear war, the destruction of the rainforests, and so on. Not only are the woods that surround the village filled with other terrors, some villagers may be actively involved in terrorising others – even while others terrify them.

Harvey understands these complexities, and takes some time to show that political identities come about in relation to people's, often contradictory, positions within various class, race, gender relations. Nevertheless, Harvey believes that it is the shared experience of class oppression that will unite radical politics, and so both the fragmentation of class politics and also political mobilisations around other injustices are a problem. And, by the way, the failure to mobilise a coherent working-class movement is blamed on the rise of other (new) social movements. However, for other radical thinkers, the new social movements represent resistance on a different terrain, grounded in a different understanding of the structural relations of power. Instead of new social movements being a threat to the unity and organisation of class struggle, they are instead a responsive and vital articulation of people's experiences and demands of multiple, interacting fields of power.

The terrain of opposition

If political opposition is to be mobilised on grounds which may not privilege class, then questions arise about why other forms of oppression are more significant and what happens when people prioritise other, sometimes highly circumscribed, (op)positions. In his outstanding study of urban social movements (1983), Castells begins by asking a deceptively simple question: 'how do structurally defined actors produce and reproduce cities through their conflicts, domination, alliances, and compromises?' For Castells, the structural position of people within capitalist relations does not necessarily mean that they must be predisposed towards class mobilisation – partly because the terrain of political struggle is defined by the state, partly because people may privilege their cultural identities, and partly because of the ways in which urban space is (re)produced.

Castells draws on a wide range of examples, including the *comunidades* of Castilla (1520–1522), the Paris Commune (1871) and the urban revolts of US inner cities (in the 1960s), but what is of interest is the way in which he uncovers basic themes in urban protests – whether they be found in squatter communities, youth organisations, tenant struggles, urban uprisings, neighbourhood associations and so on. He argues that resistance takes place as a result of demands around three basic structural issues: collective consumption, such as housing, schools, welfare provision, and so on; the defence/expression of cultural identities; and, the workings of the state and/or local government. And it is these issues of local culture, marginalisation and the attempt to create new spaces which Hasson, this volume, pursues in his detailed study of urban protest in Israel. Urban protests, for Castells, cannot be understood purely in terms of structural imperatives or as an arbitrary expression of anger or as the mimicking of protests elsewhere. Nor can the people involved in protest be easily categorised, for example, as 'the poor' or 'locals' or 'malcontents'. Indeed, acts of resistance such as inner city riots often combine many kinds of protest and different protest groups with different reasons for participating. It is instructive to look at Castells' analysis of one city's politics – San Francisco in the late 1960s and early 1970s – because it illuminates not only how he sees resistance forming, but also how he interprets the terrain on which resistance takes place. This analysis of San Francisco deals with two case studies of urban social movements: Latino community organisation in the Mission District and the gay community in the Castro.

Despite some gentrification, the Mission District – an area of Latino in-migration – was an impoverished area in the late 1960s, with poor social services and housing. Yet, according to Castells, this neighbourhood nevertheless exhibited all the vibrancy of Latin street life (pages 107–109). Moreover, the district had a large number of agencies and neighbourhood groups doing day-to-day social work. As a result of the activities of these groups, as well as conflicts and arguments or alliances and compromises between them, a dense pattern of community interactions was already in place when, in 1966, the neighbourhood was designated for redevelopment as part of an urban renewal programme. Proposals to redevelop the area in 1966, rather than being greeted with either apathy or glee, were taken as a threat to the Latino cultural identity of the area. The interconnections between the various groups meant that the community *effectively* put pressure on San Francisco's board of supervisors and the programme was cancelled. Nevertheless, the community was divided on whether urban renewal programmes were such a bad idea, and in-fighting amongst neighbourhood organisations broke out. From late 1968 onwards, neighbourhood organisations held a series of conventions in order to articulate their demands, including the need for housing and employment as well as the right to veto any further proposals for urban renewal. Despite hostile confrontations amongst neighbourhood organisations, a fragile unity was for a while maintained, but the injection of funds into the neighbourhood further exacerbated internal conflicts between neighbourhood organisations as they scrambled to gain and control scarce resources. And so, by the mid-1970s, the neighbourhood organisations had become a battlefield, rather than the grounds on which resistance to urban renewal programmes could be mounted.

In the course of this analysis of the responses of the Latino community to their economic, political and cultural marginalisation, Castells makes the following observation.

> Urban movements, and indeed all social mobilizations, happen when, in their collective action and at the initiative of a conscious and organized operator, they address one or more structural issues that differentiate contradictory social interests ... These issues, or their combination, define the movement, the people they may mobilize, the interests likely to oppose the movement, and the attitude of institutions according to their political orientation.
>
> (1983: 123)

This observation requires some unpacking. Castells is arguing that the Latino protest movement needed to define itself as 'Latino' in order to provide its own foundation, and in order to maintain that Latino identity so as to have a clear sense of what the movement was all about.[8] Further, the Latino movement needed to address one or more structural issues, where structural issues are conflicts arising from the opposed interests of different social groups. These social groups are differentiated by their location in unequal power relations. The goals of any social movement would be, then, to redress inequalities around social services, housing, employment, cultural identity, political authority and so on. Since mobilisation is goal-oriented, effective politics can be formed around any axis of power – whether the poor versus the rich, or blacks versus whites, or homosexuals versus heterosexuals – though Castells is clear that resistance will be most effective when urban movements fight along only one axis of domination. For both Harvey and Castells, then, political identity is most effective when it is singular, though for Castells this singularity can vary.

There is another aspect of Castells' remark that is worth teasing out. Castells is not only interested in the structural determinations working on political actors, he also introduces both the collective actions of groups and the self-reflexive initiatives of individuals into the equation of domination and resistance. In his analysis of gay mobilisation, for example, he takes great care to talk about the bravery of gays who came 'out of the closet'[9] and of gay political leaders, such as Harvey Milk (see Castells: 1983, chapter 14), especially in the face of homophobic backlashes, during one of which Harvey Milk was murdered. Any gains that gays have made have been won at great cost – and studies in this volume by Brown and Knopp show how these struggles continue and how pervasive homophobias still oppress sexualities that exceed the limits of a tightly circumscribed heterosexuality.

For Castells, resistance around issues of sexual liberation could only come about where gays and lesbians faced up to 'differentiation' and addressed the 'social interests' and 'institutions' which marginalised and oppressed them. It was only when gays came out of the closet, that it was possible for them to mobilise collective actions and initiate a self-aware gay liberation movement. It was at this point, Castells claims, that gays could make their 'territorial' claims on city space and begin to create a district in San Francisco where they could live out their cultural (sexual) lives – much as the Latinos did in the nearby Mission District. So, Castells argues, 'gays have succeeded in building up a powerful, though complex, independent community at spatial, economic, cultural, and political levels' (page 158). Ultimately, gays have been more successful than Latinos, for Castells, not only because they were able to articulate clearer political identities and establish clearer demands, but also because many gay activists occupied privileged positions in relation to class and race (1983: 170): in other words, they were white, male and middle class. Nevertheless, Castells warns, racism and homophobia have not gone away, and the attitudes and orientations of political institutions require constant vigilance on the part of urban social movements.

Summary

Some attention has been given, by geographers, to Castells' analysis of gay participation in the making of the city (for example, Jackson, 1989), and criticisms have been levelled at his failure to recognise lesbian participation in the production of urban space (see Bell and Valentine, 1995). Nevertheless, this work does flag some issues for understanding geographies of resistance. Like Harvey, Castells recognises the need to distinguish between power relations and to attend to issues of injustice in the formation of communities of resistance. Like Harvey, Castells demonstrates that space is constitutive not only of relations of power, but also of the needs, demands and actions of urban protest movements (see also Hasson, this volume). Like Harvey, Castells sees multiple axes of power in the making of political identities and also recognises the problems that this can present for the formation of alliances between marginalised groups. Significantly, both Harvey and Castells agree that the production and reproduction of capitalist space underpins contemporary political and economic inequalities and injustices – across the board.

Where the analyses of Harvey and Castells begin to diverge, however, is on the issue of 'resistance'. For Harvey, resistance is to be grounded in opposition to labour exploitation: the working class must oppose the strategies and tactics of vampire capitalists. In Castells' analysis, there is an analytical distinction between 'civil society' (where social – and political – identities are produced through multiple determinations of race, gender, class, sexuality and so on) and 'the state'. This distinction leads Castells to conclude that communities of resistance have to act on terms – and in the spaces – defined by the state.[10] In San Francisco, for example, the state was able to incorporate and buy off sections within the different social movements in order to reduce, marginalise or defer their demands and potential conflicts. Moreover, resources were selectively used in ways which tended to destabilise the fragile unities of the protest movements. Further, the state was able to define which 'political actions' it would recognise and also level the political playing field in ways which refused to recognise significant differences between groups (a situation which heightens the importance of Harvey's demand that radical epistemologies tell the difference). Significantly, Harvey and Castells differ on the source of political identities: for Harvey, they arise out of the objective interests of structurally opposed groups; for Castells, they are formed through the mobilisation of social groups with opposed interests which will vary between contexts.

Despite disagreements, both Harvey and Castells are concerned with the political actions of structurally located groups and, while the nature of the terrain of struggle differs for Harvey and Castells, resistance can only take place on the terms defined by structures of power. Though Castells sees resistance as starting in civil society, nevertheless battles are fought on the battlefields determined by the state; Harvey sees struggles arising out of the internal contradictions of economic exploitation, operating through uneven and unequal development across space. It is, however, not clear that resistance takes place on grounds determined by the state, if only because the state is not a coherent and unified entity. Moreover, these analyses of oppositional politics tend only to consider political those actions which are public and/or overt and/or collective – thus rendering private and/or covert and/or personal forms of resistance, which are not considered 'Political' or part of 'The Movement', invisible (following Tonkiss, forthcoming). It is not clear, further, that struggles organised exclusively in terms of a singular political identity would necessarily lead to a coherent set of demands and a unified set of tactics to achieve those demands, if only because the geographies of these demands and tactics are far from obvious, from the global to the local, from the included to the excluded. So, for example, it is hardly a foregone conclusion that a one-dimensional, old-time class politics would have been enough to save the lives of the workers in Hamlet when their lives were also subject to racism and sexism.

These difficulties are compounded when the question of the spatialities of domination and resistance is considered. Castells assuredly states that the 'territories' of lesbian resistances were more subversive than gay men's:

> So where gay men try to liberate themselves from cultural and sexual oppression, they need a physical space from which to strike out. Lesbians on the other hand tend to create their own rich, inner world and a political relationship with higher, societal levels. Thus they are 'placeless' and much more radical in their struggle.
>
> (Castells, 1983: 140)

The question of whether certain communities of struggle create richer or more impoverished inner worlds will have to wait until later, but what is of concern now is the question of 'scale' and the production of 'scale' through power relations and struggle. There is an implicit assumption in Castells' argument that the creation and occupation of a bounded physical space is not as progressive as that of 'placelessness'. The point here appears to be that political actions that are delimited by local concerns are not as subversive as those which are global (since the global is the scale of the placeless). The lesson is that effective resistance would come about only as the result of thinking global and acting local. It is hard to appreciate the limits of this well-worn political cliché until it is realised that different power relations spatialise in distinct ways. Thus, the local and the global are not natural scales, but formed precisely out of the struggles that seemingly they only contain – as Slater, this volume, demonstrates (see also Massey, 1993a, 1993b, forthcoming).

In his discussion of the struggle between the homeless, city authorities and property interests for control over Tompkins Square park in New York's lower east side,[11] Smith persuasively shows that 'scales' were not just imposed on homeless people (such that they were not allowed to move around the city or to stay in certain places), but that the scales of the city – from the body, through the region and the nation, to the globe – were constituted by the spatialisations through which struggles took place (Smith, 1993).

> Tompkins Square expresses not just the spatialization of struggle in the abstract, but the social and political inscription of the geography of the city, through which urban space comes to represent and define the meaning of these struggles.
>
> (page 93)

Scales (such as the body, locality, region, nation and so forth), boundaries, inner worlds, positions of opposition, movement, physical space all speak of the production of space, but not simply as an echo of domination – for, there are other possibilities for resistance in the dislocations through, for example, frictions of distance, the blurring of boundaries, and hiding and coming out. It can be argued that different power relations may produce different spatialisations and, further, that resistance may well operate between the spaces authorised by authority, rather than simply scratching itself into the deadly spaces of oppression and exploitation. For example, Smith (1993) describes the Homeless Vehicle Project deployed in Tompkins Square as a political intervention that was subversive in its 'jumping' between scales, in its ability to transgress the imposition of boundaries and scales, and thereby to challenge not only those scales, but also their imposition.[12] From this perspective, it is necessary to question any privileging of the global over the local or of any global–local relation prior to an understanding of the specific spatialisations of domination and resistance involved in any conflict. I will return to this point about the privileging of certain spatialities of resistance in the final subsection.

Structural analyses of power tend to assume that there are no spaces outside of power relations from which it is possible to resist the injuries and injustices of oppression. However, the suggestion that power relations might produce discontinuous spaces, which resistance might transgress or move between, implies that there could be other places in the map of resistance. One possible way to remap resistance, then, is to think about the ways in which power relations are incomplete, fluid, liable to rupture, inconsistent, awkward and ambiguous. Now, spaces of resistance can be seen as not only partially connected to, but also partially dislocated from, spaces of domination. However, even if the interactions within and between relations of power produce simultaneously fixed and fluid spatialities, and that resistance occurs as a rupture in the fabric of these interactions, consideration will still have to be given to the ways in which people take up, and practise, political identities. In the next section, it is not so much the act – and its structural determinations – that defines resistance (as has been the case so far), but the meanings that social actions take on in the practice of everyday life. In some senses, then, the narrative about resistance shifts from one about vampire killing to one about the deceptive spaces of costume and trickery, through which people blur the edges of political identity.

RESISTANCE, SPACES OF DECEPTION AND THE BLUR OF POLITICAL IDENTITIES

Potentially, the list of acts of resistance is endless – everything from foot-dragging to walking, from sit-ins to outings, from chaining oneself up in treetops to dancing the night away, from parody to passing, from bombs to hoaxes, from graffiti tags on New York trains to stealing pens from employers, from not voting to releasing laboratory animals, from mugging yuppies to buying shares, from cheating to dropping out, from tattoos to body piercing, from pink hair to pink triangles, from loud music to loud T-shirts, from memories to dreams – and the reason for this seems to be that definitions of resistance have become bound up with the ways that people are understood to have capacities to change things (see Mani, 1990), through giving their own (resistant) meanings to things, through finding their own tactics for avoiding, taunting, attacking, undermining, enduring, hindering, mocking the everyday exercise of power. That people can create their own ways of living – their own meanings and capacities – has forced a recognition that resistance can be found in everything. Here, of course, lies a problem:

if resistance can be found in the tiniest act – a single look, a scratch on a desk – then how is resistance to be recognised as a distinctive practice? For many, the answer lies in thinking through the context in which acts of resistance take place – and, significantly, the position of people within networks of power. De Certeau puts it this way:

> Innumerable ways of playing and foiling the other's game, that is, the space instituted by others, characterize the subtle, stubborn, resistant activity of groups which, since they lack their own space, have to get along in a network of already established forces and representations … Like the skill of a driver in the streets of Rome or Naples, there is a skill that has its connoisseurs and its aesthetics exercised in any labyrinth of powers, a skill ceaselessly recreating opacities and ambiguities – spaces of darkness and trickery – in the universe of technocratic transparency, a skill that disappears into them and reappears again, taking no responsibility for the administration of a totality. Even the field of misfortune is refashioned by this combination of manipulation and enjoyment.
>
> (de Certeau, 1984a: 18)

According to de Certeau, the central strategy of authority is to force people to play its game, to make sure that the game is played by its rules, then people find innumerable ways round this … they continually seek to find their own places: they rat run through the labyrinths of powers. From this perspective, resistance is less about particular acts, than about the desire to find a place in a power-geography where space is denied, circumscribed and/or totally administered. The implication is that resistance comes from a place outside of the practices of domination.

In his discussion of the practice of everyday life, de Certeau makes a distinction between the strategies of power and the tactics of resistance (both page xix; also pages 34–39):

> A strategy assumes a place that can be circumscribed as proper and thus serves as the basis for generating relations with an exterior distinct from it.

While, a tactic

> cannot count on a 'proper' (a spatial or institutional) localisation, nor thus on a borderline distinguishing the other as a visible totality. The place of a tactic belongs to the other. A tactic insinuates into the other's place.

Unhelpfully, de Certeau argues that resistance has no place of its own, but this nevertheless serves to underline that the spaces of domination and the spaces of resistance are not flattened out, made interchangeable and reversible. While strategies define a territory marked by an inside and outside, resistances *cross* these spaces with 'other interests and desires that are neither determined nor captured by the systems in which they develop' (page xviii).

Like Neapolitan drivers, who use their prosaic skills to find their way around the streets, who follow routes that trace out their interests and desires, resistances map themselves opaquely and ambiguously through geographies of power.

It may, at first glance,[13] appear that de Certeau is suggesting that the powerful control space and that resistance can do no more than act out of place, but it can also be argued that tactics of resistance have at least two 'surfaces': one facing towards the map of power, the other facing in another direction, towards intangible, invisible, unconscious desires, pleasures, enjoyments, fears, angers and hopes – the very stuff of politics (and not just radical politics, it has to be well noted). Spaces of resistance can, therefore, be seen as dis-located from those of the powerful. From this perspective, the spaces of resistance are multiple, dynamic and weak (in their effectiveness, but also because resistance is often dangerous), and only ever in part controlled by the practices of domination. Meanwhile, spaces of domination are constitutively spaces of purification and exclusion (see also Sibley, 1995): the powerful are continually vigilant of the borders, which they institute for themselves, between themselves and those they oppress (see also Slater, this volume). If de Certeau is right, resistance cannot be understood as a face-to-face opposition between the powerful and the weak, nor as a fight that takes place only on grounds constituted by structural relations – because other spaces are always involved: spaces which are dimly lit, opaque, deliberately hidden, saturated with memories, that echo with lost words and the cracked sounds of pleasure and enjoyment.

The spatial practices of resistance are not just the mobilisation of a class across space, nor the mobilisation of an interest group in a particular place, but about insinuation . . . in the sinews of the body politic – a virus to kill the vampire? Thus, resistance does not just act on topographies imposed through the spatial technologies of domination, it moves across them under the noses of the enemy, seeking to create new meanings out of imposed meanings, to re-work and divert space to other ends. Thus, Castells' studies of Latino and gay mobilisation also show that other spaces can be insinuated into the production of urban space. This is not exactly an opposition, nor a separation of one space (of domination) from another (of resistance), but a *dis-located* interaction between the two, or three or more, spaces. Resistance, then, not only takes place in place, but also seeks to appropriate space, to make new spaces – and it is partly for this reason that geographers, such as Jackson (1989), have begun to be interested in the concept of resistance.[14]

However, situating and spatialising resistance has consequences for understanding not only the spatial constitution of resistance, but also the multiple, dynamic and ambiguous spaces of political identities – and it is for this reason that it is useful to look at one struggle, as described by Frantz Fanon (1959, but see also 1952), Algerian resistance against French colonialism.[15] While this description of the brutal practices of French colonialism bears a striking similarity to de Certeau's descriptions of oppression as the control and surveillance of space, the purification of space and of the élite's constant vigilance over those they define as below (external to) them, Fanon's understanding of resistance shows how it must engage with authority's capacity both to superimpose itself onto physical spaces and to make people who they are; thus, it is highly unlikely that people will feel willing to resist if they feel they are useless and powerless and have no room for manoeuvre nor the capacity to change anything. Resistance, then, cannot simply address itself to changing external physical space, but must also engage the colonised spaces of people's inner worlds (which Castells alluded to in terms of the subversive potential of lesbian identities, see above). Indeed, it could be argued that the production of 'inner spaces' marks out the real break point of political struggle . . . maybe.

Algerian anti-colonial resistance and the veil: spaces of deception

In his book, *Studies in a Dying Colonialism* (1959), Fanon describes various aspects of the Algerian struggle against French colonialism. Characteristically, Fanon describes a war in which each side attempts to outmanoeuvre the other, in a 'dialectic' where each side identifies the weaknesses of the other and attempts to protect itself from vulnerability. As in most anti-colonial struggles, the colonising power used brutal tactics to suppress the expression of cultural identity or opposition by the colonised. However, colonialism does not simply work through the machinery of fear – soldiers on the streets, torture in prisons, arbitrary violence against any and all – but also through spatial technologies: surveillance, border guarding, controlling movement (of people, goods, information), dividing and ruling, and so on – see Routledge's study, this volume, of colonial technologies of domination and the geographies of anti-colonial resistance in Nepal. Moreover, colonial power is partly mobilised through the imposition of a system of values, that the colonised must recognise, even while they might despise it.

The coloniser and the colonised look at each other suspiciously, each needs to know where the other is. Nevertheless, though the colonisers and the colonised get bound up in an economy of 'looks',[16] neither sees the other accurately: for, as Fanon puts it, 'every contact between the occupied and the occupier is a falsehood' (1959: 65). What does this mean? The colonisers have to work furiously to figure out who it is that they are oppressing. In order to maintain control over people, they have to work out what exactly has to be suppressed and what does not, while not really understanding what is going on. Colonial authorities seek to describe what the colonised are like and then to denigrate them on the basis of those descriptions. The forms of knowledge, through which the colonisers come to know the colonised and significantly through which they become the coloniser, are fantasies – they are built out of stories, anecdotes, lies, erotica, exotica (see Young, 1995; see also Jacobs, this volume). Moreover, these fantasies are translated from place to place, from practice to practice: as Law shows (this volume), fantasies of sex and race intertwine to create a situation in which sex tourism becomes imaginable, but where the tourist and the sex worker are positioned differently in multiple relations of power – including, it should not be forgotten, those of gender and class.

Meanwhile, the oppressed are in a different position (see also Pratt, 1992). They have to recognise that the colonisers are more powerful – and this puts them in the invidious position of having to (mis)recognise themselves and their culture as having less 'value', because it is the coloniser's values that have authority, give meaning and constitute identity – however fantastic. In Fanon's experience, the colonised are battered into place by fantasms, fantasms with material effects: 'I discovered my blackness, my ethnic characteristics; and I was battered down by tom-toms, cannibalism, intellectual deficiency, fetishism, racial defects, slave-ships, and above all else, above all: "Sho' good eatin'"' (Fanon, 1952: 112). This, of course, is the familiar story of imperialism: racialised, sexualised, gendered, classed, embodied, fantasised.[17] In the case of Algeria, the intersections of these lines of subject(ificat)ion meant that the veil (the *haïk*) took on special significance in a war of manoeuvre and position between the French colonisers and the Algerian nationalist movement.

The veil was not only a visible marker of the difference between Algerian and French costume and custom, it also became a visible marker of the extent of the physical and psychic colonisation of Algeria.[18] In the exchange of looks between the coloniser and the colonised, the *haïk* demarcated not only colonised society, but also the freedom of women. As in other situations, the female body – its visibility and invisibility; its appearance and disappearance – became an intense site of struggle. Women (and their bodies) were not passive surfaces of inscription in this struggle, however. Fanon describes a situation in which women's involvement in anti-colonial resistance not only became necessary for the success of the war, but also irrevocably changed the nature and meaning of the struggle. Far from women's faces being a blank canvas on which French and Algerian men wrote their designs – military, sexual – they became the expression of resistance: resistance not only to French colonialism, but also to Algerian patriarchy.

In French eyes, what the veil said about Algerian society was that it was misogynist, sadistic and vampyric – indeed, the epitome of a mediaeval patriarchy – and this was held to be the invidious source of Algerian resistance to the French civilising project. Indeed, by seeing themselves as 'freeing' women from patriarchy – and its most visible sign, the veil – the French positioned themselves as the liberators of Algeria. Like dominoes, the fall of the veil would enable liberated Algerian women to identify with French freedoms, which would undermine the patriarchal Algerian family, which would undermine Algerian men's authority, which would undermine their resistance to French colonialism. In order to free women, the French descended on the Arab quarters of the city with charities, mutual aid societies for women and women's solidarity organisations – a kind of feminism and a kind of socialism became the techniques of liberal oppression.[19] The French also exerted other kinds of control. For example, colonial employers would invite around married male employees to see if their wives were wearing the veil – and if they were, the men would find themselves quickly out of a job. In this situation, the unveiling of the woman was a test, a test of the extent to which she had become European, free of the veil, of Algerian patriarchy, of Algerianness. At heart, French attempts to control the veil simultaneously dehumanised Algerian men, domesticated Algerian women and made femininity the stake of colonial authority and national identity – both French and Algerian (page 38).

As the veil became the sign of Algerianness for the French, so it did for Algerians, and the wearing of the veil became a sign of resistance to European colonialism. So the French were right, women in the cities took to wearing black veils – instead of the more usual white veils – as a mark of mourning. But the French were wrong, Algerian women in cities had not for a long time worn the veil, and its significance in Algerian culture had diminished over time. Indeed, colonial fascination with the veil said more about their (sexual) fantasies than anything about Algerian culture. And the French got it wrong, against their intentions, Algerian women who had stopped wearing the veil, took it up again as a sign of their hostility to French (cultural) imperialism (pages 46–47).

The wearing of the veil quickly became a highly politicised act – but its meaning also circulated and changed in a dynamic economy of sexualised exchanges. For the unveiled face meant that French men could see and leer at skin previously hidden, mysterious, forbidden. Every veil spoke of a face that they did not control, could not see, could not have. The stakes of the veil and the face were never outside of a sexualised politics in which baring and hiding women's bodies were the sight/site of fascination, of desire, of rape (see also pages 45–46).

> Every rejected veil disclosed to the eyes of the colonialists horizons until then forbidden, and revealed to them, piece by piece, the flesh of Algeria laid bare ... Every veil that fell, every body that became liberated from the traditional embrace of the *haïk*, every face that offered itself to the bold and impatient glance of the occupier, was a negative expression of the fact that Algeria was beginning to deny herself and was accepting the rape of the coloniser.
>
> (page 42)

There are scales of domination and resistance, here, for the face stood for the female body and the female body stood for the nation – in a chain of meaning that sought to unbind and bind Algeria into modernity, into a French modern, into lascivious French arms. The female body and the Algerian nation are bound together through a spatiality which is simultaneously sexualised, gendered and racialised. Further, this situation is also constituted through the spatialities of visibility and invisibility, hiding and uncovering (surveillance): whether the veil is worn or not, there is a 'play' (the word does not do justice to the brutality and humiliation of this grotesque colonial situation) between visible and invisible. If the veil is worn, then the body of the Algerian woman is covered and hidden, yet becomes the heightened object of the (male and female) colonists' sexualised gaze, and the differences amongst Algerian women and between different ethnic groups in Algeria are erased. Meanwhile, for Algerian men, the veiled woman is de-sexualised, stripped of her sexuality: according to Fanon, 'he does not see her' (page 44). On the other hand, if the veil is shed, then the woman's body is exposed to all, yet she becomes invisible as an individual, even while she is also sexualised – both by the French, and by Algerian men.

Unveiled, the Algerian woman can pass unseen as a European; veiled, the woman becomes the subject of intense scrutiny. Yet it was the naked invisibility of the colonists' costumes and customs that the Algerian resistance movement exploited, initially with great reluctance. So far, Algerian women have been seen to be subject to – and become subjects through – the 'looks' of the colonialists and Algerian men, but they were not passive objects, nor simply subject to imposed meanings that told them what they had to wear, how they had to behave, who they were supposed to be and the places they were allowed to occupy.

There were significant changes in the meaning and use of the veil throughout the resistance, just as women's roles and significance to the struggle changed. Prior to 1955, Fanon asserts, only men were involved in the struggle, but thereafter women became increasingly active. Though men resisted women's induction into active service, the demands of war meant that women were needed. At the same time, resistance leaders began to use more and more violent and terrorist tactics as French brutality and terrorism escalated. As the struggle became more and more violent, more and more dangerous, women became more and more important, taking on increasingly dangerous missions. Indeed, the bravery and commitment of women, for Fanon, belied both French and Algerian men's assumptions of what women were like and what they could do – although, it should be noted, Fanon also constructs his own version of white and Algerian femininity (more on this later).

Initially and unsurprisingly, women acted as nurses and as carers for freedom-fighting men. However, it was the geography of the Arab city that made women increasingly important to the activities of the resistance movement. The use of cloisters in the Arab city had once made it easy for Algerian men to monitor the comings and goings of women, to confine them; now this traditional architectural style enabled the French to surveil men, to watch for suspicious movements, to catch men they wanted to stop, search, arrest, take away and torture. As the French became increasingly brutal and arbitrary in their operations, so women had to get used to moving through the city, without men, into spaces that they were not used to going, and to doing so with ease.[20] The restrictions on men and the increasing mobility of women (that is, to be sure, the changing geographies of men and women) meant that it was now necessary for women to carry messages, to carry out surveillance, to carry arms and ammunition. Sometimes carrying weapons for men, until the last minute, when the men would take the gun from the women. Women had moved from being mothers and nurses, to being weapons carriers and messengers (pages 50 and 58).

During operations in the city, it was found that women's mobility through the streets was, ironically, further enhanced when they took off their veils. The only Arab women who were allowed access to the European part of the colonial city were servants. Otherwise, Algerian women were excluded and confined to the Arab quarters, and even here women's mobility was severely restricted. In order to gain access to the European city and to move freely across it, women had to become 'invisible' by making their flesh visible (pages 51–53; see also page 59, footnote 14): they had to wear western dress. Because the French assumed that an unveiled woman had assimilated to Europeanness, they assumed that they could have no part in the struggle. In this way, Algerian women could insinuate themselves into the spaces of the oppressor, under their noses, crossing across the city with a degree of ease, enabling the exchange of arms and information, establishing connections across the conquered city.

In wearing western dress, however, the women had to learn to feel at ease in these clothes – despite feeling improperly dressed, and even naked.[21] Whereas the kasbah had once shielded the women from the eyes of the French and the veil had shielded them from the eyes of disapproving Algerians, now, in western dress, they were exposed to the glare of both the French and the Algerians. The woman were made to feel ashamed – both by the sexual scrutiny of the French, and by the disapproving and insulting comments of Algerians. Men openly accused women involved in the struggle of having no shame, of being immoral. In taking off the veil, the women had stepped into a double-bind of colonial and patriarchal authority, but they openly challenged Algerian men's attitudes, forcing them to recognise their significance in the struggle, to give women greater freedom and responsibility in operations, and to change the very nature of the struggle itself. For Fanon, a significant (though unintended consequence) of the armed struggle against the French was the undermining of the father's authority within the traditional Algerian family: first, sons challenged fathers who were not committed to the struggle; then, daughters challenged men's values, but especially fathers. Ironically, the French were right about the transformative capacities of women, but they were wrong to assume that this would undermine the struggle for liberation; instead, it renewed it, gave it added purpose.

In the traditional Algerian family, Fanon suggests, girls learnt very quickly that they were not as important as boys, and learnt that their roles and place in the family were highly circumscribed and tightly controlled. There were restrictions on mobility, association, marriage, sexuality and so on.

> All these restrictions were to be knocked over and challenged by the national liberation struggle. The unveiled Algerian woman, who assumed an increasingly important place in revolutionary action, developed her personality, discovered the exacting realm of responsibility. The freedom of the Algerian people from then on became identified

> with women's liberation, with her entry into history. This woman who, in the avenues of Algiers or of Constantine, would carry the grenades or the submachine-gun chargers, this woman who tomorrow would be outraged, violated, tortured, could not put herself back into her former state of mind and relive her behaviour of the past; this woman who was writing the heroic pages of Algerian history was, in so doing, bursting the bounds of the narrow world in which she had lived without responsibility, and was at the same time participating in the destruction of colonialism and in the birth of a new woman.
>
> (page 107)

No doubt Algerian women were not as passive and carefree as Fanon intimates, but the point should not be lost – and it is also Rich's (see below) – that to write the pages of history is also to make a new map of the world. These relationships of confinement, passing, boundedness, unbinding, birth are profoundly spatial as much as they are also temporal. In occupying space, women changed that space. A situation which resonates strongly with Lefebvre's assertion that 'to change life . . . we must first change space' (1974: 190). Women changed life – they changed the rules governing gender and sexual relationships as well as those imposed by colonial administrators. As Fanon concludes, 'she literally forged a new place for herself by her sheer strength' (page 109).

Nevertheless, while women were revolutionising the revolution, the colonists were continually adapting to the changing tactics of the liberation struggle. By 1957, the French realised that resistance fighters were 'passing' as Europeans, women who had discarded the veil became suspect. In another twist, the veil once more became a way of passing under the noses of the enemy. While the colonists fantasised that beneath the veil lay a hidden beauty, a mysterious other, in fact women modified traditional costumes to conceal weapons and operational information. The women ceased to 'occupy' the bodies of Europeanness, unveiled and open to inspection, they suddenly became large and shapeless (according to Fanon) – the better to hide with. In this way, Algerian women once again learnt to use their clothes, their bodies, as camouflage – to redesign geographies of resistance top to toe. Even so, the French soon realised that European women were also participating in the struggle on the side of the Algerians. Eventually, French patrols challenged everyone – and used metal detectors to find hidden weapons. It had become clear, however, that the values of the occupier were being played with, rejected, used against them, in a continual guerrilla war of repositioning.

The struggle for liberation was not fixed in an unchanging opposition, chained to an age-old tradition, but was continually modernising, moving forward, occupying new positions as old ones became untenable. During this repositioning, women took up a place and changed the course of the struggle. Fanon draws a lesson from this: 'Colonialism must accept the fact that things happen without its control, without its direction' (page 63).

Fanon shows that, while the colonial authorities did attempt to designate, control and purify space, their reterritorialisations of Algerian spaces were never more than partial. Against de Certeau, then, colonial strategies were noticeably improper (at least in the sense of being incomplete, fragmentary and discontinuous), while the Algerian resistance had its own strategies, capable of defining what was exterior to it; meanwhile both occupier and occupied attempted, in their tactics, to work in the space of the other – through the veil, for example. De Certeau's terms become much more useful when wrenched from a narrowly dichotomised account, in which the powerful occupy and control space, and the powerless do the best they can. In this sense, strategies (for the production of space) and tactics (under the noses of the enemy) are two spatialities in the repertoire of struggle. Thus, the use of the veil was guided by the spatialities of struggle: strategies for the control of space, the definition of boundaries and exteriority; tactics for moving through spaces, (in)visibly, (un)noticed; into the veil was woven a whole universe of resistances.

For Fanon, authority must accept that things happen without its control, without its direction. Resistance can take place through the differences (even distances) between systems of oppression as much as within them. Further, the meanings and intentions of acts of resistance can have unintended, but nevertheless far-reaching consequences. What Fanon shows is how the strategies and tactics of domination and resistance are intended to gain advantage over one another, but how moves in this war continually serve to re-contextualise and re-symbolise acts of tyranny and liberation.[22] There is never one geography of authority and there is never one geography of resistance. Further, the map of resistance is not simply the underside of the map of domination – if only because each is a lie to the other, and each gives the lie to the other.

The blurring of political identities

Thus far, resistance has figured as a set of political practices, which are more and less obviously political, and political identities have been closely associated with the politics of resistance. However, Fanon's work demands a recognition that political identities are not just made out of siding either with the oppressor or against the oppressor. Subjectivities are about feelings – fears, desires, repulsions – which are not so easily contained within a narrowly structural analysis of politics. The war of repositioning is fought not only in overt political spaces, nor only through self-consciously manipulated deceits, but also in inner spaces too. Remember, Fanon plausibly stated that Algerian women felt shame and guilt when taking off their veils in the course of military activities. He argues that

> the Algerian woman must achieve a victory over herself, over her childish fears. She must consider the image of the occupier lodged somewhere in her mind and in her body, remodel it, initiate the

> essential work of eroding it, make it inessential, remove something of the shame that is attached to it, devalidate it.
>
> (page 52)

Fanon is arguing that it may be necessary to overcome resistance in order to achieve resistance. The colonised must conquer parts of themselves in order to liberate themselves: that is, power colonises internally as well as externally, and achieving the overthrow of external power is more easily conceived of than the idea of shedding the guilt and shame induced by internal colonisation – the setting up of a garrison within the conquered city of the mind.[23] Alternatively, in order to achieve progress, it may be necessary to conquer other people's resistances. The implication is this: if the situation is to be changed, then people who resist change will need to have their fears 'devalidated', their guilt, shame and embarrassment overcome, and their desires given the resources of hope. On the other hand, not all fears are 'childish' – and, indeed, given French brutality and torture, these women's fears were at least as real as childish – and not all desires are progressive. In the case of Algeria, awkwardly, it is possible to think not only of Algerian men's resistance to women's liberation, but also of the racialised desires of the French colonisers.[24]

In order to think through political identities, it is necessary to achieve an understanding of this double sense of resistance: resistance to power, resistance for power. The most developed account of resistance's role in the maintenance of repression, arguably, is articulated by Freud. Freud's development of his understanding of the term resistance developed from his patients' attempts to ignore or deny or contradict (i.e. resist) his intended-to-be therapeutic, liberating interventions – though many would not trust Freud's intentions, of course. Jacobs (this volume) also deploys Freud's understanding of the term to interpret the role of 'backlash' (psychic resistance) in conservativism; further, Dear (this volume) unpacks the invidious ways in which certain intellectual projects are 'resisted' by academics.

In one way, the two sides of resistance are the same: the individual resists the imposition of a meaning, category or custom(e) of behaviour by a powerful person or group – thus, the Algerians resisted the French just as Dora resisted Freud. Nevertheless, resistance in the political sense and resistance in the psychoanalytic sense can work in opposite directions. Political resistance seeks to overthrow the perceived dangers in the practices of the powerful, while psychic resistance seeks unconsciously to maintain the repression of traumatic or potentially dangerous memories, feelings or impulses. It should give us pause for thought that resistance seeks to maintain repression in Freud's understanding.[25] Indeed, for Freud, patients have to 'combat' five different kinds of resistance emanating from three different directions (1926: 318–320),[26] if they are to be able to work through their problems.

Outside the consulting room, however, it could well be the way in which people deal both with threats to their sense of who they are and what their lives are about and also with their desires, fears, repulsions, and so on, that will be constitutive of their politics, of the things they are prepared to undertake politically. In this sense, it could be psychic resistance that compels the most injurious and barbaric of 'political' acts, such as the bombing of the federal building in Oklahoma City or of the Olympic Games in Atlanta (as American white supremacists act out their obsessive phobias) or the horrific systematic rape of Muslim women by Serb soldiers in the former Yugoslavia (an echo of the rape of Algerian women by French soldiers) or the summary executions carried out by the (seemingly paranoid and unashamedly violent) Irish Republican Army or the seemingly non-violent spiking of trees with nails by eco-terrorists that nevertheless endanger the lives of lumberjacks or the (at least emotionally damaging) 'outing' of supposedly hypocritical people by gay activists. Indeed, psychic resistance is likely to be bound up in all political activities, whether these are believed to be progressive or reactionary.

Psychic resistance is not only unconscious, it is also highly dynamic. It can work in many ways to defend people against interventions – whether therapeutic or political – that seek to persuade, to move things on, to enable people to draw new conclusions about the reasons for things. To simply reinstate 'resistance' as an act of the powerless in unambiguous opposition to the powerful is once again to marginalise the problem of how people become involved in political struggles, why most don't get involved, why many people are reactionary, why they would prefer to see things, either stay as they are or return to some nostalgic fantasised heyday, rather than participate in progressive, radical politics – even when it is in their so-called objective interests to do so. It is to this question that the concluding subsection turns. Before this, however, it is useful to return briefly to Fanon's discussion of the mobilisation of women in the Algerian nationalist struggle – this time bearing in mind the alarming ambiguities of resistance.

This discussion of psychic resistance is not an idle speculation on opposing meanings of the word: the consequences of this interpretation of psychic resistance can be far-reaching – turning resistance into something altogether more dangerous. Since psychic resistance acts unconsciously to prevent certain, dangerous, things from coming to the surface, from being worked through, it is unsurprising that resistance in this sense can be found even in the most radical of arguments. Even as Fanon argues that women's liberation became synonymous with the freedom of Algeria, his own account of femininity nevertheless represses Algerian women's evident capacity to 'pass', 'masquerade' and 'play the game'. Fanon thoroughly identifies the ingenuity of Algerian women with the purity of the nation: he thereby dissolves women into the soil.

> It must be constantly borne in mind that the committed Algerian woman learns both her role as a 'woman on the street' and her revolutionary mission instinctively. The Algerian woman is not a secret agent. It is without apprenticeship, without briefing, without fuss, that

> she goes into the street with three grenades in her handbag or the activity report of an area in her bodice. She does not have the sensation of playing a role she has read about ever so many times in novels, or seen in motion pictures. There is not that coefficient of play, of imitation, almost always present in this form of action when we are dealing with a Western woman.
>
> (page 50)

For Fanon, the Algerian woman's actions are pure and untainted by influences other than an authentic commitment to Algerianness. The revolutionary identity of the Algerian woman is instinctive, in a complete contrast to the white woman who only ever sees herself as if in a film, playing a role. This is necessary for his analysis, because Fanon is seeking to purify the Algerian cause of French modernist or liberal values. The consequence is that the birth of a new nation is seen as inextricably bound up with the birth of a new woman – but this woman is simultaneously liberating and threatening. So, Fanon states at one point that 'the Algerian woman penetrates a little further into the flesh of the Revolution' (page 54): that this metaphor is sexualised should not need pointing out, but intriguingly it works to 'phallicise' women and 'feminise' the revolution (see Fuss, 1994).

The pernicious result is that the purity of Algerian femininity remains the essential and irreducible ingredient of a true struggle: geography (the nation, soil, land) and the woman (blood and soul) are lost one in the other, such that women are placed on the map of struggle only to be removed from history.[27] Even though Fanon wrests the struggle from an essentialised and romanticised Algerian tradition, even though he provides an account of a revolution which seeks to overthrow not only French colonialism but also an oppressive patriarchal institution, even though Fanon articulates a uniquely Algerian modern, he nevertheless anchors his analysis to an essentialised femininity, in order to evoke outrage at the rape of Algerian soil (see above). The revolution, at once unbounded and creating new spaces of action and possibility, still tethers women to the ground of the nation's struggle.

Conclusion

So far, it has been shown how spatialities are constitutive not only of domination, but also of resistance; that struggles for power are spatialised and constituted in space in specific ways – from opposition to repositioning; that power relations intersect in specific ways, that resistance in one direction can be oppression in another, and that resistance occurs in spaces beyond those defined by power relations; that acts of resistance have to be understood not only in terms of their location in power relations but also through their intended and received meanings. It has been shown that political subjectivities are constituted through political struggles, but also that there are many spaces of struggle through which people become political – as Moore's analysis, this volume, of the grounds of struggle in agrarian

Zimbabwe demonstrates. The effects of these geographies of resistance are multiple, fluid, dynamic and in some ways uncontainable or at least unintended. All of these summary points suggest that the next step is to look at the spatialities of resistance, where this permits the discussion to go beyond both the fixed grids of the latitudes and longitudes of power relations and the fluid exchanges of free-floating meanings. Spatialities of resistance imply a subject of change that is not necessarily coherent nor has an essential constitution nor a fixed goal, with a clear direction. Awkwardly, this suggests that there are no cast-iron guarantees that any one form of spatiality of resistance will be any freer than others.

THE POLITICS OF LOCATION AND GEOGRAPHIES OF RESISTANCE; OR, ON STRAYING FROM THE BEATEN TRACK

The material effects of power are everywhere. Or, maybe, one of power's most pervasive effects is that it seems to be everywhere – for better or for worse. It matters that power *seems* to be everywhere, but wherever we look, power is open to gaps, tears, inconsistencies, ambivalences, possibilities for inversion, mimicry, parody and so on; open, that is, to more than one geography of resistance – as the chapters gathered in this book convincingly demonstrate.

It has been a key feature of the arguments presented so far that resistance is analysed and theorised as a diagnosis of, and a reaction to, the injurious effects of power relations, although the idea that resistance is merely an oppositional stance has been interrupted by a sense that political identities are strategic, tactical, mobile, multifaceted, blurred, awkward, ambivalent. Further, once passive, inert and singular notions of spatiality and identity are abandoned, then it becomes clear that resistance is as much defined through the struggle to define liberation, space and subjectivity as through the élite's attempts to defeat, prevent and oppress those who threaten their authority. At the heart of questions of resistance lie questions of spatiality – the politics of lived spaces. The key sites in this political landscape of lived space need to be clarified – and this task is pursued throughout this book – but it will be useful, at this point, to discuss some significant features so far encountered because these enable a critical engagement with the 'spatialities of struggle'. They are location, boundaries, movement and territorialisation.

In her influential article, Mohanty refers to a 'politics of location' (1987). By this expression, she evokes 'the historical, geographical, cultural, psychic and imaginative boundaries which provide the ground for political definition and self-definition for contemporary US feminists' (page 74).

While the idea of a politics of location is drawn from the imaginative work of Rich, the expression takes on added meaning in Mohanty's understanding.[28] The politics of location involves not only a sense of where one is in the world – a sense gained from the experiences of history, geography, culture, self and imagination – mapped through the simultaneously spatial and temporal interconnections between people, but also the political definition of the grounds on which struggles are to be fought. In this sense, location has more to do with the active constitution of the grounds on which political struggles are to be fought and the identities through which people come to adopt political stances, than with the latitude and longitude of experiences of circumscription, marginalisation and exclusion. The idea of location does not suggest, therefore, that the grounds on which struggles are defined are permanent, fixed and universal. Instead, underlying this idea is a sense that space is discrete and discontinuous but also relational and close. The imagined political geography of location is intended to resist a politics where the spaces of difference and differentiation are erased, where the experiences of power relations are universalised, where struggles are organised only through one experience of injustice, injury and inequality.

Moreover, location in Mohanty's work conveys a sense of having to take up (often dangerous) positions in activism and struggle in order to change the plot of power. These locations cannot, however, be presumed in advance of activism and struggle. If politics is about making history, then it is also about changing space: political locations are constituted through the struggles that are supposedly fixed in them. Location is both the ground which defines struggle and a highly contested terrain, which cannot provide any secure grounding for struggle. Thus, 'gender is *produced* as well as uncovered in feminist discourse, and definitions of experience, with attendant notions of unity and difference, form the basis of this production' (Mohanty, 1987: 76) – a view also echoed by white queer theorists such as Butler (1990a). Location is simultaneously about unity and difference, about definitions of who occupies the same or a similar place and who does not, which do not presume – and, further, undermine the presumption – that there is a sameness to people's location in particular oppressive power relations and a consequent sameness of their struggle.

For Mohanty, 'the unity of women is best understood not as a given, on the basis of a natural/psychological commonality; it is something that has to be worked for, struggled towards – *in history*' (page 84). And, it can be added, in geography. Location is a spatiality of resistance in the sense that it has to be struggled for and towards. The unity of communities of resistance is formed through the production of location as much as through the uncovering of location within the fantasms of multiple power relations. Engagements in the politics of location, further, involve the definition of boundaries – but once more, these are not to be seen as fixed, impermeable and permanent. In the struggle to define alternative ways of living, people will occupy strategic locations, but these will be bounded and unbounded in ways which are designed to chalk out a place on the map of politics. For example, as Jacobs argues,

> Specific land rights struggles can influence pan-national and global indigenous rights movements. Neighbourhood territorialisations can form the basis of re-forging identity across the national borders of

> global diasporas. The politics of identity is undeniably also a politics of place. But this is not the proper place of bounded, pre-given essences, it is an unbound geography of difference and context.
>
> (1996: 36; see also Massey, forthcoming)

If locations disrupt any sense of singular, isotropic, universal experience of power relations in space and time, then boundaries are about the definition of resistant spaces, without 'insisting on *a* history or *a* geography' (page 87). While Mohanty uses the term 'temporality of struggle' to define these spaces, it may be more appropriate to think about a spatiality of struggle, in which boundaries are constantly being emplaced, repositioned, opened up and closed down. The tearing down of boundaries between people is certainly no universal panacea to the production of injurious categories of difference. Indeed, it is more likely that resistance should seek out appropriate and progressive ways of thinking about boundaries between people which enable difference to be accepted, recognised and even enjoyed (similarly, see Young, 1990). And this could enable the 'problem' of the boundaries between inner psychic resistance and outer political resistance to be addressed.

Resistance involves the spatialities of location and boundary formation, but it is also constituted through the idea of movement – a change from one place on the map to another, or possibly many others. More often than not, it is mobility that has been seen as radical and transformative. And it is demonstrably true that many oppressive practices of authority seek to control and regulate people's use of space (see Allen, forthcoming), ultimately confining people to highly circumscribed spaces – whether the cloister, the prison, the concentration camp, the housing estate, the township and so on. Indeed, it is no coincidence that communities of resistance are termed movements in much political analysis. The point seems to be that social movements move because they have an origin, a projected destination and a path to travel, over an overt public political terrain. Yet, movement does not have to be so 'big': there are tiny micro-movements of resistance, barely perceptible, even invisible or covert – quiet stealthy masquerades resistant to categorisation and definition. It should also be noted that movement also implies a change in location – if locations are multiple and defined through struggle as much as they are the grounds on which struggle takes place, then movement will not be so much a question of origins and destinations as the paths which are adopted in the course of any struggle, without insisting on *an* origin and *a* final destination. There is unlikely, therefore, to be one movement in any social movement, no one reason why people mobilise or become mobilised, no single aim, no single demand, no one path of struggle – and, far from disabling resistance, this enables it to move strategically, tactically, resourcefully from place to place.

If location is about places defined and places taken up through experience, identity and power, then these locations have distinct territorialisations. If the logic of certain kinds of power relations is to produce and uncover space in particular ways – such as the struggles to impose colonial definitions of the nation-state (see Said, 1993) or financiers' interpretations of global markets (see Gibson-Graham, 1996) – then power relations involve particular notions of space-as-territory, to be conquered, administered and regulated. In one sense, power is the power to have control over space, to occupy it and guarantee that hegemonic ideas about that space coincide with those which maintain power's authority – and this can best be seen in the coincidence of the nation and national identity (see above). In another sense, power can be mobilised through the reterritorialisation – the resymbolisation – of space, and this can be as oppressive as it can be subversive. Territories need not necessarily be spaces of exclusion, where people defined as marginal or outside the dominant values are denigrated, abjected and persecuted (see Sibley, 1995). Instead, resistance may reterritorialise space in various ways, in order to transform its meanings, undermine territory as a natural source of power, and enable territory to become a space of citizenship, democracy and freedom – within limits. Territories involve location, boundary and movement – and they will therefore overlap, be discontinuous and shifting, as people seek alternative ways of living, alternative connections to the world.

The spatialities of struggle allow what Mohanty calls a 'paradoxical continuity of self, mapping and transforming' (page 89), since the subjects of resistance are neither fixed nor fluid, but both and more. And this 'more' involves a sense that resistance is resistance to both fixity and to fluidity, as an intensely knowingly ambivalent location, as a way through the Scylla and Charybdis of too rigid political identities and entirely free-floating values. Resistance may take place as a reaction against unfairness and injustice, as a desire to survive intolerable conditions, but it may also involve a sense of remembering and of dreaming of something better. If there is a beaten track, a track laid down through the spatial technologies of power configurations, then resistance will stray from the track, find new ways, elaborate new spatialities, new futures.

To rephrase Rich's aphorism which set up this introduction, a place in history is also a place on the map; or, better, it is through places on the map that histories are made; or, making new maps will help us make new histories. When history is spatialised, it is possible to realise that people occupy many spaces, and that resistance might exist in any of them, but also 'beyond' or 'in-between' them (to evoke Bhabha's phrases, 1990a[29]). So, to continue with Rich's metaphor, we occupy many places on many maps, with different scales, with different cartographies, and it is because we both occupy highly circumscribed places on maps drawn through power cartographies and also exceed these confinements, that it is possible to imagine new places, new histories – to dream that resistance is possible. Maybe resistance is already a place on the map, but – more likely, from the discussion above – maybe it is about throwing away imposed maps, unfolding new spaces, making alternative places, creating new geographies of resistance.

ACKNOWLEDGEMENTS

I have benefited from highly entertaining and challenging conversations with John Allen, Allan Cochrane, Hilary Cottam, Lisa Law, Michael Keith, Gail Lewis, Doreen Massey, Jenny Robinson, Paul Routledge, Shaminder Takhar and Nigel Thrift. Lisa, Doreen and Jenny have been burdened further by reading a somewhat shoddy draft of this piece: if only their ideas could shine through the gloom of my understanding ... I also give my heart-felt thanks to all the contributors for their patience, insights and enthusiasm throughout this project.

NOTES

1 This maxim is used by Mohanty as a section heading (1987: 77). See also Moore, this volume.
2 I am well aware that 'power' has many different meanings – from domination and force through authority and hegemony through manipulation and seduction to ability and capacity. I am also aware that it is likely that these different forms of 'power' are likely to have distinct spatialities – see Allen (forthcoming) for a thought-provoking analysis of alternative ways of thinking about power relations and space. These issues are important, but the point of this chapter is to think through 'the spaces of resistance' rather than 'the spaces of power' so they will have to be put to one side for now.
3 It would hardly be enlightening to provide a list of radical geographers (whether Marxist, anarchist, feminist, and so on), or of representative works, since even in highly abbreviated form this would run to several pages and it would still be guilty of the sin of omission. Nevertheless, as a first step, it would be useful to look at Peet, 1977, on early radical geography; Gregory, 1978, and Soja, 1989, on critical theory; Castells, 1983, on social movements; Rose, 1993a, on feminism; Keith and Pile, 1993a, on the politics of identity; Bell and Valentine, 1995, and Duncan, 1996, on sexuality and gender; Sibley, 1995, on geographies of exclusion; Cresswell, 1996, on the politics of transgression; and Peet and Watts, 1996, on liberation ecologies.
4 The logic of capitalist investment produces uneven development across space at different geographic scales (see Smith, 1984).
5 See Painter, 1995, for a recent summary.
6 In this discussion, Harvey draws heavily on Iris Marion Young's influential discussion of oppression, justice and city life (Young, 1990, especially chapter 8). Unfortunately, it would be too much of a digression to enter into a discussion of her insightful and influential work. While justice and difference remain integral components of resistance, as commonly conceived, Young does not address the issue of resistance directly.
7 That is, it seems, marginalising already marginal groups away from the centre of oppositional politics.
8 This argument is strikingly reminiscent of discussions of the category of 'Woman' in feminism (see, for example, Butler, 1990a) – and by extension, might beg questions about the category of 'Class' in working class organisation.
9 See Sedgwick, 1990, and Brown, 1997.
10 For an introductory review of the state–civil society distinction in political and sociological thought, see Tonkiss, forthcoming – see also Slater, this volume.
11 See Abu-Lughod, 1994, and Smith, 1996.
12 See also Hebdige, 1993. According to Smith, one slogan read 'Tompkins Square Everywhere' (1993: 93). Meanwhile, the cover of Abu-Lughod's book shows a banner proclaiming 'This Land is Ours' – which echoes a rallying cry of communities in London's docklands, 'This Land is Our Land' (see Keith and Pile, 1993b: 11–16). See also Moore, this volume.
13 Other glances offer further problems, see Massey, forthcoming, and Moore, this volume.
14 It can be argued that Jackson's symbolic interactionism was particularly sensitive to the ways in which different actions take on different meanings and the ways in which dominant meanings are resisted through symbolic work (Jackson, 1989; following an analysis

pioneered by the Centre for Contemporary Cultural Studies in Birmingham, see for example Hall and Jefferson, 1976; for a recent description of cultures of resistance in Britain, see McKay, 1996). Geographers who have taken up this agenda include Cresswell, 1996a, and Jacobs, 1996.

15 There is now a vast literature on Fanon's work, for recent interventions see Gordon, Sharpley-Whiting and White, 1996, and Read, 1996.

16 'Looks' is in scare quotes because looking is not innocent power and the social production of 'looks' – the averted eyes, hiding, showing, visible or secret signs, the stare, and so on – for different takes on this, see Bhabha, 1994, Kirby, 1996, and Nast and Kobayashi, 1996.

17 For recent interventions along these lines, see Said, 1993, McClintock, 1995, and Low, 1996. Works cited in notes 15, 16 and 17 form the implicit backdrop to the discussion that follows.

18 Algeria certainly is not the only context in which dress codes become a site of struggle, see for example Radcliffe's study of Ecuador (1996).

19 Similarly, Robinson has shown how a kind of friendship was used as a technique of government in colonial and apartheid South Africa (1996).

20 For a related argument, see Jacobs, 1996: 21 and 162.

21 Fanon provides a fascinating, and it should be warned fascinated, study of the use of the body by women to pass as European (see pages 58–59). These issues deserve greater elaboration, but there is not space here to do so.

22 This does, of course, happen in peace too – see Rose, this volume, and Jacobs, this volume.

23 For a discussion of this metaphor, see Pile, 1996: 242.

24 On the difficulties of an easy valuation of resistance as good, see also Cresswell, 1996b.

25 See Freud, 1915 and 1923: 353.

26 In psychoanalysis, resistance comes from three directions – from the ego, the id and the super-ego; while it takes five different forms – repression, transference, illness, the compulsion to repeat and guilt or the need for punishment.

27 Monk, in a different context, argues that women can draw on the meanings of landscape to form their own resistances both to patriarchal domination and to create a ground for their own political identities (Monk, 1992: 124).

28 See also Mani, 1990, and Frankenberg and Mani, 1993.

29 On Bhabha's notion of 'third space', see Moore, this volume, and Law, this volume.

2

BLACK GOLD, WHITE HEAT

state violence, local resistance and the national question in Nigeria

Michael Watts

This letter is from the Prophet Mohammed. It came from Medina to Mecca ... to Wadi ... to Kukawa ... to Bornoland ... From there to Hadejia ... to Kano and from Kano to the whole world. The Prophet said there will be disaster of the wind, poverty, death of sheep and illhealth of women and men. The illhealth of women and men will affect their womenhood and manhood respectively. It would effect the old and the young of women and men East and West, South and North.

(Letter circulated by the self-proclaimed prophet Maitatsine to his Muslim followers in his Kano city community in the late 1970s, and published subsequently in *The New Nigerian* [Kaduna], March 3rd, 1984)

My Lord, we all stand before history ... I and my colleagues are not the only ones on trial. Shell is on trial here ... [and] the ecological war the company has waged in the [Niger] delta will be called into question sooner or later ... On trial also is the Nigerian nation ... I call upon the Ogoni people, the peoples of the Niger Delta, and the oppressed minorities of Nigeria to stand up and fight fearlessly and peacefully for their rights ... For the Holy Quran says in Sura 42, verse 41: 'All those who fight when oppressed incur no guilt.'

(Closing Statement prepared by Ken Saro-Wiwa leader of *MOSOP* [Movement for the Survival of the Ogoni People] on September 1st, 1995 for presentation to a military appointed special tribunal in Port Harcourt, Nigeria which subsequently hanged him on November 17th, 1995)

Early in the morning of December 29th, 1980, in the wake of two weeks of escalating violence between local state authorities and an unorthodox local Muslim preacher and his followers, federal Nigerian armed forces began a massive military assault on the sleepy residential and commercial quarter of

'Yan Awaki in the walled city of Kano, in northern Nigeria. Under ferocious aerial and ground bombardment, somewhere between five and ten thousand people were slaughtered, and another fifteen thousand injured. The body of the self-proclaimed leader, *Maitatsine* – a rough rendering from the Hausa would be 'he who damns or condemns' – who was killed in the conflagration, was exhumed by the military authorities from a shallow grave outside the city walls where he had been laid to rest by his followers, and placed on display at the local city police station. Vilified in the leading newspapers, and charged by Muslim intelligentsia with a multitude of sins, including fanaticism, hereticism and cannibalism (Yusuf, 1988), Maitatsine's project to build a renewed Muslim community (*umma*) within the heartland of Nigerian Islam had been brutally crushed by state forces. Those federal authorities who engaged in the bloodletting and who self-consciously sought to quite literally obliterate a leader and a community which saw itself as incontrovertibly Muslim in character and constitution, were acting on the instructions of a civilian administration presided over by an aristocratic northern Muslim (President Shagari) and a government that was widely held to be dominated by northern Muslim elites (the so-called 'Kaduna Mafia'). Almost one thousand followers of Maitatsine were imprisoned and in the subsequent official investigation (the Aniagolu Tribunal), blame was squarely placed on the shoulders of 'a band of fanatics'. Throughout the 1980s these 'fanatics' reappeared in various cities across the north of Nigeria (Figure 2.1).

Fifteen years later, on November 10th, 1995, Ken Saro-Wiwa and eight other prisoners – all residents of oil-rich Ogoniland in Rivers State in the southeast of Nigeria (Figure 2.1) – were awakened at dawn, shackled at their ankles, and transported from Bori military camp where they had been held during their murder trial, to Port Harcourt Central Prison. Saro-Wiwa, an internationally recognized novelist, environmental activist and leader of MOSOP (Movement for the Survival of Ogoni People) was granted his last rites by a sobbing priest, and surrendered his remaining property including his trademark pipe. In a moment of darkest farce, the executioners had presented themselves at the prison only to be turned away because their papers were not in order. Dressed in a loose gown and a black headcloth, Saro-Wiwa was, after this interregnum, led to the gallows. The pit into which Saro-Wiwa fell was shallow and the fall failed to break his neck. It took him twenty minutes to die. A videotape of the hanging was sent by courier to General Abacha, head of the Nigerian military junta, as proof of Saro-Wiwa's death. Seven others, who were also found guilty of the murder of four prominent Ogoni leaders by a kangaroo court hastily convened by the military government, suffered a similar fate. The executioners were said to have poured acid on the corpses to speed decomposition and to discourage Ogoni activists from taking possession of the bodies. Within hours of the hanging, 4,000 troops were deployed throughout Ogoniland – a Lilliputian area of 400 square miles containing half a million people and almost 100 oil wells. Nigeria's Kuwait was in effect under military occupation. Special

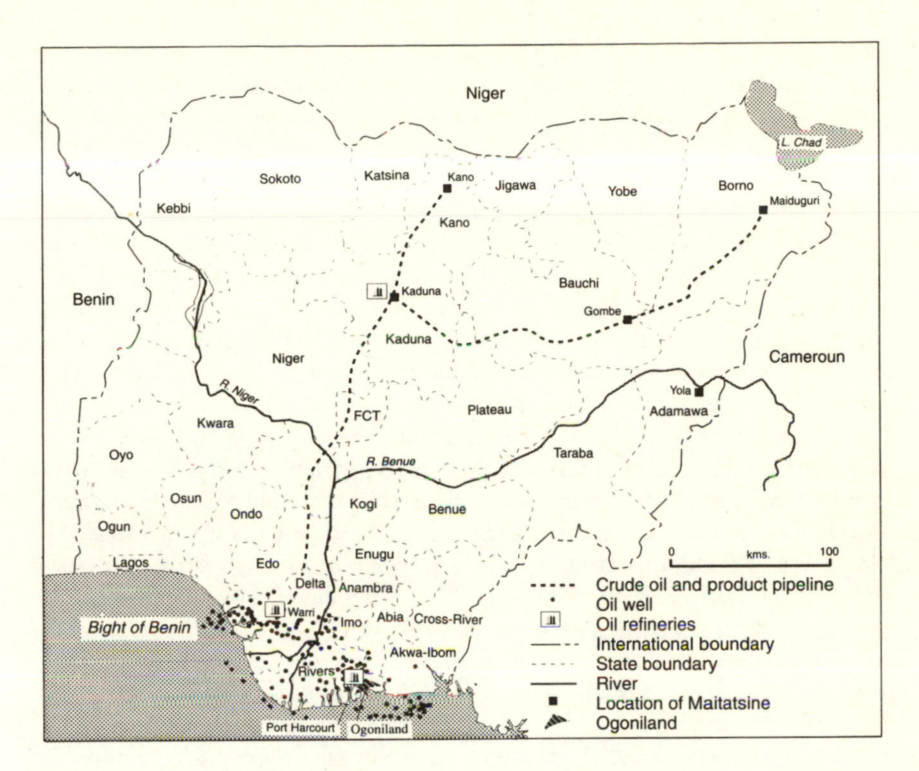

Figure 2.1 The geography of oil, Ogoni and Maitatsine

military forces beat any person caught mourning MOSOP's deceased leadership in public and embarked upon a systematic attempt to erase any trace of Saro-Wiwa's influence.

On the face of it, these two political murders are radically dissimilar; or rather the two movements are a study in contrasts. One is an admittedly curious and unconventional expression of what one might call 'political Islam' (or Islamic fundamentalism to employ the language of the popular press); and the other a striking instance of identity politics in which a resurgent ethno-nationalism confronts the slick alliance between a rapacious transnational oil company resistant to any commitment to local community development or environmental health and a state hostile to political decentralization. One reflects a worldwide debate within Islam over orthodoxy and the very essence of Muslim identity; the other is a textbook example of the sort of cultural movement – self-organizing and participatory – described by Arturo Escobar (1995) in his account of the new 'anti-development' movements proliferating throughout the South in the wake of the failure of the modernist project articulated by President Truman in his famous speech in 1949 on 'fair dealing' and 'development'. Both movements are rooted, of course, in post-colonial Nigerian political economy, and in the cataclysmic changes wrought by the discovery and exploitation of petroleum. Maitatsine stands at the apex of the *oil boom*, while the Ogoni crisis is framed by the *oil bust*. As I shall endeavor to show, Islamic reformism and Ogoni self-assertion both predate the petroleum boom and trace their origins to a period prior to the first oil explorations in the Niger Delta in the 1950s. Yet each is inexplicable outside of the maelstrom ignited by Nigerian petrolic capitalism.

But I wish to focus on another axis of contiguity which links Muslim footsoldiers in Kano with oil-stained Ogoniland, namely state violence and its relation to locality politics. What is striking about the deaths of Maitatsine and Ken Saro-Wiwa, is not that each individual perished at the hands of corrupt, venal and authoritarian state forces (in one case civilian, the other military), but rather that the movements which they built elicited such a ferociously violent assault from state authorities in which the intention was to obliterate any trace of a local community vision. It is this sense of 'overreaction', and its counter-intuitive nature, which concerns me here: not so much the *fact* of state-sponsored violence but the scale of its mobilization in relation to the perceived 'threat' of the movements themselves. My suggestion of a counter-intuitive state response refers to two things. First, the scale and escalation of state violence. Both movements could plausibly be seen as minor irritants in a country so large, wealthy and powerful as post-colonial Nigeria. Maitatsine was little more than an ornery and troublesome preacher who had after all resided in and around Kano for almost two decades prior to the events of 1980. Ogoniland is home to roughly half a million people – perhaps 0.5 per cent of the Nigerian population – and while an oil producing region was a minority backwater which, by the 1990s accounted for less than 1.2 per cent of Nigerian oil.

State violence is counter-intuitive in a second sense, namely as regards state interests, or what one could call the antinomies of ruling class hegemony. Any history of independent (post-1960) Nigeria is the history of northern (and hence Muslim) hegemony in a complex multi-ethnic state. Since 1966 when the military seized power from a First Republic tottering on the brink of chaos, northern Hausa-Fulani Muslims have ruled the country without interruption, principally under military auspices with the exception of a four year civilian hiatus between 1979 and 1983. It is, in light of this history, curious to say the very least that a federal government – in this case a civilian administration clearly dominated by northern Muslim elites – should launch a violent assault on a small community which self-identified as Muslim. That the bombardment and the casualties should occur within the traditional walled city of Kano – the *locus classicus* of traditional emirate power and authority and of Muslim sanctity – renders the violence even more paradoxical.

Saro-Wiwa was a much more visible figure than Maitatsine but it is precisely his visibility which lends state violence its particular curiosity. As an internationally and nationally recognized figure – Saro-Wiwa certainly ranked with Wole Soyinka or Chinua Achebe as one of Nigeria's global figures – his arrest garnered enormous attention and as a consequence the military junta under President Abacha felt the press of international opprobrium from all manner of political and human rights organizations, including Presidents Mandela and Clinton and the Commonwealth powers. Furthermore, Saro-Wiwa was hardly a subaltern radical on the margins of Nigerian politics. The son of a local Ogoni chief, he had been a well-connected state bureaucrat and former state commissioner who (like the Ogoni in general) sided with the federal government against their Ibo neighbors during the bloody civil war (1967–70) in which the eastern region had attempted to secede from the federation. Why then should the Abacha regime – already facing threats of international sanctions as a result of its deteriorating human rights record – hang Saro-Wiwa and his compatriots in what was clearly an inept show trial?

It would be all too easy (if not facile) to see these egregious cases of extrajudicial violence as the desparate measures of a weak and intimidated military that had degenerated into personalized and arbitrary rule. The transition from military to civilian rule had been aborted by the military in June 1993 after the presidential elections were won by a southerner. In the subsequent seizure of power by Sani Abacha, an influential Kano general, in November 1993, the military spared no mercy on internal opposition (including the arrest of democratically elected Moshood Abiola on treason charges following the annulment of the 1993 elections). Military authoritarianism fed on the social unrest precipitated by a deepening economic crisis after the oil bust in 1985 and the debt-induced austerity programs of the 1990s. And in some regards this explanation has much to commend it. But I wish to highlight another argument which turns on the fact that the Ogoni and Maitatsine movements attempted to *construct local communities and identities* at odds both with the ideology of a Nigerian federation and with northern hegemony over a fragile and contested polity without a robust sense of national identity. One was an attempt to create from below an 'orthodox' Muslim identity built around a particular reading of key Islamic texts, the other a case of ethno-nationalism triggered by the destructive consequences of oil and the transnational oil companies and an inequitable federal revenue allocation process. Both these movements presented alternative modes of citizenship and while their political horizons were in many respects local and parochial each in quite different ways, held up the specter of civil war and of political fragmentation. Each movement, as I shall show, starkly revealed how the oil boom had simply papered over the profound, structural fractures within a fissiparous multi-ethnic Nigerian state. These movements stood, in other words, as the antipodes of state legitimacy. They were also irreducibly products of, and struggles over, geography.

In this story of state fission and fictions of national identity (that is to say actual and imaginary geographies), petroleum figures centrally, because it was in one sense the resource over which the civil war was fought, and because it remains *the* central asset in Nigeria's monocultural economy. It necessarily becomes the lightning rod for political interests in general, whether issues of minority self-determination, federal hegemony, Muslim law or state corruption. In a highly regionalized and fragmented polity in which the national question is a sort of buried child, Maitatsine's and Ken Saro-Wiwa's aspirations however irreducibly local in horizon, represented a direct assault on central authority, on a particular model of petrolic capitalism, and on the ideological foundations of the post-colonial state itself. And it was precisely the weakness of central authority – the political vacuum left by decolonization (see Forrest, 1995: 39) – and the robustness of various regional subnationalisms which rendered the minority claims by MOSOP and Maitatsine so threatening to the fiction of Nigerian sovereignty and indeed to the post-colonial imaginary of Nigerian national identity. It is, in my view, only in this way that one can grasp why each movement paid so dearly for its appeal to such seemingly parochial hopes and aspirations.

NIGERIAN EL DORADO: OIL, OIL POLITICS AND UNEVEN DEVELOPMENT IN POST-COLONIAL NIGERIA

> Just as none of us is outside or beyond geography, none of us is completely free from the struggle over geography.
>
> (Edward Said, 1993)

Prospecting for oil began in Nigeria in 1908 when a short-lived German corporation commenced drilling along the southeastern coast, but the first commercial oilfield was not discovered until 1956 in tertiary sediments in the Niger Delta basin at Oloibiri, 90 km west of Port Harcourt. Eight years later the first offshore oilfield was located off the Bendel state coast. By 1996 Nigeria was producing just less than two million barrels per day, about 2.9 per cent of the world's total from 176 oilfields and some 2,000 producing wells (Figure 2.2). Sixteen of the oilfields are 'giant' (that is to say holding reserves of 500 million barrels or more) but the vast majority are small and dispersed, forty-five of which are offshore. It is however the Niger Delta basin containing 78 fields (including the largest field at Forcados) and spanning 75,000 sq. km across Rivers, Edo, Imo, Abia, Akwa-Ibom, Cross River and Ondo states, which remains the most prolific and important oil producing area in the country (see Figure 2.1). While Nigerian petroleum is typically seen to be relatively high cost (roughly 2–3 times higher than the Middle East), the quality of its marker crude, so-called Bonny Light, is high due largely to its low sulphur content.

As of January 1996 there were fourteen oil producing companies in Nigeria. All production has Nigerian participation (usually through the state oil company founded in 1971 as the Nigerian National Oil Company) in which eleven foreign oil companies operate through a complex concession system. During the Shell–BP concession era (1921–55), oil companies determined price and production in the industry (Ikein, 1990). Until the mid-1960s, the state confined its activities to the collection of taxes and rents and royalties. In effect, throughout the 1970s there was a gradual process of nationalization undertaken via the state petroleum corporation (subsequently renamed the Nigerian National Petroleum Corporation, NNPC, in

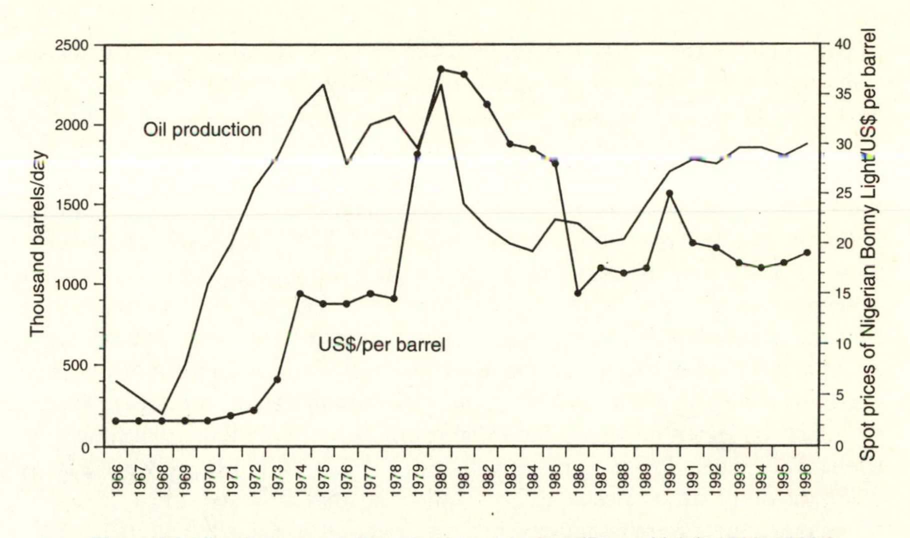

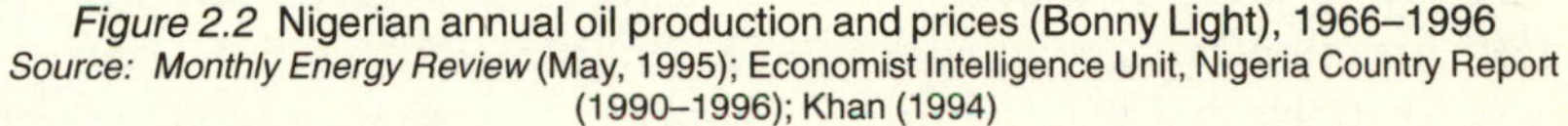
Figure 2.2 Nigerian annual oil production and prices (Bonny Light), 1966–1996
Source: Monthly Energy Review (May, 1995); Economist Intelligence Unit, Nigeria Country Report (1990–1996); Khan (1994)

1977) which increased its equity stake to 35 per cent in 1971 and finally to 60 per cent in 1979. The current leasing situation (Table 2.1), in which oil prospecting licenses and oil mining releases are dispensed to foreign companies, primarily takes the form of joint-venture contracts in which NNPC and the company share the costs of investment, exploration and production in direct proportion to their participation stakes.

Table 2.1 Oil companies holding Nigerian concessions, 1969 and 1993

Company	*Area (km²)* *1969*	*1993*
Shell–BP Petroleum Development Company	48,959	43,243
Safrap–Elf (Nigeria)	24,187	11,113
Gulf Oil Company	17,759	14,138
Nigeria Agip Oil Company (NAOC)	5,262	9,966
Mobil Producing Nigeria Ltd	5,246	4,928
Texas Overseas (Nigeria) Petroleum	5,003	2,570
Phillips Oil Company (Dubri)	3,630	232
Tenneco Oil Company	3,575	—
Union Oil Nigeria (AENR)	2,591	360
NNPC	—	40,440
Pan Ocean	—	1,005
Nigus	—	1,025
Exxon	—	2,200
Du Pont	—	2,422
Statoil	—	5,698
Ashland	—	2,346

Source: Adapted from Khan, 1994: 21

Throughout the 1980s and 1990s fiscal incentives and equity considerations have been relaxed with regard to the foreign oil companies. The 1986 Memorandum of Understanding and its 1991 revisions have substantially amended incentives and improved the terms for the oil companies, and in the last few years government has actually moved toward partial divestment of the state's share in some joint ventures as a result of acute financial constraints. The largest joint venture between NNPC and Shell accounts for 47.4 per cent of total crude production (EIU, 1996). Elf, Agip, Mobil and Chevron account collectively for another 49.2 per cent. For the most part there have been few changes in the foreign concession holders in the Nigerian oil industry, and with the exception of the civil war, production has largely been insulated from the volatile political climate (seven military governments, two elected civilian regimes, and one interim civilian government in the last 30 years). From a peak production of 2.3 million barrels per day in 1979, the output and price levels have fallen: dramatically so in the mid-1980s when oil exports were about one-fifth the level of oil export receipts in 1980. In 1995 the oil sector accounted for $9.15 billion, roughly 96.6 per cent of all Nigerian exports by value and close to 80 per cent of all state revenues (EIU, 1996; African Research Bulletin, 1996: 12476).

Oil has produced a clear centralizing dynamic in the Nigerian political economy – indeed it converted Nigeria into a strikingly mono-commodity economy on which the state directly depended. But oil's wider consequences can only be grasped in terms of the complex political geography into which the first export revenues were inserted. Nigeria is of course a huge and heterogeneous country. Among its 96 million inhabitants, there are an estimated 370 ethnic groups. Three ethno-linguistic groups – the Hausa-Fulani in the North, the Ibo in the East and the Yoruba in the West – each with distinctive religious loyalties and identifications, are disproportionately dominant, and their relative size, autonomy and resource base contribute to the instabilities of Nigeria's federal structure. As Tom Forrest (1995: 17) notes, colonial Nigeria was ruled by the British less as a single political unit than as distinct regional entities each with local authority structures and highly regionalized judiciaries, fiscal institutions (the marketing boards which taxed export commodities) and legislative assemblies with responsibility for education, health and local government. Colonial indirect rule – the policy of divide and rule in which local systems of authority and community sentiments were upheld and encouraged – entrenched a regional tripartite structure (the North, the West and the East) in which ethnic loyalties were created, refashioned and deepened. Within each region a single ethnic group predominated while federal authority (established formally in 1954) and nationalist sentiment were weak in the face of strong and fissiparous regional subnationalisms. In the 1950s, for example, political delegates remained in the regions and simply sent their representatives to Lagos.

If the regions were the source of identification and political loyalty, they were also marked by striking patterns of unequal development. The Northern Region, while larger in population and area than the other regions combined, was the poorest and least exposed to Western education. The West conversely was by virtue of cocoa, coastal access, industrial development and early education, the wealthiest region which captured 38.3 per cent of the statutory (i.e. federal) revenue allocation by 1954/5. Educational inequities contributed to regional tensions as southerners (Yoruba and Ibos) dominated federal posts and attempted to penetrate northern government. As a consequence, northerners attempted to slow down the transition to independence in the 1950s and promoted a northernization policy to limit Yoruba and Ibo incursions.

In such a highly charged, competitively regionalized polity, mobilization built upon ethnic and religious loyalties from below. As public institutions within the regions grew in the run-up to independence, patronage was distributed through contracts, public office, loans and so on. Patronage and clientelism were, in the context of intense regionalism, the legitimate perquisites of political and public office. Not surprisingly, in the wake of the federation in 1954, the establishment of political parties assumed an irreducible regional cast,[1] and state patronage developed apace. Furthermore, in the transition from the British to the first generation of Nigerian leaders, a number of key issues – public service employment, the national census – were cast and contested in strongly regional/ethnic terms. But in the political competition for power in post-colonial Nigeria, it was the economically weak and 'backward' northerners, under the careful orchestration of the British, who assumed gradual federal supremacy in an atmosphere of deepening animosity and parochialism.

The emergence of petroleum as the centerpiece of the Nigerian export economy and the mainstay of state revenues had enormous consequences for the political development of post-colonial Nigeria. First, the geography of oil mattered. Close to 80 per cent of the petroleum was located in the eastern region – more precisely in the delta which represents roughly 8 per cent of the country – and not infrequently in the territories of ethnic minorities (i.e. non-Ibo). While the civil war – the attempt by the Ibo to secede from the federation and to establish the independent state of Biafra – was not in any simple sense caused by the discovery of oil, the control of oil revenues was the central issue which precipitated the crisis of February 1967. The governor of the Eastern Region, Colonel Ojukwu, passed the Revenue Collection Edict number 11 in 1967 by which all revenues collected by the federal government would be paid to the treasury of the Eastern government. The federal (Gowon) government in response created three new states within the Eastern Region in an effort to gain support from oil producing minorities who would be awarded new-found autonomy[2] and a share of oil revenues. On May 30th Ojukwu proclaimed Biafra (the old Eastern Region) independent and war was declared on July 6th, 1967. Petroleum as a source of wealth, in other words, from its very inception was the medium through which deeply sedimented regionalisms and competitive federal politics were expressed and struggled over.

Second, petroleum underwrote a new political dynamic in the relations between the regions and the federal center. Growing nationalization of the petroleum sector and the establishment of a national oil company in 1970 channeled petroleum rents directly to federal coffers. Centrally controlled oil revenues superseded the regionally based revenues derived from the commodity marketing boards. As a consequence of the stunning growth of state revenues in the 1970s, the political center possessed a new-found fiscal capacity by which petrodollars could be used to manufacture a sort of political compliance, and conversely the regions discovered a new interest in gaining access to the seemingly infinite wealth provided by centrally controlled black gold. Petroleum enhanced the capacities of the historically weak center. But the military-bureaucratic alliance which held the political reins during the halcyon boom years was itself held hostage to the system of patronage and regional clientelism which were the hallmarks of Nigerian politics since the 1950s.

Petroleum is the key to understanding the two fundamental dimensions of Nigerian politics in the period following the defeat of Biafra in the civil war: state creation and revenue allocation. One of the first acts of the post-war military government under Gowon was to create twelve new states in 1967 from the existing four regions. Designed to balance north and south with six states, and thereby break the power and pathological competitiveness of the large regional blocs, the new state system had the effect of increasing minority access to federal funds while simultaneously making the entire state structure dependent on central (oil) revenues. Of course the demand for new states to meet the local needs for access to government resources, especially in deprived areas, was in practice difficult to halt. More states were created in 1976 and in 1991, while the number of local government areas (LGAs) within each state also proliferated, and for similar reasons. The result was the genesis of small states with little or no fiscal basis, totally dependent on what each state saw as 'their share' of the national cake (i.e. the oil monies), and a profusion (there are 589!) of corrupt, ineffective and hugely expensive LGAs driven by the logic of patronage politics. Ironically, this massive edifice effectively stymied any sense of Nigerian federalism – the dialectics of oil once more! – pointing to the ways in which vast oil revenues could not create a more robust sense of Nigerian identity and federal authority.

Nonetheless, the multiplication of states from twelve in 1967 to thirty in 1996 *did* have the effect of irrevocably breaking some aspects of the old pattern of regional power, and accordingly increased the power of minorities who came to hold some form of political representation and economic autonomy. To this extent petrodollars permitted a certain degree of political cohesion within the federation to be quite literally *purchased*. The cost of course has been an undisciplined federal structure driven by massive inflationary costs – the proliferation of state bureaucracies driven by prebendal politics – largely without a robust material base. As Khan (1994: 32) notes, the states have abandoned any pretense of a productive identity and rely unashamedly on federal handouts. The result is 'power untempered by responsibility ... [the states are] ... miniature versions of their free-spending federal paymasters' (*Economist*, 1993: 12). Oil revenues moreover did not require taxation of personal income or property, and reduced the economic and political significance of taxpayers (Forrest, 1995: 68–9) thereby removing another potential break on inflated state and federal expenditures.

Embedded in this complex multi-state military federalism, however, the obdurate tensions of old remained. First, the old regional blocs have certainly not disappeared[3] because major ethnic groups broke the old regions into smaller entities in which they could claim oil revenues from minority states (the Hausa for example created eight new states in which they became ethnically dominant). And second, the multiplication of states generated new demands by minorities and other political entities on a federal center dependent on a single and volatile source of revenue.

State creation could not be separated of course from the revenue allocation question. New states created more and complex demands on the distribution of government revenues – what from the vantage of the states and local communities was their cut of the national oil cake. In the period between the end of the Second World War and Biafran secession, the revenues received by the old Northern and Eastern regions fell (respectively from 40.7 per cent to 35.3 per cent and from 34.6 per cent to 29.3 per cent); conversely the Western State grew sharply from 24.7 to 34.9 per cent (Forrest, 1995: 22). After 1967 however the allocation process changed six times over a thirteen year period, an instability which reflects precisely the struggles over a new political order in the context of the oil boom. Indeed, revenue allocation was central to the growing conflicts between federal leaders and the Eastern Region prior to secession (over the so-called Binns allocation scheme), and in particular the manipulation of the 'rules of the game' by the northerners and westerners (Rupley, 1981). In practice, these rules have always been hotly contested. Fifteen commissions and decrees have refashioned the criteria for allocation over the period 1945 to 1996, in effect defining the relative rights of federal and local authorities to petroleum resources, and relatedly the criteria by which petroleum-based government revenues are to be allocated to the states (Furro, 1992; Ikporupko, 1996).

To simplify an enormously complex picture, prior to 1959 statutory revenue was allocated on the basis of a 'derivation principle' by which states received allocations from the federal pool in strict proportion to their contribution to these revenues (Rupley, 1981). This generally benefited northern and western regions but in the face of growing oil revenues in the 1960s they sought to change the principles of allocation. Monies to be allocated to the states came to be deposited in a Federation Account (formerly the Distributable Pool Account), the vast proportion of which was (and is) derived from oil. As the size of this account grew, new criteria were developed, largely to amend and supplant the derivation principle. By the 1960s population, need and equity principles were invoked; by the 1980s social development and internal revenue were added. In the 1990s the weighting of criteria for allocation has been as follows: population 30 per cent, equity 40 per cent, land area 10 per cent, social development 10 per cent and internal revenue 10 per cent. This new horizontal allocation system obviously privileges more populous and larger states. Hence the five oil producing states which account for 90 per cent of the oil receive 19.3 per cent of the allocated revenues (Ikporukpu, 1996: 168). Five northern non-oil producing states conversely absorb 26 per cent of allocated revenue.

Amidst the shifting sands of revenue politics and allocative criteria, several patterns are clearly evident. First, the proportion of revenues flowing to the north increased substantially from 35 per cent of the total in 1966/7 to 52 per cent in 1985. Second, the proportion of statutory revenues as a proportion of the local states' budget grew disproportionately. By 1979 state governments budgeted over 80 per cent of their revenues from federal sources (a dependency which created new competitive pressures among states to tap central oil revenues, and further deepened pressures for the creation of new states). And third, the change in the derivation principle meant that oil producing states in particular saw their share of statutory revenues fall; Bendel and Rivers States' share fell from 23.1 per cent and 17.1 per cent in 1974/5 to 6.4 per cent and 6.2 per cent respectively in 1989/90. To paint the revenue allocation picture, in short, is to depict northern hegemony in a weak and fissiparous federal system in which the oil-producing states in particular have experienced a sort of fiscal (and political) deprivation.

Running through the recent political history of petroleum in Nigeria is a strong dialectical current. Petrodollars on the one hand became the medium by which a historically frail federal center weakened regional power and manufactured a degree of political consent between center and region while simultaneously preserving northern hegemony (and reaffirming ethnic and regional loyalties). This is the heart of the Nigerian centralized patrimonial (redistributive) state (Naanen, 1995: 53). Yet on the other, petroleum revenues vastly expanded patronage politics in a way which incapacitated federal and state governments and which pulverized (to employ Guillermo O'Donnell's (1993) felicitous phrase) the social and civic life of the polity. As Khan (1994: 6) says, there is a 'vicious circle' linking the lack of effective governance to regional, ethnic and parochial interests which in turn links these special interests to the vacuum of central political authority. Unregulated patrimonialism stands as the antithesis of effective, legal–rational state action. Three decades of petroleum rewarded inefficiency, corruption and despotism[4] in a way that has torn asunder the very social, political and economic fabric of contemporary Nigeria.

A TALE OF TWO PLACES: PETROLIC DESPOTISM IN KANO AND OGONILAND

> In a paradoxical sense, it may have been the very promise of actual or even future access to oil wealth which has, to a very large extent, kept the various separatist, conflictual and destructive influences in Nigeria at bay. In other words, while the promise of the piece of the 'oil' pie

> keeps Nigeria together as a nation, the very nature of its distribution has destroyed the social, economic and political fabric from within.
>
> (Sara Ahmad Khan, 1994: 8)

An independent Nigeria inherited in 1960 an archetypical agrarian export economy dominated by groundnuts, cotton, palm oil, cocoa, and rubber. Starkly regional in character, the economy consisted of three semi-autonomous regions, each associated with a primary export commodity and a marketing board. This tripod structure, tightly circumscribed the power of the federal center. But the advent of a turbo-charged petrolic capitalism reconfigured Nigerian national space. By 1980 the country had become a monocultural economy, more so than it had ever been during sixty years of colonial rule. Oil moreover unleashed an unprecedented boom. Federal revenues grew at 26 per cent per annum during the 1970s and expanded state activity (it invested heavily in industrial development and infrastructure and manufacturing output increased by 13 per cent per annum between 1970 and 1980) unleashed a torrent of imports (capital goods increased from N422 million in 1971 to N3.6 billion in 1979) and urban construction (the construction industry grew at over 20 per cent per annum in the mid 1970s). Average annual growth rates for credit, money supply and state expenditures were stunning: respectively 45 per cent, 66 per cent and 91 per cent between 1973 and 1980! Nigerian merchandise imports increased from N1.1 billion in 1973 to N14.6 billion in 1981. The proliferation of everything from stallions to stereos produced a commodity fetishism that approximated to what one commentator aptly described as a Nigerian 'cargo cult'. Petroleum ushered in an era of public investment which radically transfigured the built landscape and deepened what one might call the institutions of modernity (hospitals, schools, roads, electrification and so on). Cities such as Warri, Port Harcourt and Lagos doubled (and in some cases tripled) in size during the boom. Consumer prices leapt upward, agriculture collapsed and the real exchange rate rose steadily, further feeding the import boom.

A revenue and investment boom of this sort occurring under highly centralized (state) auspices and framed by a highly competitive regionalized polity, vastly expanded the opportunities for patronage and clientelism that had been the hallmark of Nigerian politics throughout the colonial period. Novelist Achinua Achebe in *The Trouble With Nigeria* put the matter starkly: 'Nigeria is without shadow of a doubt one of the most corrupt nations in the world' (1983: 42). Not only did Nigerian public servants become 'more reckless and blatant' (ibid.: 43) as oil revenues continued to roll in, but in addition there was a conspicuous failure to apprehend, prosecute and punish perpetrators. *The Economist* described the culture of corruption this way:

> Contracts might be perfectly innocent ... but from the first, money told. Contracts, by convention, were inflated ... by an item called 'public relations'. This extra amount was divided in an orderly way among the officials or ministers ... In the early years it was all quite

> gentlemanly. But [with oil] the pay offs grew ... to involve literally millions on big construction contracts. The quality of work was very often skimped to allow for the payments. The beauty of it was that everybody could bribe or steal without feeling particularly dishonest ... Theft from the state brought mysterious largesse to extended families, to towns, to local political factions.
>
> (May 3rd, 1986: 18)

During the Second Republic (1979–83), the second oil boom in 1979 unleashed a frenzied feeding on public resources in which the state was mercilessly pillaged. Nigeria 'lost' $16.7 billion in oil income ('Oilgate' so-called) due to fraudulent activities and smuggling of petroleum between 1979 and 1983 (*New African*, April 1984: 11); $2 billion (over 10 per cent of GDP) was 'discovered' in a private Swiss bank account; and government ministries regularly went up in flames, the product of arson immediately prior to federal audits. Special military tribunals set up after the December 1983 coup, prosecuted governors and high ranking politicians for spectacular feats of corruption. Governor Lar accumulated N32.6 million in four years; Governor Attah made N2 million through illegal activities associated with a single state security vote in the Kwara state legislature (*West Africa*, July 2nd, 1984: 1373; July 23rd: 1511). Military officers who raided the home of the Governor of Kano State, Bakin Zuwo, discovered millions of naira in cardboard boxes piled up in his bedroom. What Czar Nicholas II once said of Romania could apply to Nigeria: 'It's not a country it's a profession.'

Fifteen years later, however, the luster of the oil boom had tarnished, and Nigeria's economic future appeared by contrast to be quite bleak, if not altogether austere. The collapse of the boom in 1981 exposed Nigeria to the structural weaknesses of oil-based rentier capitalism. The foreign exchange crisis – the visible trade balance in 1981 was N12 billion – was compounded by mounting external debt obligations. By 1982 the President and his advisers talked of the need for sacrifice and denial; the 1970s, they said, had been a time of illusion. They were promptly ousted in a military coup in 1983. By 1984 the 'boom was over' and the watchword was austerity. By 1986 Nigeria had signed a structural adjustment program (SAP) with the IMF and the World Bank and the medicine was bitter. The economy contracted, the naira collapsed from $1.12 to ten cents, and the real wage of industrial workers was savaged (Lewis, 1996). Money was scarce and popular discontent widespread, especially in the anti-SAP riots in 1988 and 1989. In the 1990s debt service ran on average at 25 per cent of export revenues, the naira collapsed in value, inflation was 57 per cent per annum over the period 1992–1995. Manufacturing output fell from an index of 100 in 1990 to 89 in 1995; average industrial capacity stood at 29 per cent (*West Africa*, January 9th, 1995: 15). The 'oil fortress' had, as *Le Monde* put it (October 2nd, 1994: 15), been rocked. As *The Economist* (June 8th, 1996: 48) recently noted: 'People complain fiercely, even by Nigerian standards, that their wealth and their future have been stolen'.[5] At least half the population lives below the poverty line and almost one-third of children die before the age of five.

What began with petro-euphoria and bountiful money in 1973 ended two decades later with scarcity, a huge debt, the deepening of economic inequalities, urban looting and bodies in the streets. Oil had vastly increased the national appetite and the capacity to consume and yet ingesting petroleum only served to contaminate everything. This roller-coaster economy, whose volatility was apparent even by the mid-1970s, had fundamentally shaped the everyday life of all sectors of Nigerian society. Seemingly for the worst.[6]

In short, oil revenues had engendered a striking contradiction within the Nigerian polity. On the one hand it became the singular means by which a military government purchased a degree of consent from the regions which preserved the fiction of Nigeria as a national entity. On the other, it created a Hobbesian state at the expense of both civil society which it systematically destroyed and of organized capitalist growth which it effectively blocked. How, then, were these contradictions displayed and experienced in Kano and Ogoniland?

Kano

A bustling, kinetic city forged in the crucible of seventeenth- and eighteenth-century trans-Saharan trade, Kano is the economic fulcrum of Nigeria's most populous state, a densely settled agricultural landscape of well over ten million people (see Figure 2.1). An ancient city whose iron workings date back to the seventh century, Kano was by the mid-sixteenth century comparable in size and significance to Cairo and Fez, and by the eighteenth century an Islamic city of international repute. In the course of sixty years of British colonial rule (1902–60), Kano emerged as West Africa's pre-eminent entrepot, a self-consciously Muslim city whose prospects were wrapped up with the fortunes of one world commodity – the peanut – used in the manufacture of vegetable oil. But it was oil of an altogether different sort that ushered in a new, and in some respects a more radical, integration of the Nigerian polity and economy into the world market in the wake of the first oil boom in 1973.

Awash in petrodollars, urban Kano was transformed during the 1970s from a traditional Muslim mercantile center of some 400,000 at the end of the Civil War (ca. 1970), into a sprawling, anarchic metropolis of over 1.7 million by 1980. Seemingly overnight Kano emerged as a fully fledged industrial periphery. The heart of Kano, the old walled city (*birni*), was engulfed by the new suburbs sprouting up outside the walls (*waje*), a process which suggested an important erosion in the autonomy and stature of the traditional Hausa core of the city.[7] At the zenith of the petroleum boom, new industrial estates sprang up at Challawa and Sharada in the city periphery, armies of migrants poured into the city, and the icons of modernity – the massive state-sponsored building and infrastructural projects – dotted the city skyline.[8] Kano was, on the one hand, a huge construction site and, on the other, a theater of wildly unregulated consumption. Murray Last has captured this ethos perfectly: Kano had become, he says, a modern city in which 'dishes for satellite television dominate the compounds of the rich whose wealth is growing ever more disproportionate as austerity ... hits the poor. A watch for $17,000; a house for $2 million; private jets to London or the Friday prayer at Mecca – this is the style to which Kano's guilded youth are the heirs' (1991: 4).

According to a World Bank study by the early 1980s, after a decade of oil prosperity, 52–67 per cent of Kano's urban population existed at the 'absolute poverty level'. This amorphous class embracing small traders, workers, informal sector workers, and the unemployed was (and remains) diverse but possessed a social unity in terms of a popular self-identity in Hausa society as commoners or *talakawa*. A *talaka* is 'a person who holds no official position ... a man in the street ... a poor person' (Bargery, 1934). As an indigenous social category it is of considerable antiquity, emerging from the social division of labor between town and countryside associated with the genesis of political kingdoms (the *sarauta* system) in the fifteenth century, and subsequently the emirate system under the Sokoto Caliphate (1806–1902), in which a lineage-based office holding class (*masu sarauta*) exercised political authority over subject populations. *Talakawa* refers, then, to a class relationship of a pre-capitalist sort but also a political relationship among status honor-groups with distinctive cultural identities and lifestyles. Naturally, the *talakawa* have been differentiated in all sorts of complex ways through the unevenness of proletarianization – the industrial working class (*leburori*) for example constituting an important social segment of the *talakawa* as such.

If the *talakawa* as a class category (and the *'yan cin rani* and *gardi* as segments of it) represent a key set of structural pre-conditions through which many Kano residents experienced the oil boom, then what were the immediate or proximate qualities of that experience in urban Kano? I shall focus on three sets of social relations: state mediation in the form of corruption, violent but undisciplined security forces, and urban social processes. I have already referred to the growing presence of the local and federal states in social and economic life. To the extent that the state was rendered much more visible in society by virtue of its expanded activities including road and office construction, contracting of various sorts, its visibility became synonymous with a total lack of moral responsibility and state accountability. In popular discourse, the state and government meant massive corruption on a Hobbesian scale. Graft through local contracts, import speculation, foreign exchange dealings, drug smuggling, hoarding of food and so on abounded among all levels of local and federal officialdom. It was commonplace to hear the governor referred to as 'a thief', or that '*Nigeriya ta lalachi*' ('Nigeria has been ruined'). State ownership even implicated the Muslim brotherhoods who owned shares in companies. The government was also centrally involved in the annual migration of 100,000 pilgrims who participated in the *hadj*, a sacred event which was widely associated with corruption and conspicuous consumption. In the return to civilian rule, state corruption reached unprecedented levels; as the *talakawa* put it, 'siyasa ta bata duniya' – 'politics has spoiled the world'.

To put the matter bluntly, state mediation of the oil boom in Kano meant corruption, chaos and bureaucratic indiscipline. The *talakawa* were systematically excluded from access to the state which they experienced as morally bankrupt, illegitimate and incompetent. The police, who had been placed under federal jurisdiction, and the internal security forces were widely held to be particularly corrupt, disorganized and violent; they embodied the moral and political degeneracy of state legitimacy. In the context of rising urban crime, it was the police who proved to be the trigger for all sorts of community violence; indeed, they were uniformly feared and loathed by Kano's urban poor. In the popular imagination the police were feared and were explicitly referred to as *daggal* – literally 'the devil'.

At another level, petrolic modernization was refracted through the lens of the urban community. The central issue here is not simply the anarchic and chaotic growth of Kano city – a terrifying prospect in itself – but the changing material basis of *talakawa* life and what it implied for the brute realities of urban living. First, urban land became a source of speculation for Kano merchants and civil merchants, reflected in the fact that the price of urban plots in the working class Tudun Wada neighborhood increased twenty times between 1970 and 1978. Land records invariably disappeared (usually through mysterious fires) and compensation for land appropriated by the state was arbitrary and a source of recurrent conflict. Second, the escalation of food prices, typically in the context of price rigging, hoarding and licensing scandals, far outstripped the growth of urban wages. The inflationary spiral in wage goods went hand in hand with the internationalization of consumption by Kano's elites – car ownership, for instance, grew by 700 per cent in six years – and by the erosion of many of the traditional occupations within the secondary labor market taken by *gardawa* and *'yan cin rani*. Kano stood at the confluence of rampant capitalist modernity and a chaotic, ramshackle state (Watts, 1996).

Ogoniland[9]

If Kano, some one thousand kilometers distant from the oil rich Niger Delta, was irresistibly drawn into the maelstrom of petrolic modernization, the oil producing communities of Ogoniland in Rivers State stood, rather curiously, at its margin. Indeed, the paradox of Ogoniland is that an accident of geological history – the location of more than ten major oilfields within its historic territory – conferred upon the Ogonis little more than massive ecological destruction and profound economic backwardness. As one of the oil representatives from British Petroleum put it: 'I have explored for oil in Venezuela ... in Kuwait [but] I have never seen an oil-rich town as completely impoverished as Oloibiri [where Shell first found oil in 1958]' (*Village Voice*, November 21st, 1995: 21). If Maitatsine railed at the flotsom and jetsom – the commodity detritus – of petrolic modernization and the moral decay which was its handmaiden, Saro-Wiwa was animated by its very absence. If some of the Islamicists took exception to the artifacts of modernity, Saro-Wiwa and the Ogoni hated modernity because *they could not get enough of it*. If Maitatsine seemed to lament the displacement of traditional transportation by the automobile, the Ogoni were angry because they could afford neither the cars nor use the roads which were the icons of petrolic success.

The Ogoni are a distinct ethnic group, consisting of three sub-groups and six clans.[10] Their population of roughly 500,000 people is distributed among 111 villages dotted over 404 square miles of creeks, waterways and tropical forest in the northeast fringes of the Niger Delta (Figure 2.3). Located administratively in Rivers State, a Louisiana-like territory of some 50,000 square kilometers, Ogoniland is one the most heavily populated zones in all of Africa. Indeed the most densely settled areas of Ogoniland – over 1500 persons per square km. – are the sites of the largest wells. Its customary productive base was provided by fishing and agricultural pursuits until the discovery of petroleum, including the huge Bomu field, immediately prior to independence. Part of an enormously complex regional ethnic mosaic (Figure 2.4), the Ogoni were drawn into internecine conflicts within the delta region – largely as a consequence of the slave trade and its aftermath – in the period prior to arrival of colonial forces at Kano in 1901. The Ogoni resisted the British until 1908 (Naanen, 1995) but thereafter were left to stagnate as part of the Opopo Division within Calabar Province. As Ogoniland was gradually incorporated during the 1930s, the clamor for a separate political division grew at the hands of the first pan-Ogoni organization, the Ogoni Central Union, which bore fruit with the establishment of the Ogoni Native Authority in 1947. In 1951, however, the authority was forcibly integrated into the Eastern Region. Experiencing tremendous neglect and discrimination (Okpu, 1977), integration raised longstanding fears among the Ogoni of Ibo domination. As constitutional preparations were made for the transition to home rule, non-Igbo minorities throughout the Eastern Region appealed to the colonial government for a separate Rivers State. Ogoni representatives lobbied the Willink Commission in 1958 to avert the threat of exclusion within an Ibo dominated regional government which had assumed self-governing status in 1957. Minority claims were ignored, however, and instead the newly independent Nigerian government passed the Niger Development Board Act in 1961 to look into the special problems of the delta and to improve social services and infrastructure. But as Okilo notes (1980: 14), the federal government in keeping with its history of bad faith toward the delta, 'failed to develop one square inch of the territory'.

Politically marginalized and economically neglected, the delta minorities feared the growing secessionist rhetoric of the Ibo and consequently led an ill-fated secession of their own in February 1966. Isaac Boro, Sam Owonaro and Nottingham Dick declared an illegal Delta People's Republic but were crushed and were subsequently, in a trial which is only too reminiscent of the

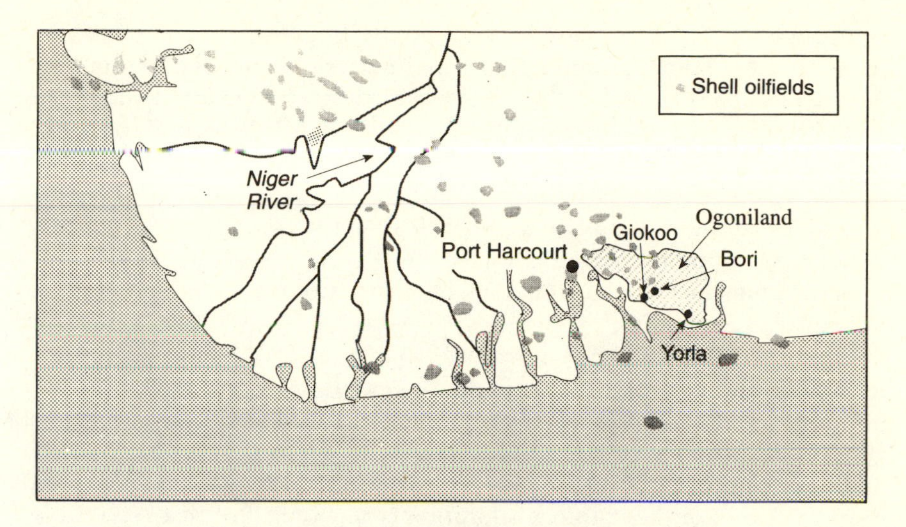

Figure 2.3 Location of Ogoniland and Shell oilfields

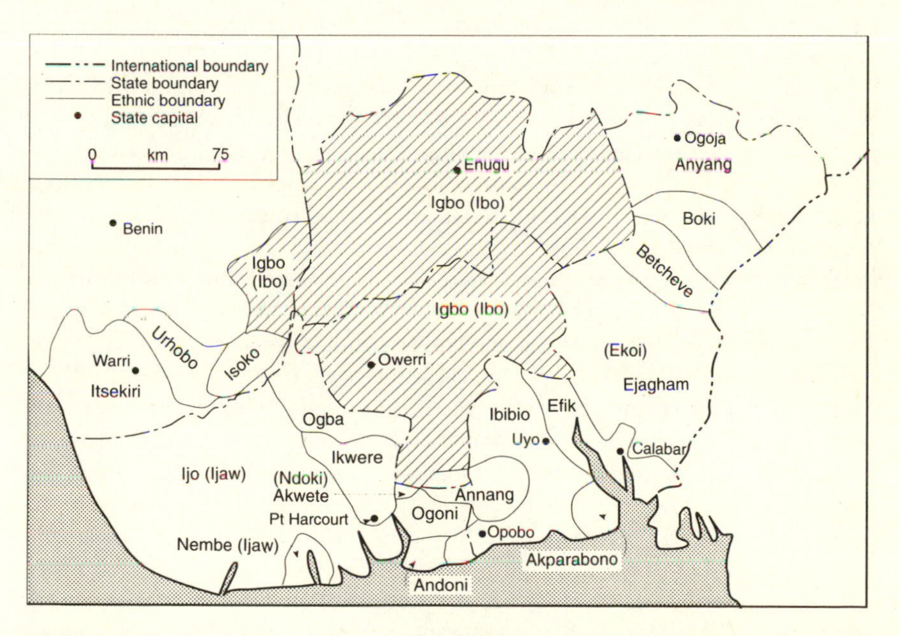

Figure 2.4 Ethnic geography of Eastern Nigeria (Rivers, Delta, Imo, Abia, Cross-River and Akwa-Ibom States)
Source: Saro-Wiwa, 1992: 13

Ogoni tribunal in 1995, condemned to death for treason. Nonetheless, Ogoni antipathy to what they saw as a sort of internal colonialism at the hands of the Ibo, continued in their support of the federal forces during the civil war. While Gowon did indeed finally establish a Rivers State in 1967 – which compensated in some measure for enormous Ogoni losses during the war – the new state recapitulated in microcosm the larger 'national question'. The new Rivers State was multi-ethnic but presided over by the locally dominant Ijaw, for whom the minorities felt little but contempt.[11] In Saro-Wiwa's view (1992), the loss of 10 per cent of the Ogoni people in the civil war was ultimately for nought as federal authorities provided no post-war relief, seized new onshore and offshore oil fields, and subsequently sold out the minorities to dominant Ijaw interests.

Something like Ogoni nationalism, long predated the oil boom but it was deepened as a result of it. Ogoni fears of what Saro-Wiwa calls 'monstrous domestic colonialism' (1992), were exacerbated further by federal resistance to dealing with minority issues[12] in the wake of the civil war and by the new politics of post-oil boom revenue allocation. Rivers State saw its federal allocation fall dramatically in absolute and relative terms. At the height of the oil boom, 60 per cent of oil production came from Rivers State but it received only 5 per cent of the statutory allocation (roughly half of that received by Kano, Northeastern States and the Ibo heartland, East Central State). Between 1970 and 1980 it received in revenues one-fiftieth of the value of the oil it produced. In what was seen by the Rivers minorities as a particularly egregious case of ethnic treachery, the civilian Shagari regime reduced the derivation component to only 2 per cent of revenues in 1982, after Rivers State had voted overwhelmingly for Shagari's northern dominated National Party of Nigeria. The subsequent military government of General Buhari cut the derivation component even further at a time when the state accounted for 44.3 per cent of Nigeria's oil production. In 1992 the Oil Mineral Producing Areas Commission (OMPADEC) was established to develop projects in oil producing communities (funded by 1.5 per cent of oil revenues) but it has been marred by corruption, inefficiency and fierce conflicts among oil producing communities (Osaghae, 1995: 333; Welch, 1995).

Standing at the margin of the margin, Ogoniland appears (like Chiapas in Mexico) as a socio-economic paradox. Home to six oilfields, half of Nigeria's oil refineries, the country's only fertilizer plant, a large petrochemical plant, Ogoniland is wracked by unthinkable misery and deprivation. During the first oil boom Ogoniland's fifty-six wells accounted for almost 15 per cent of Nigerian oil production[13] and in the past three decades an estimated $30 billion in petroleum revenues have flowed from this Lilliputian territory; it was, as local opinion had it, Nigeria's Kuwait. Yet according to a government commission, Oloibiri, where the first oil was pumped in 1958, has no single kilometer of all-season road and remains 'one of the most backward areas in the country' (cited in Furro, 1992: 282). Few Ogoni households have electricity, there is one doctor per 100,000 people, child mortality rates are the highest in the nation, unemployment is 85 per cent, 80 per cent of the population is illiterate and close to half of Ogoni youth have left the region in search of work. Life expectancy is barely fifty years, substantially below the national average. In Furro's survey of two minority oil producing communities, over 80 per cent of respondents felt that economic conditions

Table 2.2 Reasons for the neglect of the oil producing areas in Rivers State

Item	*Obagi*		*Obrikom*		*Oboburu*		*Port Harcourt*	
	Yes	*No*	*Yes*	*No*	*Yes*	*No*	*Yes*	*No*
Ethnic/tribal factors	23.5	75.5	89.5	10.5	—	100	40.5	59.5
Lack of representation	83.2	11.8	100	—	72.2	27.8	45.9	54.1
Minority status	53.8	41.2	100	—	33.3	66.7	54.1	45.9
Religious considerations	11.8	85.2	5.3	94.7	—	100	8.1	91.9

Source: Furro, 1992: 313

had deteriorated since the onset of oil production, and over two-thirds believed that there had been no progress in local development since 1960. No wonder that the systematic reduction of federal allocations and the lack of concern by the Rivers government was, for Ogoniland, part of a long history of 'the politics of minority suffocation' (cited in Ikporukpo, 1996: 171). Petroleum was, in the local vernacular, being pumped from the veins of the Ogoni people. Systematic neglect not only deepened the sense of possession over local oil resources – 'Leave us our oil for us . . . Rivers State oil for Rivers State only' read an editorial in *The African Guardian* (1991: 12) – but affirmed the belief that exclusion was primarily a result of ethnic and minority discrimination (Table 2.2).

If Ogoniland failed to see the material benefits from oil, what it *did* experience was its devastating ecological costs – what the European Parliament has called 'an environmental nightmare'.[14] The heart of the ecological harm stems from oil spills – either from the pipelines which criss-cross Ogoniland (often passing directly through villages) or from blowouts at the wellheads – and gas flaring. As regards the latter, a staggering 76 per cent of natural gas in the oil producing areas is flared (compared to 0.6 per cent in the US). As a visiting environmentalist noted in 1993 in the delta, 'some children have never known a dark night even though they have no electricity' (*Village Voice* Nov. 21st, 1995: 21). Burning 24 hours per day at temperatures of 13–14,000 degrees Celsius, Nigerian natural gas produces 35 million tons of CO2 and 12 million tons of methane, more than the rest of the world (and rendering Nigeria probably the biggest single cause of global warming). The oil spillage record is even worse. According to Claude Ake there are roughly 300 spills per year in the delta and in the 1970s alone the spillage was four times than the much publicized *Exxon Valdez* spill in Alaska (Table 2.3). In one year alone almost 700,000 barrels were soiled according to a government commission. Ogoniland itself suffered 111 spills between 1985 and 1994 (Hammer, 1996: 61). Figures provided by the NNPC document 2,676 spills between 1976 and 1990, 59 per cent of which occurred in Rivers State (Ikein, 1990: 171), 38 per cent of which were due to equipment malfunction.[15] Between 1982 and 1992 Shell alone accounted for 1.6 million gallons of spilled oil, 37 per cent of the company's spills worldwide.

Table 2.3 Spillage of crude oil in Nigeria, 1972–1979

Year	*Crude oil production*	*Recorded number of spills*	*Approximate quantity unrecovered spills oil (Billion barrels)*
1972	665,293,292	5	39,000
1973	719,376,760	7	2,619
1974	823,317,843	24	23,368
1975	660,146,040	16	3,544
1976	758,055,728	26	3,133
1977	766,052,636	61	4,374
1978	597,719,241	143	101,211
1979	513,193,653 (July)	71 (October)	638,235

Source: Report of the Presidential Commission on Revenue Allocation, Vol. III, Apapa, Federal Government of Nigeria, 1980: 400

The consequences of flaring, spillage and waste for Ogoni fisheries and farming have been devastating. A recent spill in 1993 flowed for 40 days without repair, contaminating large areas of Ogoni farmland. Petroleum residues appear in the rivers at levels of 60 ppm and in the sediments around the Bonny terminal they reach lethal levels of 12,000 ppm. In the ecologically delicate mangrove and estuarine regions of the delta, oil pollution has produced large scale eutrophication, depletion of aquatic resources and loss of traditional fishing grounds (see NEST, 1991: 44; Benka-Cocker and Ekundayo, 1995) which now threaten customary livelihoods. The loss of farmland, dating back to massive blowouts at Bomu in 1970 and in Ibobu in 1973, has further eroded the subsistence capabilities of Ogoni communities[16] (*Newswatch*, Dec. 18th, 1995: 10). Indeed, it is the direct threat to the means of subsistence by petrolic destruction that led Saro-Wiwa to talk of Ogoni genocide.

In almost four decades of oil drilling, then, the experience of petrolic modernization in Ogoniland has been a tale of terror and tears. It had brought home the worst fears of ethnic marginalization and minority neglect: of northern hegemony, of Ibo neglect, and of Ijaw local dominance. The euphoria of oil wealth after the civil war has brought ecological catastrophe, social deprivation, political marginalization, and a rapacious company capitalism in which unaccountable foreign transnationals are granted a sort of immunity by the state.

THE MORAL ECONOMY OF ISLAM AND ETHNIC COMMUNALISM: COMMUNITY, IDENTITY, RESISTANCE

Money, modernism, millenarianism: the Maitatsine insurrection[17]

[W]hat the 'ulama [divines] are doing is to attempt a definition of orthodoxy – a (re)ordering of knowledge that governs the 'correct'

> form of Islamic practices ... This like all practical criticism seeks to construct a relation of discursive dominance ... [But] 'Orthodoxy' is not easy to secure in conditions of radical change.
>
> (Talal Asad, 1993)

The body of a self-proclaimed Muslim prophet, Maitatsine, is exhumed from a shallow grave near Rigiyar Zaki village on the outskirts of Kano, a massive, sprawling metropolitan area in the commercial and religious heart of Hausa-speaking northern Nigeria. Concerned to prevent any sort of martyrdom after ten days of insurrectionary struggle between the prophet's followers ('*Yantatsine*) and the Nigerian military, the authorities embarked upon a campaign to discredit the movement as the brainchild of a psychopath, a 'fanatic' and 'heretic'. According to one Muslim intellectual, Maitatsine was 'witchcraft married to cannibalism' (Yusuf, 1988). He was, in fact, a brilliant Qu'ranic student schooled in the science of Qu'ranic exegesis, and had lived and taught in Kano city for almost twenty years in the 'Yan Awaki quarter. However, throughout the 1970s his public teaching became increasingly idiosyncratic and unorthodox; anti-modernist and syncretic in style, damning all those who read any text other than the Qu'ran, who carried money, who rode on bicycles or in cars, who smoked cigarettes, wore buttons or watched TV. Highly critical of established Sunni practice, Maitatsine vilified the local Muslim clergy and gradually built a community of followers within the traditional walled city of Kano. Radical and literalist in thrust, Maitatsine articulated an alternative Muslim project. By 1980 he had mobilized some 10,000 foot soldiers to advance his millenarian vision.

Maitatsine's community lived in a sort of liminal space between the old and the new Kano, a transition zone of motor parks, markets, cinemas and low income houses, a sort of urban jungle. Followers and students scavenged for work and alms in this informal sector, living in makeshift dwellings, preaching and begging at major intersections. By 1977 the followers were increasingly visible, vigorously attacking corrupt and unjust leaders and clergy (*ulema*), and more generally denouncing materialism. Several clashes between 1974 and 1979 had escalated tensions between Maitatsine and the police. On December 18th, 1980, four police units were sent to arrest some of the preachers after rumors of conflagrations and attacks on the central mosque. Disorganized police forces were attacked by Maitatsine's followers armed with bows and arrows and daggers. Police arms were seized, vehicles burned and by late afternoon a plume of smoke hung over the city. Amidst growing chaos and confusion, the police were unable to control the situation, and by December 22nd large numbers of Maitatsine supporters were reportedly entering Kano while trucks full of corpses were leaving the city. After five days of escalating violence, the Nigerian army intervened on December 29th with ten hours of mortar barrage. According to the official tribunal 4,177 were killed but the figure is a huge underestimate; 15,000 were injured and 100,000 rendered homeless. Maitatsine was killed in the fighting and buried by his followers as they retreated from the city. Close to 1,000 of the *'yantatsine* were subsequently arrested and imprisoned.

The *'yantatsine* constituted a sort of archetypical Fanonite class. For the most part relatively recent migrants to Kano (*'yan cin rani* and *gardi* in Hausa), drawn into the city through longstanding Qu'ranic networks by the urban construction boom unleashed by state invested petrodollars. They brought with them a sort of jacobin or populist reading of Islam acquired through a vital rural Muslim educational system, and an experience of social relations rooted in the moral economy of the peasantry. Their consciousness was very much that of the 'commoner' class, *talaka* in Hausa, a complex and dense term which originated in relation to the genesis and growth of the tributary Hausa states, and subsequently the Sokoto Caliphate, in the eighteenth century (Watts, 1983; Lubeck, 1985). But the 'yantatsine's experience of urban life and of proletarianization was directly mediated by the particular form of petroleum-capitalism that I sketched previously. Corruption, illicit gain of magic money, rapid inflation, escalating land prices and rentals, and a sort of city life that combines commodity fetishism (the consumer boom of the 1970s) with Hobbesian anarchy. The state becomes *the* expression of corruption and moral degeneracy; its local representatives, the police, were despised as described, as *daggal*, literally 'the devil'.

The historic role of Maitatsine was, in a sense, to interpret this experience in cultural terms in ways which resonated with everyday life. Maitatsine articulated a particular experience through what one might call the local hermeneutics of Islam. Rather than seeing Islam as prescriptive in simple ways, it is a text based religion which is made socially relevant through enunciation, performance, citation, reading and interpretation. It is this interpretive and dialogic tradition within Islam which points to how texts are used, by whom, and with what authority (Fischer and Abedi, 1990; Zabaida, 1995). Maitatsine drew upon a legitimate Muslim tradition at a moment when Islam as a whole was embroiled in debates over modernity and change. In Nigeria, as elsewhere in the Muslim world, Islam is not a monolith but contains important institutional, ideological and social organization tensions within its circumference. There have been struggles between and against the Sufi brotherhoods in Nigeria since the 1950s, and the effort by Ahmadu Bello to create the *Jama'atu Nasril Islam* (the Organization for the Victory of Islam) in 1962 can be understood as an attempt to unite a *divided* Muslim community as a way of securing northern political hegemony in the newly independent Nigerian federation. But over the past thirty years the fissiparous nature of the Muslim community has deepened. A primary axis of difference and debate links the Darika brotherhoods – Quadriyya and Tijaniyya – to which many Nigerian Muslims are aligned, and newer so-called fundamentalist groups such as the powerful *Jama'atu Izalat Al-Bidah Wa Iqamat Al Sunnah* or Izala (roughly translated as the Removal of False Innovations and the Establishment of Orthodox Tradition) headed by Abubakar Gumi with strong connections to King Faisal of Saudi Arabia, and the Muslim Students Society itself split between pro-Iranian and pro-Saudi factions (Ibrahim, 1991). These local divisions and debates are not unrelated to the extraordinary ferment throughout the Muslim world during the last two decades and as Muslims, in other words, 'their differences are fought out on the ground of [orthodoxy]' (Asad, 1993: 210) in an attempt to establish a discursive dominance.

In light of my emphasis on flexibility, diversity and moral economy, how might one characterize the Muslim landscape in northern Nigeria which existed on the eve of the oil boom when Maitatsine was already active within Kano? I want to draw attention to the shifting influence of the Sufi brotherhoods, most especially the Qadriyya which is associated with the holy war (*jihad*) of Usman dan Fodio in the early nineteenth century and more generally with the traditional northern Nigerian Hausa-Fulani aristocracy, and the Tijaniyya which had grown and flourished in the fertile soil of colonial politics and merchant-capitalism. Unlike the Qadriyya which was part of a class alliance between the local Muslim ruling class (*sarauta*) and the colonial state, Tijaniyya contained specific radical, anti-colonial beliefs which appealed to a traditional merchant class operating in the interstices of the colonial economy (Paden, 1973). Since the 1960s, however, the hegemony of both brotherhoods has been contested by a number of anti-Sufi reformist sects in northern Nigeria – notably the Izala founded in 1978 by Mallam Idris in Jos in central Nigeria – whose aim is to abolish innovation and to practice Islam in strict accordance with the Qu'ran and the *Sunna*. Anti-Sufi sentiment in Nigeria can be traced to the 1930s but proliferated during and after the Second World War with the growth of a class of new elites who articulated the virtues of the modern social transformations in contra-distinction to the vices of tradition (Umar, 1993). The anti-Sufi movement has accelerated largely through the leadership of Sheikh Abubakar Gumi who has employed the Nigerian Broadcasting Corporation and the Kaduna-based Hausa language newspaper *Gaskiya ta fi Kwabo* to great effect in promoting his message that Sufism is not part of Islam because it emerged long after Islam had been completed. Referred to as the new jihadists, the doctrines of Izala are identical with Gumi's views but while critical of the West's decadence they are neither anti-modern nor anti-materialist in any simple sense, and hence are widely supported by northern Nigerian intelligentsia and by aggressive new modernists and businessmen. In some senses, Izala ideology 'is to the Islamic brotherhoods what Protestantism is to Catholicism' (Gregoire, 1993: 114). But it is also a form of protest against traditional non-capitalist values and against the perceived corruption of religious values and practices in the context of what Umar (1993: 178) calls a 'shift from communal to an individual mode of religiosity . . . in tune with the rugged individualism of capitalist social relations'. Whether Izala is simply a cathartic response to the tensions and frustration of Nigerian capitalism[18] is less relevant for my concerns than the fact its *tafsir* broadcasts and its polemical pleas for the separation of mosques contributed to a growing sense of cleavage and division within Nigerian Islam, and to the wider legitimacy of anti-Sufi sentiment. Indeed, the reappearance of Maitatsine (and related movements) in Maiduguri, Yola and other cities in the north during the 1980s (see Figure 2.1) pays eloquent testimony to the continuity of this debate and cleavage.

To put the matter succinctly, what seems like a fundamentalist irruption led by a fanatical heretic is more about a particular class, and cultural, reading of petrolic despotism. That is to say, a reading of the speeding up of modern life and an acceleration of commodities in circulation against a backdrop of wide ranging debates within Islam over modernity itself. The appeal of Marwa, and the refashioning of local identity that he represented, must be located in a series of complex articulations: between Islam and capitalism, between pre-capitalist and capitalist institutions, and between class and culture. It would be much too facile to see the millenarian qualities of the movement as a lumpen insurrection plain and simple. The *'yantatsine* were uniformly poor; 80 per cent had income well below the minimum wage. But the 'fanatical' qualities of the social explosions which occurred between 1980 and 1985 can only be fully comprehended in terms of the material and status deprivation of *'yantatsine* recruits scrambling to survive in an increasingly chaotic and Hobbesian urban environment, and the unprecedented ill-gotten wealth and corruption of the dominant classes in urban Kano.

Marwa was, first of all, a long time resident of Kano and was witness to the extraordinary transformation in the political and cultural economy of urban Kano during the oil years. As a charismatic preacher with a compelling, if idiosyncratic, reading of the Qu'ran, Marwa recruited followers from the influx of migrants and students into the city and from the marginal underclass of Kano. His disciples enlisted at the lorry parks and railway stations in Kano where they typically sustained themselves by selling tea and bread (Federal Government of Nigeria, 1981). His followers (migrants, itinerant workers, gardawa) were products of the same Qu'ranic system as Marwa himself; most of his Kano followers were educated in the *makarantar allo* system. Furthermore, Marwa's use of syncretist and pre-Islamic powers resonated strongly with migrants from the countryside where vestiges of ancient Hausa metaphysical belief remained quite influential. Slowly Marwa was able to build up an enclave in 'Yan Awaki fashioned around a disciplined, and self-consciously austere, egalitarian community of Muslim brothers (*'yan'uwa*) who supported themselves largely through alms though land and urban gardens were appropriated to support the devotees.[19] Marwa's unorthodox and literalist interpretation of the Qu'ran focused specifically on the icons of modernity: bicycles, watches, cars, money and so on. But it would be mistaken to view Maitatsine as anti-modern or simply harking back to some earlier uncontaminated tradition of Islam. The movement employed modern arms when necessary and Marwa emphasized the ill-gotten quality of goods not their inherent illegitimacy. Likewise his desire to seize key institutions in Kano city – the radio, the state electricity company – is hardly anti-modernist. Kano merchants, bureaucrats and elites were implicated in his critique,[20] but it was the state that was the embodiment of moral bankruptcy, and the police its quintessential representatives. Any affiliation to the state naturally contaminated Islamic practice.

In this way, one can begin to see why an idiosyncratic Muslim movement represented such a threat to the newly elected Shagari government which itself represented Northern Muslim interests. Maitatsine attempted, unsuccessfully to create a new orthodoxy in times of radical, oil-induced change but in doing so he raised questions which resonated among large sections of a complex and fissured Northern Muslim community. His assault on the moral decay of Islam was not only an attack on a notoriously corrupt government, but on the ruling Muslim elites who had overseen the decay of the moral fabric of society at large. Not only did his critique have wide appeal within the Muslim community – as the activities of the Izala suggest – but his ragtag movement represented a potential wedge within the northern ruling bloc and an appeal to civil society as the repository of justice and morality. In responding to the challenge, state authorities had to deny that Maitatsine and his followers were Muslim at all and that his self-proclaimed prophethood was illegitimate. In this sense, Islamism – in whatever form – presented a political crisis for the northern ruling elites, suggesting new tensions and fractures within the old bloc (*Africa Confidential*, September 9th, 1994: 1–2). Maitatsine held a mirror to the waning power and cohesiveness of the postcolonial northern order. The ruthless way in which he was killed – and his followers imprisoned – was the very measure of President Shagari's tenuous hold over a crumbling political regime.

Drilling fields: oil, self-determination and the minority question in Ogoniland

> On the oilfields of Ogoni
> exploring black gold efficiently
> is the maxim of the mad moghul
> and human life but cold obstacle
> in the frenzied path of a search
> guided by blood and flow charts
>
> (Ken Saro-Wiwa, no date)

The hanging of the Ogoni nine in November 1995 – accused of murdering four prominent Ogoni leaders who professed opposition to MOSOP tactics – and the subsequent arrest of nineteen others on treason charges, represented the summit of a process of mass mobilization and radical militancy which had commenced in 1989. The civil war had, as I have previously suggested, hardened the sense of external dominance among Ogonis. A 'supreme cultural organization' called Kagote which consisted largely of traditional rulers and high ranking functionaries, was established at the war's end and in turn gave birth in 1990 to MOSOP. A new strategic phase began in 1989 with a program of mass action and passive resistance on the one hand (the language is from MOSOP's first President, Garrick Leton) and a renewed effort to focus on the environmental consequences of oil (and Shell's role in particular) and on group rights within the federal structure. Animating the entire struggle was, in

Leton's words, the 'genocide being committed in the dying years of the twentieth century by multinational companies under the supervision of the Government' (cited in Naanen, 1995: 66).

A watershed moment in MOSOP's history was the drafting in 1990 of an Ogoni Bill of Rights (Saro-Wiwa, 1992). Documenting a history of neglect and local misery, the Ogoni Bill took head-on the question of Nigerian federalism and minority rights. Calling for participation in the affairs of the republic as 'a distinct and separate entity', the Bill outlined a plan for autonomy and self-determination in which there would be guaranteed 'political control of Ogoni affairs by Ogoni people ... the right to control and use a fair proportion of Ogoni economic resources ... [and] adequate representation as of right in all Nigerian national institutions' (Saro-Wiwa, 1992: 11). In short the Bill of Rights addressed the question of the *unit* to which revenues should be allocated – and derivatively the rights of minorities. Largely under Saro-Wiwa's direction, the Bill was employed as part of an international mobilization campaign. Presented at the UN Sub-committee on Human Rights, at the Working Group on Indigenous Populations in Geneva in 1992 and at UNPO in The Hague in 1993, Ogoni became – with the help of Rainforest Action Network and Greenpeace – a *cause célèbre*.

Ken Saro-Wiwa played a central role in the tactical and organizational transformations of MOSOP during the 1990s. Born in Bori as part of a traditional ruling family, Saro-Wiwa was already, prior to 1990, an internationally recognized author, a successful writer of Nigerian soap operas, a well connected former Rivers State commissioner, and a wealthy businessman. Saro-Wiwa was also President of the Ethnic Minorities Rights Organization of Africa (EMIROAF) which had called for a restructuring of the Nigerian federation into a confederation of autonomous ethnic states in which a federal center was radically decentralized and states were granted property rights over on-shore mineral resources (Osaghae, 1995: 327). Under Saro-Wiwa, MOSOP focused in 1991 on links to pro-democracy groups in Nigeria (the transition to civilian rule had begun under heavy-handed military direction) and on direct action around Shell and Chevron installations. It was precisely because of the absence of state commitment and the deterioration of the environment that local Ogoni communities, perhaps understandably, had great expectations of Shell (the largest producer in the region) and directed their activity against the oil companies after three decades of betrayal. There was a sense in which Shell *was* the local government[21] (*Guardian*, July 14th, 1996: 11) but the company's record had in practice, been appalling. In 1970, Ogoni representatives had already asked Rivers State government to approach Shell – what they then called 'a Shylock of a company' – for compensation and direct assistance (a plea which elicited a shockingly irresponsible response documented in Saro-Wiwa, 1992). Compensation by the companies for land appropriation and for spillage have been minimal and are constant sources of tension between company and community. Shell, which was deemed the world's most profitable corporation in 1996 by *Business Week* (July 8th, 1996: 46) and which nets roughly $200 million profit from Nigeria each year, by its own admission has only provided $2 million to Ogoniland in 40 years of pumping. Ogoni historian Loolo (1981) points out that Shell has built one road and awarded 96 school scholarships in 30 years; according to the *Wall Street Journal*, Shell employs 88 Ogonis (less than 2 per cent) in a workforce of over 5,000 Nigerian employees. Furthermore, the oft-cited community development schemes of the oil companies, only began in earnest in the 1980s and have met with minimal success (Ikporukpo, 1993). In some communities, Shell only began community efforts in 1992 after 25 years of pumping, and then providing a water project of 5000 gallons capacity for a constituency of 100,000 (*Newswatch*, Dec. 18th, 1995: 13).[22]

In an atmosphere of growing violence and insecurity, MOSOP wrote to the three oil companies operating in Ogoniland in December 1992 demanding $6.2 billion in back rents and royalties, $4 billion for damages, the immediate stoppage of degradation, flaring and exposed pipelines, and negotiations with Ogonis to establish conditions for further exploration (Osaghae, 1995: 336). The companies responded with tightened security while the military government sent in troops to the oil installations, banned all public gatherings, and declared as treasonable any claims for self-determination.[23] Strengthening Ogoni resolve, these responses prompted MOSOP to organize a massive rally – an estimated 300,000 participated – in January 1993. As harassment of MOSOP leadership and Ogoni communities by state forces escalated, the highpoint of the struggle came with the decision to boycott the Nigerian presidential election on June 12th, 1993.[24]

In the wake of the annulment of the presidential elections, the arrest of democratically elected Mashood Abiola and the subsequent military coup by General Abacha, state security forces vastly expanded their activities in Ogoniland. Military units were moved into the area in June 1993 and Saro-Wiwa was charged with, among other things, sedition.[25] More critically inter-ethnic conflicts exploded between Ogoni and other groups in late 1993, amidst accusations of military involvement and ethnic warmongering by Rivers State leadership.[26] A new and aggressively anti-Ogoni military governor took over Rivers State in 1994 and a ferocious assault by the Rivers State Internal Security Task Force commenced. Saro-Wiwa was placed under house arrest, and subsequently fifteen Ogoni leaders were detained in April 1994. A series of brutal attacks left 750 Ogoni dead and 30,000 homeless; in total, almost 2,000 Ogonis have perished since 1990 at the hands of police and security forces.[27] Ogoniland was in effect sealed off by the military. Amidst growing chaos, Saro-Wiwa was arrested on May 22nd, 1994, and several months later was charged with the deaths of four Ogoni leaders.

In evidence that has come to light during and after the military tribunal which charged Saro-Wiwa and eight others, the trial was marked by massive irregularities, including witnesses paid by the government to falsely implicate MOSOP leaders. The defence team, faced with a kangaroo court, withdrew in June 1995, and Saro-Wiwa and the other defendants were sentenced to death on October 31st, 1995. In spite of international outcry and enormous political pressure on the Abacha regime by the Commonwealth, Nelson Mandela and the human rights community, Abacha refused to rescind the death sentence. Throughout 1996, military operations in Ogoniland resulted in more deaths (in January 1996) and the arrest of 18 MOSOP leaders in March to prevent them from meeting with representatives of a UN Mission which investigated the murder trials in the first two weeks of April. As I write, it has become clear that the Abacha regime's military occupation of Ogoniland is determined to obliterate Saro-Wiwa's legacy at all costs and to send a clear signal to the other politically active oil-producing minorities in the delta.

In spite of the remarkable history of MOSOP between 1990 and 1996, its ability to represent itself as a unified pan-Ogoni organization remained a knotty issue, particularly for Saro-Wiwa. There is no pan-Ogoni myth of origin (characteristic of many delta minorities), and a number of the Ogoni subgroups engender stronger local loyalties than any affiliation to Ogoni nationalism. The Eleme subgroup has even argued, on occasion, that they are not Ogoni. Furthermore, the MOSOP leaders were actively opposed by elements of the traditional clan leadership, by prominent leaders and civil servants in state government,[28] and by some critics who felt Saro-Wiwa was out to gain 'cheap popularity' (Osgahae, 1995: 334). And not least the youth wing of MOSOP, which Saro-Wiwa had made use of, had a radical vigilante constituency which the leadership were incapable of controlling. In the same way that so-called orthodox Islam had tensions within it – there are several moral economies of reformism of which Maitatsine is one – so also was a sense of pan-Ogoni identity and self-interest fractured, in this case along lines of generation, tradition, and political strategy in regard to the federation. What Saro-Wiwa did was to build upon over fifty years of Ogoni organizing and upon three decades of resentment against the oil companies, to provide a mass base and a youth driven radicalism – and it must be said an international visibility – capable of challenging state power.

How then can one understand the ferocity with which state forces entered Ogoniland, engaged in extrajudicial executions, and harassed and detained MOSOP leadership? First, the escalation of MOSOP activities occurred at a delicate moment in the military transition process in which Ogonis had made the unprecedented move of boycotting the elections. The fact that a non-northerner (Abiola) had in all likelihood won the election – thereby threatening the post-1960 political order – further exacerbated the delicacy of the moment. Second, MOSOP explicitly challenged the very notion of Nigerian federalism and articulated a series of claims (a new ethnic confederation) which stood at odds with the power structure on which the military rested. Third, the MOSOP campaign had effectively threatened the production of oil itself upon which the entire ruling order rested. By the end of January 1994 the eight major oil companies estimated their losses at $200 million. The oil fortress had been shaken in an unprecedented fashion, and both companies and government called for urgent measures to safeguard the economic lifeline of the country. Fourth, MOSOP converted a *local* minority and green movement into a *global* human rights cause linked to corporate irresponsibility (with, it should be said, global implications through gas flaring). And last but not least, Saro-Wiwa had intended MOSOP to have a 'ripple effect'. Despite the abject poverty of the Ogoni, other oil producing communities had almost certainly fared worse – in terms of ecological costs, economic deprivation and political marginality – and it was perhaps no surprise that other delta communities responded to the initiatives of MOSOP. Following the MOSOP precedent, a number of southeastern minorities pressured local and state authorities for expanded resources and political autonomy: the Movement for the Survival of Izon/Ijaw Ethnic Nationality was established in 1994, the Council for Ekwerre Nationality in 1993 and the Southern Minorities Movement (28 ethnic groups from five delta states) has been active since 1992. The Movement for Reparation to Ogbia (MORETO) produced a charter explicitly modeled on the Ogoni Bill of Rights in 1992.[29] These groups directly confronted Shell and Chevron installations (Human Rights Watch, 1995; Greenpeace, 1994) and in turn have felt the press of military violence over the last four years. The point is simply that MOSOP was a flagship movement for a vast number of oil producing communities and threatened to ignite a blaze throughout the oil producing delta. In so doing MOSOP presented a vision of, indeed an alternative to, post-colonial Nigerian politics.

MOSOP exposed, in sum, the explicit collusion between rapacious transnational capital and a corrupt military state on the one hand, and on the other presented an alternative to the so-called 'tripod model' of Nigerian federalism dominated by the three ethnic majorities – over which the Abacha regime held an increasingly tenuous sway. While Saro-Wiwa endlessly asserted that MOSOP concerns had never envisaged secession as such, this was precisely how the Abacha regime expressed its concerns over Ogoni self-determination.[30] For Abacha, faced with coup plots, an increasingly divided north, popular discontent over the transition process, and economic recession, Ogoni claims threatened the very existence of post-colonial Nigerian federalism and the petrolic patrimonialism which supported the whole rotten structure. In losing control over oil, the Federal government by definition – irrespective of the specificities of Saro-Wiwa's model of ethnic federalism – would *de facto* commit political suicide.

STATE VIOLENCE, THE NATIONAL QUESTION AND LOCALITY POLITICS

> Nigeria is not a Nation. It is a mere geographical expression. The word 'Nigeria' is merely a distinctive appellation to distinguish those who live within the boundaries of Nigeria from those who do not.
>
> (Obafemi Awolowo, 1960)

Two movements, two charismatic leaders, two murders. In one case the idiom was Islam, the moral decay of the Muslim community and the call for renewal; in the other, minority rights, self-determination and confederal politics. Both were, as I have tried to argue, fundamentally shaped by two vectors: a despotic petrolic capitalism on the one hand, and the national question on the other. These two vectors were themselves intimately linked because, as Paul Lubeck (1995: 18) notes, the globalizing impact of the oil boom 'set the stage for the crisis of the secular nationalist state once the [oil] rents were spent'. Oil provided the political cement with which held the Nigerian federation together after the civil war, and yet it was also the political dynamite which threatened to blow the entire post-colonial federal order apart. Petroleum was both Scylla and Charybdis.

Both the Maitatsine insurrection and MOSOP can also be seen as political defeats, and it is the nature of this defeat at the hands of ferocious state violence which I have tried to explain. My argument has turned on the fact that each movement challenged the fiction of Nigeria as nation, a fiction that Nigerian politician Awolowo referred to as a 'mere geographical expression'. In his new memoire, Nobel Laureate Wole Soyinka – himself a political refugee currently residing in Europe – sees Nigeria's crisis as precisely the lack of a moral community synonymous with nationhood. What, he says, has 'a geographical space such as … Nigeria come to mean … when a leader speaks of national sovereignty'? (Soyinka, 1996: 45). It is the fact that there is no organic connection between geographical space and national identity but only an invention of a purported unity by a handful of leaders at the top, that renders Nigeria close, to employ his own language, to the end of its own history. And it was to this contradiction of geography which both these movements spoke. Maitatsine made such a challenge, at a moment when the frailty of oil revenues was already apparent, by challenging the ideological bedrock of northern hegemony. Maitatsine unmasked the northern ruling oligarchy. A north without cohesion not only undercut the Hausa-Muslim dominated post-1960 order, but carried the prospects of the collapse of the federation itself. Saro-Wiwa and MOSOP, at a moment when the bankruptcy of the petro-economy was crystal clear, both exposed the complicity of the oil companies and the military state (an unmasking of another sort) and explicitly provided another model of revenue allocation and by extension of political confederation. Islam and ethnicity in their own ways challenged the rickety structure on which Nigerian nationhood was built.

In his brilliant new book *Citizen and Subject*, Mamood Mamdani (1996) has traced the crisis of the contemporary African state to the political legacy of colonialism which left what he calls a system of decentralized despotism. This despotism in the name of native authorities and customary law, was animated by ethnicity and religion. And in Mamdani's view any reaction against this despotism in the post-colonial era has typically borne the institutional imprint of that mode of rule. In this sense Maitatsine and MOSOP are exemplary cases. But it is also true that the impoverished sense of citizenship and subject which is the hallmark of decentralized despotism was precisely what Maitatsine and Ken Saro-Wiwa sought to address. It is on this larger field of citizenship and subject that, in my view, Nigerian state violence against local alternatives can best be understood. While it is often said that discussions of resistance may romanticize its politics or may exaggerate its capacities, the two cases I have presented show vividly how two movements which are fundamentally local can strike to the very heart of state power. But the obverse is also true. In providing alternative visions, Maitatsine and the Ogoni nine came to represent a murderous confirmation of the state's monopoly of the means of organized violence.

ACKNOWLEDGEMENTS

I am grateful for the research assistance of Chitra Ayar and James McCarthy and for information provided by various activist groups including Greenpeace, Human Rights Watch, the Association of Concerned African Scholars and the Rainforest Action Network (San Francisco), and Mr Odigha Odigha. The research was supported by the SSRC/MacArthur Foundation Peace and Security Program.

NOTES

1 The Northern People's Congress under Ahmadu Bello in the North, the Action Group under Awolowo in the West, and the National Convention of Nigerian Citizens under Azikwe in the East.

2 In the new Gowon scheme which created three new states in the east, the Ibo received 14 per cent of total oil revenues; under Ojukwu's secessionist plan, Biafra would receive 67 per cent of oil revenues (Forrest, 1995: 32).

3 In the 1990s roughly 12 of the 19 positions in the Armed Forces Ruling Council (the former Supreme Military Council) have been held by northerners (Osaghae, 1991: 257).

4 These antinomies of oil are no better captured than in the astonishing entity which is the state petroleum company, the NNPC. The Irikefe Report in 1980 which discovered 'vast irregularities' in the awarding of contracts, contributed to the NNPC's history of personnel turnover which resembles Italian politics. An estimated $14 billion disappeared from NNPC between 1979 and 1983; according to the 1994 Okigbo Report, $12.5 billion was unaccounted for between 1988 and 1994 (*Africa Confidential*, November 4th, 1994: 6). In 1990–1991 when oil prices leapt by $12.5 billion as a result of the Gulf crisis, there was no recorded increase in NNPC or government revenue! In 1992, the World Bank estimated this gap between official and unofficial oil earnings was $2.7 billion, roughly 10 per cent of GDP (*The Economist*, 1993: 8).

5 Watching a minister sweep through Lagos in a 12-vehicle convoy, complete with an armoured car, a journalist asked bystanders what would happen if they got out and walked: 'We'd kill them' they replied (*The Economist*, June 8th, 1996: 8).

6 According to Pat Uyomi of Lagos Business School, personal income has grown at 0.02 per cent since the oil boom: 'most Nigerians lived better before the 1973–74 oil price rise' (*The Economist*, June 8th, 1996: 48).

7 For Kano citizens (*Kanawa*), the classical Muslim heart of the city is the *birni*. But from 1902 onwards, the suburbs outside the *birni* grew rapidly, populated in the first instance by Westerners and Nigerians from the south of the country. As such, *waje* was associated with Christianity, permissiveness and evil. As Barkindo notes, *waje* was an evil to be tolerated but 'this view radically changed from about 1970 onwards' (1993: 94).

8 The absolute number of manufacturing establishments and of industrial wage workers, the scale of direct investment by multinationals, and the shares of federal and regional state capital in industrial output, all witnessed positive growth rates throughout a boom period presided over by a succession of military governments (1972–9).

9 I have been fortunate in making use of copious information provided through the Ogoni

and other websites: www.oneworld.org/oca/ (Ogoni); ww.gem.co.za/ELA/ken.html (Earthlife Africa) and Amnesty International. I have also relied heavily on Human Rights Watch, the Unrepresented Nations and Peoples Organization (UNPO), and the United Nations Mission to Ogoniland in 1996. Information was also provided by Claude Ake and Odigha Odigha.

10 Ogoniland consists of three local government areas and six clans which speak different dialects of the Ogoni language. MOSOP is in this sense a pan-Ogoni organization.

11 The Ogoni and other minorities petitioned in 1974 for the creation of a new Port Harcourt State within the Rivers State boundary (Naanen, 1995: 63).

12 What Rivers State felt in regard to federal neglect, the Ogoni experienced in regard to Ijaw domination. While several Ogoni were influential federal and state politicians, they were incapable politically of exacting resources for the Ogoni community. In the 1980s only 6 out of 42 representatives in the state assembly were Ogoni (Naanen, 1995: 77). It needs to be said however – and it is relevant for an understanding of state violence against the Ogoni – that the Ogoni have fared *better* than many other minorities in terms of political appointments: in 1993, 30 per cent of the Commissioners in the Rivers State cabinet were Ogoni (the Ogoni represent 12 per cent of the state population) and every clan has produced at least one federal or state minister (Osaghae, 1995: 331) since the civil war. In this sense, it is precisely that the Ogoni *had* produced since 1967 a cadre of influential and well placed politicians (including Saro-Wiwa himself) that their decision to move aggressively toward self-determination and minority rights was so threatening to the Abacha regime (Welch, 1995).

13 According to the Nigerian Government, Ogoniland currently (1995) produces about 2 per cent of Nigerian oil output and is the fifth largest oil-producing community in Rivers State. Shell maintains that total Ogoni oil output is valued at $5.2 billion before costs!

14 Nigerian geographer David Ogbonna also laid out almost twenty years ago, a devastating portrait of ecological death and corporate irresponsibility in Rivers State in an important document which went relatively unnoticed (1979).

15 The oil companies claim that sabotage accounts for a large proportion (60 per cent) of the spills, since communities gain from corporate compensation. Shell claims that 77 of 111 spills in Ogoniland between 1985 and 1994 were due to sabotage (Hammer, 1996). According to the government commission, however, sabotage accounts for 30 per cent of the incidents but only 3 per cent of the quantity spilled. Furthermore, all oil producing communities claim that compensation from the companies for spills has been almost non-existent.

16 *Newswatch* reported that oil and burnt soil have rendered land unusable 25 years after a blowout at Ejama-Ebubu (December 18th, 1995: 11).

17 For a more detailed discussion of Maitatsine, see Watts, 1994, 1996.

18 The critique focuses especially on certain innovations: the folding of arms while praying, not facing Mecca while praying, collections of fees by Mallams, and the wearing of amulets. The latter is especially interesting because it speaks to the popular belief across many segments of society in non-Muslim spirits and powers. It is precisely this source of power (often associated in northern Nigeria with non-Muslim Hausa [Maguzawa] and the urban underworld) which Maitatsine drew upon; indeed Maitatsine was recognized as a sort of sorcerer. Several of Maitatsine's lieutenants carried talismans (one quoted in the *Aniagolu Report* spoke of: 'if I were cut into pieces and die I shall come back to life').

19 According to Saad (1988: 118), a survey conducted among the arrested *'yantatsine* revealed that 95 per cent believed themselves to be Muslim.

20 The *'yantatsine* did not necessarily stand in opposition to the popular classes or indeed the merchants and shopkeepers among whom they lived. In the 1980 conflicts for example, *'yantatsine* who had occupied a cinema on the Kofar Mata Road told local residents that their fight was strictly with the police, while young immigrant workers living in Fagge appeared to be just as fearful of the police and of the vigilantes as they were of *'yantatsine*. Marwa himself scrupulously returned property to their rightful owners if appropriated by his followers.

21 Prior to the cessation of operations in 1993, Shell was the principal oil company operating in Ogoniland, pumping from five major oilfields at Bomu/Dere, Yorla, Bodo West, Korokoro and Ebubu.

22 According to the survey by Furro (1992), 83 per cent of respondents were dissatisfied with the role of the oil industry in his four oil producing communities and two-thirds had received no compensation from the companies for various damages to land and property.

23 The fact that protests around Shell and other installations have resulted in the death and destruction of Ogoni persons and property (Human Rights Watch, 1995), the fact that oil is currently pumped under military protection throughout the delta, and new evidence of Shell's role in both importing arms and assisting the Nigerian Mobile Police Force and internal security forces in attacks on oil communities (*Village Voice*, Jan. 23rd, 1996: 23; *New York Times*, February 13th, 1996: A4), all lend credence to the so-called 'slick alliance' theory that Shell is complicit with the state in the assaults on MOSOP and other oil producing community organizations.

24 It is hard to underestimate the significance of this decision in view of the popular dissatisfaction with the transition process and the implication that the Ogonis had opted out of the federation (with all of its associations).

25 The history of events since June 1993 are detailed in Human Rights Watch (1995), UNPO (1995) and the UN Mission (1996).

26 It seems clear that the conflicts between Ogonis and Andonis and Okrikas were almost certainly effected by state authorities, disgruntled community elders and the oil companies (Osaghae, 1995: 337; Human Rights Watch, 1995).

27 The evidence implicating the military in widespread extrajudicial killings, rape, looting and arbitrary arrest and detention is overwhelming, and the ruthlessness of Lt Col. Paul Okuntimo in particular is especially distressing (see Human Rights Watch, 1995).

28 This was raised of course in the murder trial, where Saro-Wiwa was accused of referring to some of the elders and chiefs as 'vultures' (Human Rights Watch, 1995: 29).

29 In addition there are the following active minority groups in the Delta: the Oyigba Forum, the Bonny Indigenous Peoples Federation, the Ndoni Community Association, O'Elobo Eleme and Uzugbari Ekpeye.

30 This was the substance of Information Minister Walter Ofonagoro's address to Oxford University in 1995 (*Newswatch*, May 22nd, 1995: 17) and of the efforts by the Nigerian High Commission in the US in attempting to garner support from the African-American community (*The Nation*, December 25th, 1995).

3

A SPATIALITY OF RESISTANCES

theory and practice in Nepal's revolution of 1990

Paul Routledge

This research is based upon two visits that I made to Nepal. The first was in 1990 when the country was embroiled in a popular revolution against the autocratic rule of King Birendra. I was in the country primarily as a witness to the events that took place, having recently completed my doctoral research in India. Although I participated in several demonstrations against the king, donated some of my photographs to a human rights organization, and, as asked, wrote about the revolution in the US media (under Kala, 1990[1]), I felt that I had understood little about the dynamics of the revolution. I returned to Nepal two years later, to spend two months conducting informal interviews with the participants of the revolution, including students, political party activists, human rights workers, academics, writers, and politicians. As a result, I was able to gain some insights into how space and power had been contested within some of Nepal's urban spaces. In so doing, I gained in knowledge and theoretical insight, what I lost in the intensity of emotions experienced during those moments of lived rebellion. Hence this writing about resistance falls far short of the experience of resistance. It can only be a partial and positioned narrative. However, based upon my research and my first-hand experiences, I want to discuss some aspects of what I would term the 'spatiality of resistance' that was effected during the Nepali revolution. In order to do so, let me first explicate how I understand the term 'resistance'.

THEORIZING RESISTANCE

The theorization of resistance is itself a problematic exercise. The complex, contradictory, and lived nature of resistance is frequently erased, or at best generalized, in theoretical approaches (see Melucci, 1989; Routledge, 1993). Indeed, some, such as Hecht and Simone have argued that theorizing resistance simply nullifies acts of resistance (1994: 83). Certainly attempts to theorize resistance have been fraught with an intellectual taming that transforms the poetry and intensity of resistance into the dull prose of rationality. Such taming takes at least two forms. First, there is the process of teleological taming which confines resistance to a temporal dimension, determining in advance the path that resistance must take in order to realize certain universal principles such as Reason and Freedom.[2] Hence we might assess resistance according to its progress along consensus approved trajectories, or precalculated curves of history.

Second, there is the process of macropolitical taming whereby resistances are located within various empirical unities such as class struggle and economic contradictions in order to be considered progressive. These tamings negate the local/particular and the heterogeneity within resistance in order to search for generalized explanations. As such they fail to take account of the spatio-cultural specificity of resistance (see Routledge and Simons, 1995). Therefore, it may be more appropriate to talk of resistances rather than 'resistance'.

I use the term resistance to refer to any action, imbued with intent, that attempts to challenge, change, or retain particular circumstances relating to societal relations, processes, and/or institutions. These circumstances may involve domination, exploitation, subjection at the material, symbolic or psychological level.[3]

Resistances are assembled out of the materials and practices of everyday life, and imply some form of contestation, some juxtaposition of forces. These may involve all or any of the following: symbolic meanings, communicative processes, political discourses, religious idioms, cultural practices, social networks, physical settings, bodily practices, and envisioned desires and hopes. These actions may be open and confrontational, or hidden (see Scott 1985, 1990) and range from the individual to the collective. Their different forms of expression can be of short or long duration; metamorphic, interconnected, or hybrid; creative or self-destructive; challenging the status quo (as in Nepal) or conservative (Calderón *et al.* 1992: 23).

Resistances may be interpreted as fluid processes whose emergence and dissolution cannot be fixed as points in time. Myriad processes contribute to, overlap with, and extend beyond the dates we may wish to choose for the birth and death of resistance. Indeed, the boundaries of such resistances can be vague; there is not a clear division between past, present and future practices. Following Deleuze and Guattari (1987) we may theorize resistances as rhizomatic multiplicities of interactions, relations, and acts of becoming. A multiplicity cannot lose or gain a dimension without changing its nature, indeed it is constantly transforming itself. Any resistance synthesizes a multiplicity of elements and relations without effacing their heterogeneity or hindering their potential for future rearrangement (1987: *xiii*, 33, 249). As rhizomatic practices, resistances take diverse forms, they move in different dimensions (of the family, community, region etc.), they create unexpected networks, connections, and possibilities. They may invent new trajectories and forms of existence, articulate alternative futures and possibilities, create autonomous zones as a strategy against particular dominating power relations.

Dominating power is that which attempts to control or coerce others, impose its will upon others, or manipulate the consent of others. This dominating power can be located within the realms of the state, the economy, and civil society, and articulated within social, economic, political, and cultural relations and institutions. Patriarchy, racism, and homophobia are all faces of dominating power – that which attempts to silence, prohibit, or repress dissent, that which is intolerant of difference, that which engenders inequality, and asserts the interests of a particular class, caste, race, or political configuration at the expense of others (Routledge and Simons, 1995: 473). Having said this, practices of resistance cannot be separated from practices of domination, they are always entangled in some configuration. As such they are hybrid practices, one always bears at least a trace of the other, that contaminates or subverts it. For example, social movements frequently suppress their own internal heterogeneities and sub-groups in the interests of some broader strategy. Hence the anti-Vietnam war resistance in the United States during the 1960s has been criticized for its sexism, while the women's movement has been criticized for excluding lesbians and women of colour (see de Lauretis 1990). In addition, certain resistances are themselves a reproduction of dominating power, for example, the anti-abortion Operation Rescue mobilizations in the United States which form part of a broader attack upon women's reproductive, civil, and economic rights within the country (Faludi, 1992).

A SPATIALITY OF RESISTANCE

Different social groups endow space with amalgams of different meanings, uses, and values. Such differences can give rise to various tensions and conflicts within society over the uses of space for individual and social purposes and the domination of space by the state and other forms of class and social power.

> Socio-political contradictions are realised spatially. The contradictions of space thus make the contradictions of social relations operative. In other words, spatial contradictions 'express' conflicts between socio-political interests and forces; it is only *in* space that such conflicts come effectively into play, and in doing so they become contradictions *of* space.
>
> (Lefebvre, 1974: 365)

Within such contradictions of space, particular places frequently become sites of conflict where the social structures and relations of power, knowledge, domination and resistance intersect. In the case of Nepal, the domination of space by the autocratic state was challenged by a popular movement within the urban spaces of the Kathmandu Valley.

The practices and discourse of such resistance require some form of coordination and communication. This usually involves some form of collective action, although resistance can be at the individual level.[4] In order to effect this resistance, actants[5] must establish (however temporarily) social spaces and socio-spatial networks that are insulated from control and surveillance. Such spaces may be real, imaginary, or symbolic. Bell hooks (1991) refers to these spaces as 'homeplaces' which act as sources of self-dignity and agency, sites of solidarity in which and from which, resistance can be organized and conceptualized. As I will show below, this notion is very useful in understanding why particular urban spaces in Nepal became the principal sites of struggle. However, in addition to the notion of location imbued in the concept of homeplaces, they are also sites of difference and distance, and separation and limitation (Kirby, 1996: 208). Following Foucault (1986) we might also term them heterotopias.[6] Such 'performed spaces' contain physical and social boundaries where resources are available to some and not to others (Pile and Thrift, 1995a: 374). Such places of resistance are ambiguous in character and may be conceptualized as 'third spaces' (Bhabha, 1994). That is, they are places where resistance is never a complete, unfractured practice, but rather places where practices of resistance are always entwined in some way with practices of domination such as marginalization, segregation, or imposed exile (hooks, 1991: 33–49).

Nevertheless, in order to understand the process of resistance, it is necessary to understand how such sites are created, claimed, defended, and used (strategically or tactically). Such spatial practices imply a strategic mobility that involves the tactical interactions and communication relays within resistance collectivities, and between them and their opponents and allies (see Routledge, 1996). These practices may also effect movements of territorialization and/or deterritorialization. The former implies the (temporary or permanent) occupation of space, while the latter involves a movement across space, implying an imminent dispersion (Deleuze and Guattari, 1987). Moreover, the relationship between resistance practices and the places wherein they are articulated are mutually constitutive, albeit in different ways. Hence, while the strategic mobilities of resistance may constitute particular spaces as homeplaces, the material, symbolic, and imaginary character of places will also influence the articulation of resistance. As a multiplicity, the precise blend of these relations is always unique. As I will show below, this notion of strategic mobility, and its mutually constitutive relationship to space, is of particular use in understanding the political and cultural character of Nepali resistance.

Strategies and tactics are different, of course. In the sociological sense, tactics form the specific tools for the articulation of broader political strategies. In the spatial sense, de Certeau (1984a) defines strategies as creating places that can be delimited on their own as sites of resistance that serve as bases from which opponents can be excluded and challenged. Tactics are actions undertaken in the territory of one's opponents, within the opponent's field of vision. I will show that this spatial interpretation of strategies and tactics is more ambiguous than de Certeau would suggest.

The spatiality of resistance inquires into: (1) how spatial processes and relations across a variety of scales, as well as the particularities of specific places, influence the character and emergence of various forms of resistance; (2) how practices of resistance are constitutive of different relationships to space, via strategic mobilities, or uses of space; (3) how these relationships enable or constrain such articulations of resistance; and (4) how the character and meaning of place may change when it becomes a site of resistance. I shall explore some dimensions of these questions in this chapter, recognizing that they can only elucidate some threads of the multiplicity that was Nepal's terrain of revolutionary resistance.

A NOTE ON 'REVOLUTION'

Theoretically, the metanarrative of revolution can ignore social differences and particularities, involve reified categories, transform moral and political conflicts into technical and metatheoretical disputes and locate social movements within a purely temporal dimension (see Seidman, 1992). As such it is of questionable explanatory power. While Nepal's Movement for the Restoration of Democracy (MRD) was inspired, in part, by Eastern Europe's revolutions of 1989, and while the communist factions within the movement also drew inspiration and influence from earlier revolutions and revolutionary theory, the events of spring 1990 were spatially contextualized within the specific political, economic and cultural landscape of Nepal. As I will argue, a spatiality of resistance examines the local narrative of revolution, in this case as it was manifested in Nepal.

PLACE AND THE EMERGENCE OF RESISTANCE

A consideration of economic, political, and cultural developments within Nepali society, and the particular place-specific characteristics of the Kathmandu Valley, and the urban spaces therein, enable an understanding of some of the reasons for the emergence of Nepali resistance to King Birendra's regime.

Dominance without hegemony

The relations of power within Nepali society, include pre-modern (feudal) and modern elements, and comprise an interwoven web of caste, class, patronage and kinship. Nepali society is permeated with traditions of status and privilege, and economic and cultural differences that reflect caste and ethnic inequalities and feudal land relations. For example, 9 per cent of Nepali households own land holdings above 3 hectares and control 47 per cent of the total cultivated land, while 66 per cent of Nepali households own holdings of less than 1 hectare which accounts for 27.4 per cent of the total cultivated land (INSEC, 1991, 1992; Yadav, 1984). Social, economic and political inequalities are compounded by the caste system, which is dominated by the Brahmin, Chhetri and to a lesser extent the Newar castes. According to Shaha (1990) 80 per cent of the positions of political and religious power and economic profit (especially through the ownership of land) are held by these three groups. Meanwhile, the majority of the population are condemned to the ravages of grinding poverty, bonded labour, illiteracy, and inadequate health care. For example, over 60 per cent of Nepal's approximately 18.5 million population live beneath the official poverty line (INSEC, 1991, 1992).

The royal family, headed by the (Chhetri) king, controls the social order in alliance with a religious (Brahmin) hierarchy, the army, landed interests, and modernizing capitalist, bureaucratic, and political elites. This control has been maintained through the coercive force of the army, controlled by the king, and through the inequalities of power institutionalized in the panchayat system. Based upon a pyramidal structure of village, district, zonal and national councils, with the king at the apex, the panchayat system provided the king with autocratic power over the political, economic and cultural life of the country. This power was institutionalized through a variety of repressive laws, designed to curtail any opposition to the government.[7] Both parliament and the judiciary answered only to the king; freedom of peaceful assembly and association was denied; imprisonment without trial was legally sanctioned; and torture and deaths within prison were commonplace (Shaha, 1990; INSEC, 1991, 1992; FOPHUR,1990a, 1990b).

However, the Nepali state's grip over the subaltern population has been characterized as a 'dominance without hegemony'. It has been challenged many times throughout Nepal's history, and at times has had to compromise with such contestation from below (Parajuli, 1992). Nepal's Movement for the Restoration of Democracy (MRD) had historical antecedents in various movements and struggles for democracy that had been waged in Nepal during the previous 50 years. In 1950 the Rana oligarchy (who had ruled Nepal for 104 years) was overthrown by an armed rebellion led by the Nepali Congress party, and replaced with a constitutional monarchy and, subsequently, a parliamentary government. In 1962 the Nepali Congress launched a short-lived and unsuccessful armed struggle, from bases within India, against the (then) King Mahendra's banning of political parties and repression of political opposition. During 1971–2 Maoist factions within the communist movement launched an armed peasant insurgency in the Jhapa district of eastern Nepal which was eventually suppressed by the army. Following the crowning of King Birendra, after his father's death in 1972, the Nepali Congress again launched a short-lived and unsuccessful armed insurrection. In 1985 the Nepali Congress launched a non-violent civil disobedience movement against the government which was again short-lived and unsuccessful. It was not until 1990 that the various oppositional forces in Nepal were to unite in an attempt to overthrow the panchayat system.

Seed crystals for resistance

Landlocked between China and India, Nepal's economy is heavily dependent upon trade with India which accounts for 40 per cent of Nepal's foreign trade (Ali, 1989). During a 1989 dispute over a trade treaty between the two countries, India closed 13 out of 15 transit points along its border with Nepal, seriously curtailing the flow of imports into the country. The economic blockade precipitated the rationing of kerosene, petrol and diesel within Nepal as well as sharp price increases, and served to exacerbate economic problems within the country and fuel popular dissatisfaction with the government. The emergence of the resistance of 1990 was aided by several factors. First, communications and transport facilities (e.g. roads, telephones, television, etc.) had been improved during the 1980s, enabling improved communication between underground opposition groups. Second, improvements in education had increased the literacy rate in Nepal from 10 per cent in 1960 to 35 per cent in 1990, improving the awareness of the population (via the local press media) concerning the inequitable character of the country's political economy (Interviews, Kathmandu, 1992). Third, encouraged by the success of popular movements in Eastern Europe in 1989, the principal opposition parties, the Nepali Congress (NC) and the United Left Front (ULF) – a coalition of seven of Nepal's communist parties – decided, despite ideological differences, that it was in their mutual interest to join together to launch a non-violent Movement for the Restoration of Democracy (MRD).

Strategic non-violence

The movement's demands focused on the dismantling of the panchayat system, the restoration of parliamentary democracy, and the reduction of the king's powers to those of a constitutional monarch. To contest power, the MRD adopted a variety of non-violent sanctions, ranging from intervention, non-cooperation and protest and persuasion (see Sharp, 1973). Intervention took the form of 'liberating' urban space (as in Patan, below). Protest and persuasion took the form of demonstrations and symbolic protests such as blackouts (see below). Non-cooperation took the form of country-wide strikes (*bandhs*). An estimated 50–60,000 teachers and 30–40,000 workers were involved in the strikes (Interview, Kathmandu, 1992).[8]

The strategic use of sanctions was premised upon the withdrawal of popular consent and cooperation from the panchayat regime in an attempt to secure movement demands. This strategy was aided by the active participation of many facets of civil society in the resistance,[9] and was coordinated by an underground steering committee, established because of government arrests of party leaders. The choice of non-violent sanctions was due to several factors: (1) armed struggle was seen as having no mass base among the population; (2) armed struggle was expected to incur extreme government repression as in the past; and (3) it was believed that a commitment to non-violence would be likely to elicit international support and sympathy for the movement's demands (Interviews, Kathmandu, Patan, 1992), reflecting the movement's self-location within a global as well as local and national context.

Homeplaces of resistance

During the revolution outbreaks of resistance were widespread throughout Nepal. Demonstrations were conducted in the urban centres of Pokhara, Biratnagar, and Janakpur as well as other towns in the Terai region of the country. However, movement leaders agreed that the principal terrains of resistance against the panchayat regime were located within the Kathmandu Valley, and particularly in the capital city of Kathmandu, and the surrounding towns of Patan, Kirtipur, and Bhaktapur. The urban spaces of the Kathmandu Valley became the principal homeplaces of the resistance. Kathmandu and Patan in particular, emerged as sites where the power of the panchayat regime was contested. Their urban landscapes not only influenced the articulation of resistance, they were also consciously transformed by MRD activists into a tool of their struggle, into a terrain of resistance.

Kathmandu's location as the capital of Nepal, the home of the king (and hence the centre of royal power), and the locus of political, administrative, and economic power made it an appropriate site for the contestation of such power by the MRD: the potential impact of the movement would be greatest if directed at the heart of power. This was further facilitated by the improved communications that existed within the city, and the Valley; the fact that communications outside of the Valley were difficult[10] and the dense concentration of population that existed within the city (400,000) and the Valley (800,000). These factors facilitated mass mobilizations within the urban areas which were not possible elsewhere in the country. Kathmandu was also the location of the principal organizing headquarters for the main opposition political parties (excluding the Labour and Peasants' Organization which was located in Bhaktapur), and for the student organizations who were involved in the movement.[11] The surrounding rural villages of the Valley, were at times used as shelters for movement activists, since the rugged terrain and dispersed location of the shelters impeded detection by the authorities (Interview, Kathmandu, 1992).

Concerning social relations within the area, the Kathmandu Valley (and especially the urban centres of Kathmandu, Patan, Bhaktapur and Kirtipur) is the traditional home of the Newars. As Levy (1990) notes, after the Gorkhali conquest of the Kathmandu Valley in the late eighteenth century, the Newars ceased to be *the* people of Nepal with their own kings. Instead they were only one of approximately seventy linguistic and ethnic groups in the enlarged territory of Nepal and a conquered one at that, ruled by a Gorkhali king (pp. 14–15, 47–52). Although some Newars held positions of power in both the Rana oligarchy and the subsequent royal regimes there was widespread discontent among the Newar community at what they perceived as discrimination and repression by the predominantly non-Newar government, and a non-Newar king (Interviews, Kathmandu, Patan, 1990, 1992). By focusing the movement in the Valley, the movement was able to mobilize popular frustration and anger against the regime.

STRATEGIC MOBILITY AND THE TRANSFORMATION OF PLACE

Through a consideration of the strategic mobilities of struggle, we can gain some insights into how Nepali resistance effected different relationships to space, how these enabled the practice of resistance, and how they transformed the character of particular places. In so doing, we can understand the rhizomatic character of resistance – in that it takes diverse forms and moves within different dimensions – and how places and resistances are mutually constitutive of one another.

Packs and swarms

Two spatial practices enacted within Kathmandu and Patan were those of the *pack* (Canetti, 1962) and the *swarm* (Ross, 1988). Packs are small in number and often constitute 'crowd crystals' that may precipitate crowds or demonstrations. The pack does not openly confront dominating power, it is more secretive, utilizing underground tactics, surprise, and the unpredictability of deterritorialized movement. Packs effect a movement of deterritorialization of space – they tend to move across space, rather than occupying it. Their action always implies an imminent dispersion. The swarm, by contrast, is large in number, effecting a movement of territorialization. The swarm openly confronts dominating power by weight of numbers, by occupying space – be it physical, symbolic, political, or cultural. The relationships to space of these agencies – those of territorialization and deterritorialization – are always relative, always interwoven with one another.

Concerning the tactics of the swarm, numerous demonstrations were conducted within the urban areas, particularly Kathmandu, Patan and Bhaktapur, movement slogans calling for the end to the panchayat regime and a return to democracy. By temporarily occupying streets and squares, Nepalis articulated, both physically and symbolically, their resistance to the regime. During the demonstrations, packs of students initiated spontaneous corner demonstrations whereby small groups of students would assemble at a strategic location within the city, shouting anti-government slogans, burning effigies of the king, and distributing movement literature; then disperse if the police arrived and reassemble at another location. Often many of these corner demonstrations would be held simultaneously at various locations so as to stretch police capabilities of deployment. Various diversionary tactics were employed by activists (e.g. running through the streets with *mashals* [burning torches]) to draw police attention away from movement meeting sites. During several of these demonstrations conflicts would occur between demonstrators and the police. Deviating from the non-violent discipline of the movement, some activists battled the police with rocks and street stones, which were met by the *lathis* (wooden night-sticks), tear gas and bullets of authority.[12]

Relays and interwoven spaces

An important tactic utilised during the revolution was the blackout protest, whereby all of the households in Kathmandu and Patan were asked to turn out all of their lights as a symbol of dissatisfaction and resistance against the government. These protests were often called during the evening curfews that were imposed by the government in an attempt to quell the movement. The blackouts symbolically communicated popular resistance to the government, enabled city residents to grasp the extent of popular support for the MRD, and acted as a morale booster to the movement. They also enabled increasing numbers of people to show solidarity to the movement and to challenge the curfew and join demonstration swarms under cover of darkness with a reduced chance of being identified by security forces.

Although the blackouts were called by the movement leadership, the communication of the action was conducted by city residents. Residents relayed the message of the action from rooftop to rooftop across Kathmandu.[13] In doing so, they drew upon and utilised various spatio-cultural practices. Traditional Newar houses within the city consist of only three, four or five storeys. The upper storey opens out onto a porch (*kaisi*) which is used for various rituals. One of these – the flying of kites during the Mohani festival as a message to the deities to bring the monsoons to an end – involves symbolic communication (Levy, 1990: 188, 748n). The porches are also used for more secular activities such as the drying of clothes and talking with neighbours. By informing their neighbours of the blackout protests from their rooftop porches, residents utilised a cultural space that was already important for both community and symbolic communication. In so doing, a space of interwoven meanings was produced. The rooftops acted as a place for the performance of religious rituals, daily activities, and resistance. The latter was facilitated by the propinquity and low elevation of the city dwellings, and the fact that they were out of the purview of government forces. Once on the streets, people set fire to car tyres to act as temporary barricades across the narrow streets, and pitched battles between armed riot police and stone-throwing demonstrators ensued, the incendiary of protest lighting up the darkened city.

The public contestation of space transformed the physical character of Kathmandu and Patan, inscribing upon them a mosaic of signs which spoke of the ferment in their streets. Photograph displays were erected in the public squares depicting victims of police torture and killings, political prisoners, and activists who had disappeared. Numerous windows of government offices and shops were broken. In the middle of roads, burned-out skeletons of government buses attracted crowds of onlookers. Torn-up street stones, used in battles with the police, lay strewn across streets and pavements. Upon city buildings and temple walls were daubed pro-democracy and political party slogans.

Place as strategy: transformed space

Patan, a Newar town with a history of communist activity, 2 kilometres from the capital, served as the base of operations for the movement's underground leadership during the duration of the MRD. The movement consciously used the urban topography of Patan to accommodate the exigencies of underground activism. Within Newar towns, community relations are focused around village-like spatial segments of the town called *twa:*, which represent important loci of personal and household identification. They constitute the realms of interpersonal community beyond the extended family (Levy, 1990). Usually a *twa:* will be centred around a spacious square at the centre of a matrix of narrow winding streets and bazaars. These squares are usually paved and used for various agricultural and commercial purposes, as well as serving the immediate communities as a focal point where the inhabitants of that particular community meet.

The numerous squares within Patan were used by the MRD as meeting places for the discussion and planning of movement strategies. In so doing, the uses and meanings of these sites were transformed from places where the local communities would meet, to places where movement activists would organize resistance. Being interlinked within a labyrinthine web of streets, the squares afforded a protected space out of the purview of the government. The narrow streets of the town prevented any mass deployment of government forces, or the deployment of armed vehicles, while aiding the escape of activists from the police. The interconnected network of backstreets that traversed the town enabled activists to avoid the main streets, and to move unhindered from one end of Patan to the other, and from Patan to Kathmandu without detection. Throughout the revolution, the movement was able to consciously utilise the spatio-cultural configuration of the urban to break government curfews, and maintain communications between movement members in Patan and Kathmandu concerning actions, and to conduct meetings. Indeed, this use of urban space enabled the underground leadership of the MRD to remain uncaptured during the entire uprising. This use and transformation of the urban formed part of a conscious spatial strategy of resistance.

Place as strategy: liberated space

The most dramatic contestation of power and space occurred when Patan was defended against incursion by government forces for a period of one week during the revolution. During a blackout protest, the seven approach roads to the town were barricaded[14] and trenches were also dug to prevent the entry of government vehicles and personnel. The effectiveness of the barricades was enhanced by the fact that there is only one major entrance to Patan from Kathmandu, across the Bagmati River, via the Patan Gate.

The movement drew again upon local spatio-cultural practices to defend Patan from incursion by the government. In Newar culture, social relations are arranged in successively inclusive units – the household, the *twa:*, the town, etc. – whose boundaries must be protected at each level. Because it is the focus of face-to-face communities beyond the extended family, the *twa:*, or one of its neighbourhood subdivisions, is the realm in which individual and family reputation (or 'face') is at risk. To ensure moral integration within the wider community, individual and household reputation must be protected from (moral) incursion. Within the town, the *twa:* is a significant locus of action in itself (e.g. during household religious ceremonies, funeral processions, etc.), and in concert with the other *twa:* that constitute the town (e.g. during various religious festivals). Because of its social significance, the reputation and moral integration of the *twa:* (and thus of its inhabitants) within the wider community of the town is also to be protected. Likewise with the reputation and moral integration of the town within the country (see Levy, 1990: 182–6).

A member from every household in the *twa:* nearest each barricade, volunteered for the People's Security Committees to staff the barricades. In defending the town from incursion by the government, each member of the committee was also protecting their *twa:*, and their household. The traditional importance of ensuring moral protection within the boundaries of household, *twa:*, and town, was transcribed onto movement organization for the defence of Patan. The committees consisted of between 50 and 100 people, were armed with *khukris*, tools, and sticks, and staffed the barricades around the clock. All those entering the town were checked. To warn of the approach of government forces, temple bells were rung, whereupon people would rush to the barricades to prevent the government's entry. At night wood and tyres were burned in the streets and residents would gather *en masse* to protect the town, and conduct torchlight demonstrations.[15] During its liberation, the meanings associated with Patan as a place of home, community, and Newar culture became interwoven with those of Patan as a place of activist networks, meeting sites, and organization. Patan acted as a material and symbolic site of resistance to the royal regime, one that served to inspire the imaginations of the residents of the Kathmandu Valley during their challenge to the government (Interviews, Kathmandu, 1990, 1992).

After one week of liberation, the government announced a curfew and the army intervened in Patan. Movement leaders decided not to resist the army for several reasons: (1) the movement was unarmed (apart from *khukris*, tools, and sticks) and did not want to incur unnecessary bloodshed; (2) the movement did not want to de-legitimize its predominantly non-violent character; and (3) the movement wanted to avoid a direct confrontation with the army. The strategy of the movement was to challenge the panchayat system rather than the army or the king directly (the army being under the king's control rather than the control of the panchayat) (Interviews, Patan, 1992).

Despite the government intervention in Patan, the town continued to act as the base of underground operations of the movement. In contrast to Kathmandu, Patan possessed several characteristics that contributed to its utilization as a movement base: (1) the town had a history of communist activity and was, at the time of the MRD, a stronghold of communist support;[16] (2) unlike the mixed (migrant) ethnicities of Kathmandu's population, Patan had a relatively homogenous Newar population, which was easier to organize (especially given the strength of the communists, and the Newari resentment against the government) and more difficult for government agents to infiltrate; (3) unlike Kathmandu, Patan was neither the centre of government nor the place where government coercion was most concentrated (Interviews, Patan, 1992).

Globally placed

As stated earlier, the improvement of communications technology aided the movement in its struggle. The movement (particularly human rights and student groups) made use of a well developed underground press as a vector of dissent, to disseminate information concerning movement activities and the government's human rights abuses. Photocopying machines were used to print daily action reports, in Nepali and English, that were circulated both inside and outside Nepal. Events taking place within the urban spaces of the Kathmandu Valley were relayed to other places of struggle within Nepal, serving to inform and inspire those resistances. The availability of fax technology enabled the movement to communicate events to six international sites (including the United States, Germany, Australia and Switzerland) and keep the world watching, reflecting the conscious location of movement action within global (as well as local and national) space. Kathmandu as contested space became globally placed within media vectors. By relaying events taking place within Kathmandu and the other arenas of resistance to an international audience, the movement was able to partially globalize their resistance. In so doing, the movement was able to elicit international support from US, German and Swiss aid agencies, who threatened to withdraw aid to the Nepali government if the repression of the movement continued (Interviews, Kathmandu, Patan, 1992).

THE SPATIALITY OF DOMINATION

The government's response to the movement was predominantly coercive, ranging from arrests and detention to torture and police shootings of activists. The government predominantly utilized armed riot police during the course of the conflict although the army was mobilized in the later stages of civil unrest. Because of the spatial strategies of the MRD, the government attempted to control urban space, and its inhabitants, through a variety of methods.

Massive deployments of police and security forces were made in Kathmandu and police observation posts were erected throughout the city. Armed police set up road blocks in parts of the city, and along the roads leading out of the capital. During times of heightened unrest, curfews extending up to 22 hours in length were imposed by the government. During the course of the conflict, the offices of human rights groups were ransacked; a solidarity group meeting of various professional groups at Tribhuvan University was surrounded by police and all of the participants arrested; and eye-witness reports spoke of police throwing students from the windows of campus buildings to their death (Interviews, Kathmandu, 1990). Reports of police shootings in Kathmandu, Bhaktapur, Patan, Kirtipur and elsewhere in Nepal were widespread (FOPHUR, 1990a, 1990b). The most notorious example occurred during a country-wide *bandh* called by the movement on April 6th, when 200,000 people held a non-violent demonstration against the king outside the royal palace in Kathmandu. Following a police attack upon the crowd with *lathis*, some of the demonstrators responded by throwing bricks at the police. The security forces pumped tear gas into the crowds and then, as demonstrators broke shop windows and defaced King Mahendra's statue, the security forces opened fire on the crowd, killing at least 50 people[17] and injuring hundreds of others in what became known as the 'Massacre of Kathmandu'.[18] These attempts by the state to impose its power upon the urban space and population of Kathmandu were challenged directly by the movement – against a backdrop of burning tyres, street barricades and confrontational demonstrations.

Dominance/compromise

Two days after the 'Massacre of Kathmandu', the king announced that the ban on political parties had been lifted, and after consultations between political party leaders and the king, the movement was called off. An interim government, including representatives from the NC and ULF, was established. Across Kathmandu, the flags of the previously banned political parties appeared, hung from rooftops, and displayed in shop windows. The police observation posts were dismantled, and the constant police patrols of the city were curtailed. In the following few weeks renewed conflicts erupted on the streets, as the king attempted to circumvent the opposition's demands. Demonstrators and police clashed again on the streets, the curfew was renewed, and troops were deployed in the capital. However, by mid-May, a Constitution Committee was formed by the interim government, paving the way for elections the following year.

The king's compromise with the movement can be attributed to a variety of factors: (1) external factors such as the trade embargo with India (and the possible threat of destabilisation by India), and the threats of withdrawal of international aid to Nepal if the repression continued; (2) the strength of the opposition forces (which included Nepal's intelligentsia and professional classes) and the success of their underground networks that were difficult to locate and neutralize; (3) the movement's use of non-violent sanctions which not only imposed economic and political costs on the regime, but also legitimized the movement in foreign eyes while at the same time de-legitimized the government's use of violent repression; (4) the extensive use of the army to suppress the movement would have risked armed struggle (witness Nepal's recent history) and a possible civil war. This raised the spectre of Indian support to the opposition (as in the past) and, with the trade embargo, the possibility of an Indian invasion; (5) there were rumours that a section of the army had revolted against the use of more force against their own citizens, that army morale was low, and that the army might rebel and stage a coup; and (6) the existence of foreign pressure for a compromise solution to prevent the communists from seizing power (Interviews, Kathmandu, Patan, 1992). Clearly, international factors were as important as local ones in precipitating the king's decision, giving credence to the movement's strategy of eliciting international support and sympathy for their struggle. The result of the compromise was that the NC and ULF would contest parliamentary elections, with the king being demoted to a constitutional monarch.

Compromise/cooption

As a result of their revolution, Nepalis were able to vote (for the first time in over 30 years) in national elections for a parliamentary democracy, replete with a constitution protecting political, human and economic rights (INSEC, 1991). Despite this 'success', many of the fundamental power relations that permeate Nepali society – such as those of caste, land, gender, and ethnicity – and their institutionalisation (through the judiciary, legislation and the executive powers) have remained intact. In addition, the king retains control of the army. A compromised democratic state power has replaced an autocratic state power. Nepal's revolution of 1990 was located within the permissible terrain of the state: although the MRD sought to eliminate the panchayat system and curtail the king's absolute power, the goal of parliamentary democracy acknowledged the political and economic integrity of the Nepali state, its structural and cultural differentiations, and the elite nature of party politics. In this sense, the party-led MRD was reformist rather than revolutionary in its goals, characterized by what Cohen and Arato (1992) term 'self-limiting radicalism'. Indeed, many activists whom I interviewed felt that the movement was coopted by the political parties (Interviews, Kathmandu, Patan, 1992).

There was also a clear spatial outcome to the revolution. Within the cities people have experienced greater freedom of expression, and freedom to associate. However, according to human rights workers these freedoms continue to be seriously compromised in rural Nepal, where the feudal system continues intact, caste hierarchies continue to serve as a means of exploitation and domination, and the landlords, panchayat power brokers and elites continue to wield much of the power in the country (Interviews, Kathmandu, 1992). Those few gains elicited by the movement appear confined to the urban areas of the country, in part reflecting the spatially specific terrain of movement agency.

PLACES AND RESISTANCES

This narrative of Nepal's revolution represents but some of the web threads that constituted the event and experience of that resistance. Through a discussion of some of the practices of resistance conducted within the Kathmandu Valley, my purpose has been to argue for a spatially contextualized interpretation of collective action. Such an interpretation provides some important insights into the relationship between resistance practices and the places in which, and from which, they are articulated. As I mentioned at the beginning of this paper, there is no overarching theory that can be constructed that will hold true for all times and places. Indeed, there is no precise, stable meaning to the terms 'resistance' and 'place'. They are always negotiable and contested. Despite similarities in the strategic sanctions that are adopted for the waging of conflict, the articulation of resistance is always contingent upon the spatio-cultural conditions of its emergence and the character of its participants. As I have shown by reference to Nepal's revolution, resistances are rhizomatic multiplicities that can involve myriad interwoven processes, relations, and meanings, be they symbolic, imaginative, political, cultural, or material. Attentive to these caveats, a spatiality of resistance inquires into why resistances emerge where they do; how practices of resistance are constitutive of different relationships to space; how these relationships enable or constrain such articulations of resistance; and how the character and meaning of place may change when it becomes a site of resistance. From these questions, and their application to the articulation of resistance in Nepal, we may posit some broad theoretical approaches with which to inquire into other resistances: (1) that resistances frequently attempt to construct 'homeplaces' which act as sources of agency and solidarity in which and from which, the strategies of resistance can be organized and conceptualized; (2) that these places of resistance are ambiguous, entailing inclusions and exclusions, as well as relations of domination and can thus be understood as 'third spaces'; (3) that these places are not bounded and stable, but are interconnected with broader (regional, national, international) processes and sites; (4) that the relationship between resistance practices and the places wherein they are articulated are mutually constitutive: strategic mobilities of resistance (e.g. relays, interactions) constitute places as networks, homeplaces, targets, etc., while the material, symbolic, and imaginary character of places influences the articulation of resistance; and (5) that the meaning of places may change, temporarily or permanently when they become sites of resistance.

In Nepal, sites of resistance emerged across the country. However, the MRD created a homeplace for its resistance within the Kathmandu Valley. Although important resistances were conducted in Bhaktapur and Kirtipur, the focus of the resistance was located within the urban areas of Kathmandu and Patan. Kathmandu was the locus of political, administrative, and economic power; possessed excellent communications facilities; and dense concentration of population. It was also the headquarters of many dissident political parties and organizations. Further, both Kathmandu and Patan contained large Newar populations who could be mobilized against the non-Newar government.

These homeplaces were also sites of political, economic, and cultural difference and distance: for example, between the members and supporters of the regime and those of the MRD, and between those groups within civil society who became involved in the resistance. These places of resistance can be conceived of as third spaces: ambiguous sites wherein the forces and relations of resistance and domination were entangled. For example, not all of the residents of Kathmandu were against the royal regime. Some became suspected of being informers or *agents provocateurs* and were subjected to public humiliation, and/or execution by MRD supporters. In addition, houses that did not turn off their lights during the blackout protests were marked as enemy houses: their windows were broken and threats were made that the houses would be burned if they did not extinguish their lights the next time (Interviews, Kathmandu, 1992). Finally, while resistances are enacted in particular places they are frequently unconfined to their locality. As was noted regarding Nepal, the homeplaces of resistance emerged as a result of regional and national processes; they were part of a wider (national) struggle, articulated within many different places within Nepal; and these resistances produced effects at the regional, national, and international level.

Through an analysis of the spatio-strategic mobilities of resistance, we have gained some insights into how the MRD's resistance and the places in which it was articulated were mutually constitutive. Through the tactics of the pack and the swarm, through movement relays and interactions, places of resistance were created, claimed, defended, and used. The articulation of resistance drew upon an intimate sense of place and community of the urban residents and consciously sought to use and transform urban space into a tool of the revolution. At the same time, the various spatio-cultural practices that pertain to Newar life – the use of rooftop communications, the use of squares, and Newar notions of protection from moral incursion – crucially informed movement practice. Moreover, as places became interwoven with strategic mobilities, they became transformed by the itinerancies of struggle. Places became networks and circuits for movement relays and interactions, e.g. backstreets and rooftops became relay networks, and community squares became sites of movement interaction (e.g. for strategic planning). In addition, resistance also occurred 'out of place' in that the MRD sought to globalize its resistance in order to elicit international support.

From the experiences of Nepali resistance, we might conclude that the spatial articulation of strategies and tactics is more ambiguous than de Certeau (1984a) might suggest. Hence strategies might delimit a homeplace of resistance but as we have seen these are themselves ambiguous sites. For example, Patan became a liberated space delimited from the autocratic space of the royal regime, but contained within it 'opponent sites' such as the Mangal Bazaar police station. Also, while tactics such as the pack and the swarm certainly took place within the space of the opponent (e.g. outside the royal palace) tactics such as the blackout protest were conducted within resisters' homes and, simultaneously, within the broader, *contested* space of the city. Therefore, it is useful to think of the places of resistance as third spaces, i.e. ambiguous spaces where the marginalized are (momentarily) located at the centre of power, where certain power relations are confronted, while others are (re)inscribed.

Finally, the practice of resistance can change the meaning of particular places. Clearly, changes may be interpreted differently between and within various groups, while others may not experience a change at all. However, within Kathmandu and Patan, for the protagonists of the MRD, the everyday meanings that the squares had as community meeting places, or the rooftops had as sites for symbolic rituals, became transformed as they were adapted for the exigencies of political action and communication. While this transformation of meaning was temporary – taking place during a particular moment within the unfolding drama of resistance – more permanent changes also occurred. This is because, not least, every resistance, whatever its outcome, leaves a trace in the memory of those who have lived and witnessed the drama of collective action. These traces may be evoked in songs, poems, tactics and strategies that may inspire and inform future resistances. They may remain as stories retold across generations and cultures, such as those celebrated in Eduardo Galeano's *Memory of Fire*. They may also reside in the memories that particular places evoke. For example, the streets around the royal palace are remembered as the site of the Massacre of Kathmandu, while the town of Patan is remembered for its temporary liberation (Interviews, Kathmandu, Patan, 1992). On city buildings and temple walls, faded slogans of resistance remain as reminders of the events of 1990. While economic, political, and cultural inequalities remain in Nepal, the backstreets, squares, and rooftops of Kathmandu and Patan continue to evoke the memories of revolution, whose trace articulates them as potential places of resistance.

ACKNOWLEDGEMENTS

I would like to thank Steve Pile for his useful comments on an earlier draft of this chapter.

NOTES

1 I wrote under a pseudonym because I was unsure, at the time of writing, what political consequences might accrue to me (e.g. being denied entry into Nepal) as a result of my criticisms of Nepal's government.

2 Indeed Foweraker (1995) argues that resistance in the form of social movements is often uncritically assumed to be 'progressive'.

3 Subjection refers to a form of power which makes individuals subjects. This can mean being subject to someone else by control and dependence, as well as being tied to one's own identity by conscious self-knowledge. Domination refers to ethnic, social and religious coercions. Exploitation refers to the social and spatial relations that separate individuals from what they produce. These economic, cultural and political forces and relations are interrelated (Foucault, 1983, and see also Soja and Hooper, 1993).

4 However, it is rare to find a resisting subject who acts totally alone without some form of

interrelationship with others, e.g. those who provide some form of logistical, emotional, or material support, or those who turn a blind eye when illegal actions are performed.

5 Collective entities in action (see Haraway, 1992).

6 A real space that is 'other' to the norms of conventional space and time, a counter-site in which other real sites of culture are simultaneously represented, contested and inverted (1986: 24).

7 The Public Security Act, 1961, empowered the government to detain or imprison anyone suspected of anti-government activities, without recourse to the judiciary, on the grounds of security and peace within the country. The State Affairs (Crime and Punishment) Act, 1962, empowered the government to punish anyone showing a disrespectful attitude against the working of the government, and barred the courts from any jurisdiction over those arrested under the Act. The Organization and Association (Control) Act, 1962, proscribed political parties, demonstrations, political meetings, public expression, and the publication of articles. Offenders could be imprisoned for up to three years. The Police Organization Act provided the police with wide powers of arrest, search and detention, while the Press Act banned the independent media and brought the news media under government control (Baral, 1977).

8 These strikes were notable because of the participation of a wide cross-section of Nepal's civil society. Notable absences in the movement were rural landlords, and the *Panchas* (village, town and district administrators) whose power and privilege were dependent upon the system that was challenged by the MRD. Although the *bandhs* were most effective in the urban areas (where mass mobilizations were possible), they also took place in rural areas, as previously mentioned.

9 Resistance was conducted by trades unions, community organizations (e.g. the Informal Sector Service Centre); professional organizations (e.g. of doctors, lawyers, teachers, and journalists); artists; student organizations, particularly the All Nepal National Free Students Union (ANNFSU) affiliated to the ULF, and the Nepal Students Organization (NSO) affiliated to the NC; and human rights groups, particularly Human Rights Organization of Nepal (HURON) and Forum for Protection of Human Rights (FOPHUR).

10 There is only one major road out of the Kathmandu Valley. While this impeded communications outside of the valley, it did enable the movement to successfully blockade the road out of the city to disrupt the flow of goods and supplies to and from Kathmandu (Interview; Kathmandu, 1992).

11 Tribhuvan University campus is located approximately 6 km from the centre of Kathmandu, outside Kirtipur.

12 In Kathmandu residents would keep their house doors open to shelter activists if police attacked a demonstration. When police fired tear gas into the crowds, residents would throw water from their rooftops to dampen the tear gas.

13 According to some student activists, houses that did not turn off their lights were marked as enemy houses: their windows were broken and threats were made that the houses would be burned if they did not extinguish their lights the next time (Interviews, Kathmandu, 1992).

14 The barricades were located at Patan Gate, Kumbheswor, Pulchok, Kumari Pati, Lagankhel, Sundhara and Thysal.

15 In addition to the barricading of the town, the inhabitants *gheraoed* (surrounded) and captured the main police station, at Mangal Bazaar. Barricades were erected and trenches dug around the police station, and the 128 policemen therein were held in 'custody' for the duration of Patan's 'liberation'.

16 According to Patan activists, the town had a history of peasant movement revolts, particularly against the right-wing Gorkha Parishad during the 1950s (Interviews, Patan, 1992). During the Patan Uprising, two communist groups were particularly active in the town: the Communist Party of Nepal (CPN) (Unity Centre), a coalition of the CPN 4th Congress and CPN Mashal; and the CPN (Marxist-Leninist), now the CPN (United Marxist-Leninist) (Interview, Patan, 1992).

17 This figure is based on those taken to Bir hospital in Kathmandu after the massacre (FOPHUR, 1990a, 1990b).

18 Human rights groups have estimated that as many as 500 people were killed during the MRD and over 25,000 people arrested. It is also estimated that 75 per cent of those killed were below the age of 25 years (Interviews, Kathmandu, 1992).

4

REMAPPING RESISTANCE

'ground for struggle' and the politics of place

Donald S. Moore

> Just as none of us is outside or beyond geography, none of us is completely free from the struggle over geography. That struggle is complex and interesting because it is not only about soldiers and cannons but also about ideas, about forms, about images and imaginings.
>
> (Said, 1993: 7)

Recent approaches in cultural studies often invoke a 'tropology of resistance and hybridity' in their analyses of subaltern actors traversing landscapes of culture, power, and social contestation (Loomba and Kaul, 1994: 3). Across the less post-structural terrain of agrarian studies, 'resistance' also remains a trenchant keyword even as its multivalent meanings, theoretical deployments, and political consequences continue to animate spirited debate. Despite studies of the historically, geographically, and culturally specific struggles over territory, rarely do the politics of place occupy critical ground. Ironically, as the processual, dynamic, and power-saturated aspects of identity have become prominent features of cultural studies,[1] they threaten to become relatively naturalized fixtures of the contemporary theoretical landscape, the background against which analyses carve themselves out. One acute danger is the possibility of eclipsing the situated practices through which identities and places are contested, produced, and reworked in particular localities. '[T]he geographical metaphors of contemporary politics,' Liz Bondi asserts, 'must be informed by conceptions of space that recognize place, position, location, and so on as *created*, as *produced*' (1993: 99).[2] Similarly, Caren Kaplan (1994: 138) warns that 'any exclusive recourse to space, place, or position becomes utterly abstract and universalizing without historical specificity.' Just as the conceptual keywords of 'resistance' and 'identity' demand to be problematized, then, so do predominant notions of 'place.'

If, as Doreen Massey argues, 'we make our space/spatialities in the process of our various identities,' (1995: 285) then we need to conceive of localities not as inert, fixed backdrops for identity struggles, but rather themselves products of those contestations. Instead of viewing geographically specific sites as the stage – already fully formed constructions that serve as settings for action – for the performance of identities that are malleable (if also constrained and shaped by multiple fields of power), this vision insists on joining the cultural politics of place to those of identity.[3] As bell hooks (1990b: 8) suggests, 'Cultural critics who are committed to a radical cultural politics ... must offer theoretical paradigms in a manner that connects them to contextualized political strategies.' Her view resonates sharply with the 'politics of theory' advocated by Stuart Hall, another doyen of contemporary cultural studies: 'Not theory as the will to truth, but theory as a set of contested, localized, conjunctural knowledges which have to be debated in a dialogical way' (1992b: 286). Significantly, a politics of place and positionality – a localized and grounded site of theoretical production – animates both visions, bridging a dialogue between theorizing place and the place of theorizing in cultural studies. 'Ground for struggle,' then, becomes an advisedly chosen spatial metaphor for emphasizing the situated practices that shape, but do not necessarily determine, the formations of identities and places.

For those negotiating the ever-dwindling half-life of epistemic shifts in the Western academy, 'social theory' is itself a moving target, what Derek Gregory pegs as a 'series of overlapping, contending and contradictory discourses' (Gregory, 1994: 78). The relatively recent proliferation of spatial metaphors across that shifting expanse, as Gregory points out, 'implies that the *discourse* of geography has become much wider than the discipline' (page 81). Geographers whose work has travelled widely beyond disciplinary divides – such as Edward Soja (1989, 1993) and David Harvey (1989, 1990) – see the (re)assertion of spatial concerns as a necessary corrective for temporality getting, well, historically too much air time in Grand Theory. Yet critics warn of the dangers of 'anemic geographies' (Sparke, 1994) and caution that 'Spatial metaphors are problematic in so far as they presume that space is not' (Smith and Katz, 1993: 75).[4] Insofar as space is currently 'hot,' moreover, the seductive powers of theoretical fashion similarly invite problematization. Is it possible to enlist a spatial metaphor that 'frees conceptions of identity and landscape from a repressive fixity and solidity' (Nash, 1994: 242)?

Amidst the politics of place, how does one begin to understand sites of resistance without losing sight of the grounded struggles of women and men? As Bell and Valentine (1995: 230) suggest, 'By reading resistance as spatial practice – defying the propriety of place which keeps certain people out of place, without a home, or lost in space – we can see how contested and embattled terrains can be reinscribed, redefined, remapped.' Yet what are the dangers of romanticizing these remappings of resistance and according alternative geographies an insurrectionary, insurgent or emancipatory potential? My own strategy moves from cites to sites, first identifying related positions within the seemingly disparate fields of cultural studies and literature on Third World agrarian politics. My concern is to chart how 'sites of resistance' have been conceptualized both for individuals and social collectivities. I then turn to the particularities of people and place, attending to historical contexts, in a specific ground for struggle located in Zimbabwe's Eastern Highlands.

CITES OF RESISTANCE IN AGRARIAN AND CULTURAL STUDIES

Douglas Kellner (1995: 39) has recently warned against the 'fetishization of resistance' in cultural studies. Another prominent critic, Angela McRobbie (1994: 162), seeks to 'deconstruct resistance by removing its metapolitical status,' to 'reinsert resistance at the more mundane, micrological level of everyday practices and choices about how to live.' Pieterse (1992: 11, 13) complains that 'Resistance is the default discourse of the left' and 'its politics are opaque, they must be decoded by context.' Similarly, Benita Parry lauds Fanon and Cesaire for their 'unwillingness to abstract resistance from its moment of performance' (Parry, 1994: 179), suggesting critical re-readings of their germinal positions. Hence, prominent cultural studies practitioners stress the importance of historical, cultural, and geographical *specificity* to any understanding of 'resistance.'

In the 1980s, 'ethnographies of resistance'[5,6] had their overlapping agrarian studies counterpart in the 1980s growth-industry of 'everyday forms of peasant resistance.'[7] Despite significant differences in approach, both literatures emphasized the importance of daily cultural practices, giving flesh and blood to myriad 'local contexts.' In these studies, capitalism, colonialism, and their legacies confronted the often tactical ploys of symbolic bricoleurs who fashioned alternative meanings from powerful forces of oppression. Culture, in particular, was conceptualized as a site of struggle, a fiercely contested terrain of symbolic and material practices. But as Lila Abu-Lughod (1990: 42) notes, many of these studies had a 'tendency to romanticize resistance, to read all forms of resistance as signs of the ineffectiveness of systems of power and the resilience and creativity of the human spirit in its refusal to be dominated.'

James C. Scott's (1985) *Weapons of the Weak*,[8] a germinal text overlapping these related yet distinct discursive fields, chronicled the quotidian practices of rural Malaysians negotiating local class relations and the interventions of state officials in the countryside.[9] Scott countered an earlier emphasis in agrarian studies on revolution and large-scale, collective mobilizations expressing open defiance of state policies.[10] For Scott, 'everyday resistance' consisted of

> the ordinary weapons of relatively powerless groups: foot dragging, dissimulation, desertion, false compliance, pilfering, feigned ignorance, slander, arson, sabotage, and so on. These Brechtian – or Schweikian – forms of class struggle have certain features in common. They require little or no coordination or planning; they make use of implict

> understandings and informal networks; they often represent a form of individual self-help; they typically avoid any direct, symbolic confrontation with authority.
>
> (Scott, 1985: xvi)

Situating his analysis of resistance primarily in an analytics of *class* to the exclusion of other productive and social inequalities (notably gender, age, and ethnicity), Scott focused on peasant acts '*intended* either to mitigate or deny claims' asserted by 'superordinate classes' (1985: 290). In his view, the subaltern engages a single matrix of power structured by the social relations of production, a 'location' overdetermined by the structural grid of class. Scott foregrounded the question of consciousness by stressing the *intentions*, rather than consequences, of everyday peasant resistance. Significantly, his approach expanded the domain of 'politics' beyond formal organizations and collective mobilizations, extending to what de Certeau (1984a) calls 'the politics of ploys' embedded within 'the practice of everyday life.'[11] Rather than measuring resistance against a yardstick of widespread social and political economic transformation, the micro-politics of tactical maneuvers took center stage.

While appreciating Scott's attention to the texture of local politics, critics have directly confronted Scott's position on theoretical ground. Feminists have stressed Scott's neglect of gender at the expense of class, pointing to multiple productive and social inequalities in rural Malaysia and elsewhere.[12] Anna Tsing, among other recent critics, underscores the myriad 'intersections of power and difference – gender, ethnicity, regionalism, state control – within which the poor and the rich find their oppositional identities' (1993: 90) implicitly echoing Chandra Mohanty's (1991: 14) insistence on the 'multiple intersections of structures of power' neglected by any rigid class analysis. The 'complications' of 'decentred, multiple selves, whose lives are shot through with contradictions and creative tensions,' Dorinne Kondo argues, 'explode binary oppositions' and displace essentialized understandings of class (Kondo, 1990: 220, 224). Rather than simply suggesting an 'additive' solution to Scott's model – including gender along with class[13] – critics have called into question his entire framing device.

It is a particular strategy of 'enframing,' as Timothy Mitchell (1990) astutely points out, that spills beyond Scott's binary oppositions – 'material versus ideological, actions versus words, observable versus hidden, coerced versus free, base versus superstructure, body versus spirit' – and across a Manichaen metaphysics of power and resistance prevalent in contemporary social theory: 'we mistake [the] effects of certain coordinated practices for the existence of a distinct metaphysical realm of structure or meaning that stands apart from what we call material reality' (1990: 560–562). The metaphor of culture-as-text, most closely associated with a Geertzian interpretive hermeneutics, similarly comes under attack for conceiving of 'a world of subjects who always author their own collective narratives and whose cultural identities are thus unique and self-produced. Built into the theory, therefore, is the latent notion of a subjectivity or selfhood that pre-exists and is maintained against an objective, material world, and a corresponding conception of power as an objective force that must somehow penetrate this non-material subjectivity' (Mitchell, 1990: 562).[14] The autonomous, sovereign self, an artifact of 'western' theory, thus authorizes a metaphor of culture-as-text that scripts the perfomance of social actors. The 'text,' a fixed metaphoric 'site,' resides in a static space removed from practice, performance, power, and process. From this angle, resistance becomes the intentional thwarting of 'external' forces from an imagined space of autonomy conceived of as 'outside' or 'beyond' power. Insubordination 'cites' the oppositional cultural text which is not only spatially removed, but temporally prior, to the situated practices of women and men. The metaphysics of power, then, seduced by abstract spatial and temporal metaphors, neglects the *realpolitik* of place and practice.

Thus, in Scott's (1990) *Domination and the Arts of Resistance*, an attempt to develop a more general theory of resistance, spatial metaphors meld with those of a 'hidden transcript' underwriting ideological insubordination to produce what he terms an 'infrapolitics of the powerless.'[15] 'Social spaces of relative autonomy' and 'offstage social spaces' (1990: 118) become, for Scott, *sites* where power does not saturate or colonize the consciousness of slaves, peasants, and other subalterns. Significantly, 'offstage' is defined as 'beyond direct observation by powerholders' (1990: 4), a 'hidden' transcript to counter a 'public' one. This theatrical metaphor, underscored by the notion of a 'script' governing social action, defines the spatial parameters of Scott's model. The fixed viewpoint of a presumed audience of 'powerholders' defines separate spaces for power and resistance. 'Public' space becomes naturalized by power's optic rather than viewed as one of the discursive effects through which power works.[16] As feminists have long noted, the 'public' is a politically charged and gendered social space, contrasted with notions of 'private' and 'domestic' and freighted with histories of exclusion, subordination, and control.[17] One of the ways power works, then, is through mapping social inequalities onto spatial categories that are produced through those processes, boundaries carved out through historical struggle. Performances do not unfold in a pre-given discursive field; rather, they shape the very texture and contours of that terrain (cf. Butler, 1990a, 1993).

Rigid spatial dichotomies similarly pervade the texture of a germinal perspective within Subaltern Studies, a post-colonial academic and political intervention critically interrogating the colonial historiography of India while producing alternative histories of 'nation,' 'community,' and 'the people.' Ranajit Guha's (1982: 4) inaugural manifesto emphasizes the historiographical neglect of 'the *politics of the people*,' stressing the '*autonomous* domain' of 'subaltern classes and groups'. Significantly, subalternity is given a 'site' beyond the reach of power. The resonance with Scott's (1985: 328) 'realm' of 'autonomy' – 'a social space in which the definitions and performances imposed by domination do not prevail' – is striking. Just as Scott's intentional consciousness of a sovereign self represents a privileged locus of resistance, a site not colonized by power, so too Guha's collective consciousness of insubordination is nurtured 'beyond' the reach of power in a 'domain' or 'space' of autonomy.[18] Both understandings of 'authentic' insurgency and insubordination run counter to Foucault's insistence that 'resistance is never in a position of exteriority in relation to power' (1978: 95).

As Stuart Hall (1981: 232) argues, 'there is no whole, authentic, autonomous "popular culture" which lies outside the field of force of the relations of cultural power and domination.' Rather, the 'symbols and meanings embedded in the day-to-day practices of subordinated groups,' what Nugent and Alonso (1994: 21) term 'popular culture,' are shot through with the histories of interaction with state and other 'external' agents. For this reason, Joseph and Nugent (1994: 21, 22) criticize the literature on peasant insurgency for its 'tendency to insist on the autonomy and distinctiveness of forms of "popular" resistance, as though they were self-generating phenomena sprouting in a sociocultural terrarium.' Spatial metaphors have been crucial to this critique. Reflecting on the shifting terrain of social theory, Kaplan (1995: 48) asserts: 'A notion of links between locations and subjects deconstructs the long-standing marxist cultural hegemony model by demonstrating the impossibility of finding a pure position or site of subjectivity outside the economic and cultural dynamics that structure modernity.' Instead of conceiving of a space of subalternity, insurgency, and resistance 'outside' of power, domination, or hegemony, the challenge becomes to understand their mutual imbriction. For this reason, Gyan Prakash (1990: 407) advocates the project of 'writing history as a site of contest.'

While Subaltern Studies does not represent one unified theoretical position, practitioners have engaged their critics and more recently underscored 'relative cultural and political autonomy' (Chakrabarty, 1992: 101), pointing to the 'potential for resistance within the structure of power' (Prakash, 1992: 180). Ranajit Guha himself has recently qualified his position, writing of the 'unbounded' 'field of politics' riven with the 'flux of ... fusion and displacement' (1989: 233). Articulation, rather than autonomy, and heterogeneity, rather than communitarian primordialism, have become pronounced features of the historically textured analyses of Subaltern Studies and its interlocutors.[19] Haynes and Prakash (1991: 3) underscore that 'neither domination or resistance is autonomous.' Resistance, they argue, 'should be defined as those behaviors and cultural practices by subordinate groups that contest hegemonic social formation, that threaten to unravel the strategies of domination; "consciousness" need not be essential to its constitution' (page 3).[20]

For Scott, in contrast, 'intentions are inscribed in the acts themselves' (1985: 301) and individual consciousness becomes a site of authorship, a microcosm of the larger social 'hidden transcript' author-izing forms of resistance. Yet the tendency to map resistance onto a 'site' of consciousness is not unique to working in a fundamentally Marxian or structuralist problematic. Chela Sandoval charts the feminist features of a '*topography* of consciousness in opposition, from the Greek "topos" or place, insofar as it represents the charting of realities that occupy a specific kind of cultural region' (1990: 11). In her post-structural conceptualization, 'subject positions, once self-consciously recognized by their inhabitants, can become transformed into more effective sites of resistance to the current ordering of power relations' (Sandoval, 1990: 11). How, then, does one balance an understanding of multiple subject-positions and cross-cutting matrices of power with oppositional acts forged from historically, geographically, and culturally specific locations? If critical analysis dispenses with the notion of an authentic insurrectionary space 'outside' of power – nurturing either an originary and insubordinate individual consciousness 'offstage' or sustaining the 'autonomy' of insurgent collectivities – then how does one begin to conceptualize sites of resistance; and from what critical ground? What, in turn, do such theoretical moves imply for understanding the polyvalent politics of place?

GROUND FOR STRUGGLE IN ZIMBABWE'S EASTERN HIGHLANDS

> How are places forged from spaces, who has this power to forge, how is it contested and fought over, and what is at stake in these contestations?
>
> (Pred and Watts, 1992: 195)

Hugging the Mozambican border in Zimbabwe's Eastern Highlands, the state-administered Kaerezi Resettlement Scheme is now home to approximately 1,000 smallholder families (some 5,500 official residents) spread across 18,500 hectares of rugged mountains and steep river valleys. Under colonial rule, when Kaerezi was the white-owned Gaeresi Ranch, married black male labor tenants planted pine and wattle plantations, usually 'exchanging' six months of ranch labor for 'permission' to graze, cultivate, and establish a homestead on the property where their wives and children cultivated crops in family parcels. The cash demands of rent and taxes, as well as increasing commoditization layered colonial capitalism over pre-colonial trade and migration routes, insinuating Kaerezi into a regional and transnational political economy. Adult men often worked as migrant laborers in distant mines and urban centers as far away as Cape Town, their absence enabled by their wives' and children's daily and seasonal agricultural labor. Kaerezi has long been linked to elsewheres, underscoring that 'the global has always been part of the construction of the local' (Massey, 1994a: 116).

In the decades following the colonial occupation of Rhodesia in 1890, the pace of global–local articulations increased despite a limited white settler presence in the Eastern Highlands. In 1905, a Johannesburg-based syndicate purchased approximately 63,000 hectares, including the area now comprising the resettlement scheme, from the British South Africa Company. Land speculation from afar and the lack of white-orchestrated ranching or agricultural operations on the property accorded with the wider regional pattern of 'paper alienation, with European "farms" existing only on the surveyor-general maps' (Palmer, 1977: 227). But as Harley (1988: 283, 285) notes, 'the scramble for Africa' demonstrates the 'power effects' of maps, one of which is to serve as 'a social apparatus for legally regulating appropriated lands and exacting taxation.' Africans living on expropriated lands like Kaerezi, where 'paper alienation' underwrote the recent history of colonial conquest, endured the double burden of paying rents and taxes as well as their gendered consequences: increasingly, men sought cash through long-distance migrant wage labor; women remained on the land, enduring the increased loads of agricultural labor, managing households in the periodic absence of their husbands and kinsmen.

In the 1930s, after a group of UK shareholders bankrolled the purchase of the vast Inyanga Block encompassing Kaerezi, the two Hanmer brothers divided the property, the junior taking Gaeresi Ranch along the international border. Amidst the rising tide of African nationalism in the late 1960s, the white owner sought to turn the men living on Gaeresi Ranch into full-time workers, binding them to ranch operations with new 'agreements.' As one elder man recalled the episode in 1991: 'You will sign, certifying that forever I will work for you and you will not be allowed to go and work anywhere. There will be nothing like working for 6 months and then going away [as a labor migrant]. You have to be at work always.' Kaerezians' struggles over place in the 1960s, then, were not over its pure essence 'outside' of history nor were they arguments to incarcerate a locality 'beyond' its articulations with distant elsewheres. Rather, Kaerezians sought both local landrights for family parcels that women would manage and an *un*fixing of men from place, freeing them from effective rural proletarianization by allowing male mobility, and greater earning potential, in pursuit of wage labor in distant locales.

The period of heightened labor unrest coincided with the death of a chief in 1965 widely believed to be a colonial puppet, more loyal to the dictates of indirect rule than to his followers. Kaerezians elected a new leader, Rekayi Tangwena, who had recently returned to more permanent local residence after extended periods of work in Bulawayo, the major city in the country's south. The Rhodesian government, however, refused to recognize Rekayi's authority, targeting him instead as a catalyst to labor unrest. It declared him a 'squatter' along with 36 men singled out for eviction under the 1931 Land Apportionment Act which made it illegal for Africans to reside in areas designated 'European' without tenancy contracts. Many Kaerezians not officially named in the proclamation also experienced forced removals.

From 1967 to 1972, Chief Rekayi Tangwena and his followers fought evictions from the ranch in internationally publicized court cases and on the land, arguing that their ancestral rights pre-empted the colonial alienation of land. Hanmer, the white owner, invoked the trope of private property inscribed in the records of a deed office. Tangwena and his supporters, in opposition, argued they did not live *on* a ranch but rather *in* a chieftainship ruled by Rekayi. The government land surveyor, who had sited boundary beacons, testified to the ranch's border. 'Tribal' elders rebutted with their understandings of the chieftainship's extent, itself carved out of the previous century's conquest of the domain of a powerful rainmaking lineage, describing prominent features of the landscape. They invoked social memories of a previous Chief Tangwena and his followers who migrated to Kaerezi from Barwe territory in what is now Mozambique.

Hanmer and state officials had a radically different understanding of Tangwena's chieftainship. They invoked an administrative optic[21] mapping ethnic identities onto fixed territories isomorphic with tax constituencies and the tribal polities governed through the dictates of British indirect rule, overseen by salaried and appointed chiefs. The Tangwena, colonial logic argued, were people out of their 'proper place' since their 'true' chieftainship was rooted across the international border. Hanmer and state officials cited the chiefly residence of Rekayi's government-appointed and salaried successor, a homestead near the international boundary and clearly beyond the ranch's boundary beacons. Rekayi's citational practices counter-asserted a different politics of location, weaving ethnic identity, 'tribal' authority, and landrights into his lineage's historical residence at a homestead on property claimed by Hanmer. The chief's cultural sedimentation in the local landscape, he argued, was further buttressed by historical precedence; when the Hanmers first arrived to Kaerezi, they encountered an established chiefdom.

Tenants' assertions of ancestral rights to territory, denying the white land owner's claims to labor by virtue of their residence, invoked shades of a 'sedentarist metaphysics' linking ethnic essence to the local landscape (Malkki, 1992, 1995). During the heat of evictions, Rekayi Tangwena reasoned: 'Africans have a deep spiritual feeling for their traditional land. I will only leave this land if the graves of my ancestors, the sacred hills and the valleys are transferred to the place where they want me to go.'[22] He clarified his logic to the Rhodesian national press in 1970: 'They will have to move the land with us.'[23] One can interpret his gestures as tactical moves, a 'politics of ploys' (de Certeau, 1984a) that appropriate and turn against itself the colonial administrative discourse that mapped tribal structure on the landscape, 'fixing' essentialized ethnic identities in partitioned rural territories. Yet the prominence of 'production politics' (Burawoy, 1985) on the ranch, and concerns about men being permanently *tied* to place under Hanmer's proposed labor 'agreement,' interrupt any notion of ethnic or place-specific 'purity' or isolation. The politics of place, in this context, suggest struggles over the articulations of identity and locality, mobility and return, a nodal point of routes sedimented with fiercely held claims to historical roots.

In 1969, in the face of a government proclamation ordering his eviction, Rekayi's response remained emphatic: 'I will not be moved. This is our home and I shall die here. They can kill me if they like.'[24] The white ranch owner, in opposition, enlisted the support of the Rhodesian government and its armed 'security' forces, opting for brute domination rather than hegemony. Fearful of the sacred violation of colonial principles of racial exclusion from 'white' land, government officials ordered a road constructed, and Africans were forcibly evicted from the ranch. Rhodesian security forces destroyed huts, burned crops, and seized cattle. Displaced yet defiant, Kaerezians sought refuge in the thickly forested surrounding mountains, returning, usually under cover of nightfall, from their mountain refuges to rebuild huts, cultivate fields by hand and hoe, and plant crops. Despite government 'offers' to resettle those declared 'squatters' in a number of proposed sites, a core group of former labor tenants tenanciously clung to the land. Several hundred Kaerezians built huts in their chief's compound, by all accounts the epicenter of defiance, only to see them repeatedly destroyed. Kaerezians' actions resonated with bell hooks' suggestion that 'when a people no longer had the space to construct homeplace, [they] cannot build a meaningful community of resistance' (1990a: 47). In the 1990s, vivid memories of resistance to evictions from Gaeresi Ranch wove themselves into the discursive construction of 'community' during the post-colonial period.

Under the increasing militarization of the ranch, marked by a heliocopter raid in 1972, hundreds of the displaced fled to the thick brush of surrounding mountains, living without permanent shelter for months before eventually settling across the border in Mozambique in a FRELIMO-controlled liberated zone.[25] During the raid, police forcibly seized over a hundred Kaerezi children who had been sent for safe keeping to a small grammar school situated on Nyafaru Farm, an 800 hectare multi-racial cooperative completely encircled by Gaeresi Ranch. In 1991, a woman recalled being among the schoolchildren rounded up and loaded onto trucks bound for the capital, likening the experience to cattle bound for a slaughterhouse, a narrative Rhodesian police apparently encouraged: '[The Rhodesian District Commissioner] was telling our parents to move from this area. Rings with our names on them were put around our necks. We were told that we are being taken to butcheries to be killed.' Defiance and fear mingled, in turn, articulating with the rising tide of nationalist guerrilla activity and government aggression in rural areas. Kaerezians' struggle against evictions ultimately turned to flight under the escalation of state violence.

POSTCOLONIAL POSITIONS, 'RESISTING' RESETTLEMENT

During the 1990s, Kaerezians recalled this episode of local history, remembered as distinct from the national liberation struggle, as 'suffering for the land' (*kutambudzikira nyika*) or 'struggling for territory' (*kushingirira nyika*).[26] Yet they also recalled with pride their contributions to the nationalist struggle, particularly their role in aiding the escape of Robert Mugabe who later became leader of Zimbabwe's post-independence government. Kaerezians shepherded Mugabe, with Rhodesian security forces in hot pursuit, to guerrilla bases in Mozambique from where he orchestrated a nationalist military strategy. 'That escape,' one nationalist argues, 'guaranteed our independence' (Moyana, 1987: 42). Hyperbole aside,

Mugabe's escape and subsequent leadership of Zimbabwe's post-independence government secured crucial political capital for those displaced Kaerezians who aided his wartime flight.

When Kaerezians returned from forced exile at Zimbabwe's independence in 1980, they participated in a post-colonial struggle over place that reworked symbolic sedimentations in the landscape while carving out novel forms of territory and identity. Zimbabwe's newly independent government purchased the property, declaring it a 'resettlement scheme' where appropriately named 'settlers' would receive use rights to state property. Government officials would allocate permits to reside, cultivate, and graze livestock in functionally and spatially separate categories of land to 'household heads': married men, divorced women, and widows. The chief and his six headmen, most of whom had been prominent in the struggle against colonial evictions from the ranch, invoked their authority to 'rule the land' (*kutonga nyika*) and asserted authority over land allocation to residents not of a 'resettlement scheme' but rather a 'chiefdom.'

Declared a national hero by the post-colonial government, Rekayi's struggle for landrights on Kaerezi wove itself into the narrative of the nationalist liberation war, a struggle for the territorial sovereignty of a majority-ruled nation-state. His burial, in 1984, at Heroes Acre national monument in the capital invoked its own politics of place. Some Kaerezians protested that Rekayi should rest in the culturally appropriate chiefly graveyard in Kaerezi, ensuring his spirit's solidarity with the local landscape and the livelihoods of its inhabitants. Rekayi's burial in Harare, however, ensured that his grave's symbolic capital, inscribed in the struggle for nation rather than locality, would endure in nationalist routes, not chiefly roots.

Rekayi's successor, who had never lived in Kaerezi, was appointed acting chief by the government in 1986, but had still not been formally installed in 1990 when I began a 26 month stint of fieldwork. While state officials relied on Kaerezi's status as property that changed hands from one legal owner (the former white rancher) to another (the government), the new chief asserted that the territory was his ancestral inheritance (*nhaka*) by virtue of his birthright. At the same time, the acting chief had no objections to the government salary he continued to receive despite his objections to state resettlement policy. My own arrival negotiated both 'state' and 'traditional' authority circuits; before I held a meeting with my future neighbors, I procured letters of permission for my research from the district administrator, the highest ranking government official in the region, as well as the chief. Having secured permission to traverse a state-administered 'resettlement scheme' as well as a 'chiefdom,' I sought acceptance to share the lived landscape with its residents. Like them, but from a radically different perspective, I, too, negotiated a landscape saturated with power, competing cultural meanings, and historical struggles for landrights constituting a '"fractal" world of overlapping boundaries' (Guyer, 1994: 215). Moreover, my fieldwork's own 'politics of location' was more than metaphorical. I lived in a Kaerezi family's homestead sited, contrary to state policy, outside the government's planned residential grids, a white anthropologist among black 'squatters.'

Commoners, women and men caught in the cross-fire among state and 'traditional' authorities, invoked their own social memories of 'suffering for the land' (*kutambudzikira nyika*) during colonial evictions. Recalling historical migrations of the Tangwena people from Mozambique to Zimbabwe, long-standing gendered patterns of wage labor migration to distant mines and cities, and the comings and goings of state administrators and white ranchers, the memories of Kaerezian women and men suggested the 'accumulated history of a place, with that history itself imagined as the product of layer upon layer of different sets of linkages, both local and to the wider world' (Massey, 1994b: 156). Kaerezians' recognition of these connections speaks further to a 'need to analyze the power geometries that intersect with one another in any given place – the lines of domination, subordination, influence, power *and resistance* that connect the place to others and are therefore constitutive of its very character' (Crush, 1994b: 314).[27]

In 1991, the black, Zimbabwe-born District Administrator whose US university degree bore the traces of global routes, candidly explained: 'The Tangwena people epitomized symbolic resistance to the colonial government. At independence, it was almost impossible for the [post-colonial, black majority-rule] government not to support the chief. Tangwena was a "special case." Our problem has always been defining just how special.' After Rekayi's death, government efforts tested the political capital embedded within the resonances of that resistance. According to the government resettlement logic, functionally and spatially distinct categories of land – residential, arable, and grazing – would neatly map compartimentalized practices onto the landscape. Permits would be allocated to 'household heads': married men, divorced women, and widows. Concentrated linear settlements paralleling the scheme's dirt access roads allowed, the plan argued, easier provision of 'infrastructure' such as boreholes, grinding mills, clinics, and schools. Families would 'reside' in the densely-packed linear 'settlements' on fixed 100 m × 50 m plots within a grid. They would commute to arable holdings at the site of their past homesteads. In yet a third spatio-functional category, they would pasture cattle. The scheme eclipsed the historical sedimentations of livelihood practices inscribed on the landscape insofar as most homesteads were intentionally settled near Kaerezi's abundant springs and natural waterpoints. Moving into the concentrated linear grids, where the government promised to dig boreholes at a future, unspecified date, represented moving away from reliable waterpoints and toward uncertainty and a greater dependence on the government. Moreover, Kaerezians argued that they would not be able to protect their crops from maurading wildlife at night and protested that the crowded linear grids would encourage witchcraft, thievery, and social conflict.

Despite known opposition to resettlement policy, after the death of Rekayi, the chief who was also a nationalist hero, state functionaries tried to demarcate planned linear settlements within the scheme in 1987. But pegs were uprooted, the peggers intimidated, and the exercise postponed indefinitely. 'Resettlement is viewed as eviction,' a local schoolteacher in his twenties told me in 1991. A man in his sixties, born on the land before it became Gaeresi Ranch explained:

> it was said our resettlement was different from others. Traditional leaders were supposed to carry on their duties ... I do not agree with these planned linear settlements. It's giving us problems just like people living in the 'keeps' [the bitterly-remembered concentrated security settlements Rhodesian forces shepherded rural communities into during the liberation war]. No freedom (*rusununguko*). No one will agree to it unless they are forced ... We will be like prisoners.

Rather than 'hidden transcripts,' opposition was publicly voiced to state officials, prompting the scheme's on-site administrator to inscribe his lament in a 1989 government report: 'Basically, people hate resettlement implementors.'

Most commoners 'resisted' state villagization efforts – attempts to relocate Kaerezians from their current scattered homesteads to concentrated linear residential grids severed from historically situated livelihood practices – by simply staying put. When a few families had, in 1987, feared government evictions, they began to build huts inside their assigned residential plots in the government-imagined linear settlements. The new chief, anxious to associate himself with the legacy of his famous nationalist predecessor known for fierce assertions of territorial sovereignty, ordered those moving into the linear grids to stop, fining them and forcing them to destroy their half-built huts. The new chief frequently proclaimed: 'I will chase away anyone who builds inside the government's linear grids in my territory (*nyika yangu*).' But by 1991, shortly after the government's highly publicized burning of squatters' huts in a nearby state forest, many voiced concerns that they, too, would be forcibly evicted by state authorities if they did not follow government resettlement policy. An elder man expressed many of his neighbors' fears: 'This government is made of matches.' After a threatening letter from the District Administrator arrived, effectively informing Kaerezians that anyone *not* residing in their assigned resettlement plot would lose their landrights, several families began preparing sun-dried bricks on their government-allocated stands, ostensibly to signal plans to build residences.

Amidst the escalating controversy, I encountered Angela, a woman in her late fifties, preparing bricks inside 'the lines' at the site allocated to her husband. With her hands covered in muddy clay, she told me of her trip to the chief's court where she was summoned for violating his edict forbidding hut construction in the linear grids. 'I want to build a good house on my pegged stand and if I am "arrested," my children will live peacefully in a good house,' she explained. 'When I get back from jail I will also live in a good house.' Her 'arrest' by the chief would not have sent her to 'jail,' but it ran the risk of cultural sanctions, fines, and banishment if the chief could mobilize community opinion behind him. 'Struggles over representation,' Harvey (1993b: 23) suggests, are 'as fiercely fought and as fundamental to the activities of place construction as bricks and mortar.' Angela explained that she argued her case at the chief's court, stressing the predicament caused by her neighbor, widely believed to be a powerful sorcerer, who had made threats against her family and asserted claims to the homestead she then inhabited. After consultation with his councillors, the chief told Angela to move the site of her new homestead to a spot just outside her allocated government stand. He managed to save face, despite what appeared to be no actual move of the small house's location, by reinscribing Angela's defiance of his order with his own intentions. For good measure, he dispatched his son to oversee the 'change' in the house's siting, asserting his authority over attempts at constructing inside the government grids, and making clear his close surveillance and regulation of that space.

While Angela defied the chief's edict, she was not an advocate of government resettlement policy. She spoke frequently about her desire to live *madiro*, or freely, without unwanted intervention. She invoked her own participation in struggles against colonial evictions on Gaeresi Ranch, her life in the mountain forests surrounding Kaerezi after the Rhodesian helicopter raid in 1971, her cooking for Robert Mugabe as he fled Rhodesian security forces, and her friendship with the previous chief. She frequently reminded me that the new chief had lived outside Kaerezi during what one of her neighbors described as 'the Tangwena history of suffering.' 'During the armed struggle,' he recalled bitterly, 'the new chief was comfortably seated elsewhere.' The young man was among the Tangwena children seized by Rhodesian security forces during the major 1971 raid. Taken by truck to a fenced social welfare camp in the capital, he did not see his parents for more than two years. On many occassions, Angela, the young man and many of their neighbors spoke at length about the new chief's not sharing in their struggle for place by virtue of his historical residence outside Kaerezi, suggesting 'a place on the map is also a place in history' (Rich, 1984: 212).

Angela may have 'resisted' the chief's authority by refusing his decree, but the same situated practice – assembling sun-dried bricks on her government-allocated residential site – represented, to some extent, a capitulation to state power, a tactical move to avoid the threat of eviction. Her husband's absence at their construction site, particularly after a chief's messenger physically threatened him there, and in the chief's court appeared to be as consciously gendered as was Angela's constant presence. She recalled with pride her finesse of gendered expectations during the war: 'The colonial government didn't suspect women involving themselves in politics. One day Rhodesian soliders arrived at our home and we covered our chief, Rekayi, with grass. When the soldiers saw that there were only women present, they saw us as "politically helpless" and they went away.' Perhaps she wagered that her presence at the disputed site would depoliticize defiance of chiefly authority, domesticating the risk of violence. Angela's legal rights to land, according to government resettlement policy, were necessarily mediated through her husband. Patriarchy, 'traditional' authority, and state power all converged on her residential site. Suspended in competing matrices of power and authority, a cross-fire of claims and counter-claims, Angela's hope for a 'homeplace,' a specific ground for struggle, sought to carve out a precarious livelihood from an uncertain future.

'THIRD SPACES' AND DISCURSIVE COMPLICITY

> As for politics ... space, like identity, is contingent, differentiated, and relational, and ... it thus makes little sense to conceive of any space as stabilized, fixed, and therefore outside of the possibility of counter-hegemony. In this view, all space-identity formations are imbued with oppositional potential. And thus a practical task for politics is to activate this potential through denaturalization, exposure, and contestation so as to achieve new appropriations and articulations of space and identity.
>
> (Jones and Moss, 1995: 256)

Angela's actions and Kaerezians' historical 'suffering for the land' suggest a 'dialectic of cultural struggle ... in the complex lines of resistance and acceptance, refusal and capitulation, which make the field of culture a sort of constant battlefield. A battlefield where no once-for-all victories are obtained but where there are always strategic positions to be won and lost' (Hall, 1981: 233).[28] Cultural practices caught in the 'entanglements of power and resistance' (Haynes and Prakash, 1991), Kaerezians livelihood struggles offer critical ground for interrogating contemporary social theory. They encourage me, moreover, to reflect on the traffic between the politics of theory and theorizing the politics of place.

To counter static representations of what Foucault recognized as a prominent trend in 'western' social theory – to treat space as 'dead, the fixed, the undialectical, the immobile' (Foucault, 1980: 70) – critics have advocated 'retriev[ing] from habitual invisibility the spatiality of local politics' (Smith, 1992: 60). To enliven this space, then, is to move its discursive 'site' from that of an assumed inert backdrop against which social practices unfold to the foreground of analyses of resistance and the cultural politics of identity. This move connects the metaphoric site 'in theory' with a politics of place 'on the ground.' The complex traffic between those discursive sites, however, suggests a need for a critical 'third space' to mediate while simultaneously problematizing those binary oppositions. As Bammiker (1994: xvii) asserts, 'identities are always constructed and lived out on the historical terrain between necessity and choice, the place where oppression and resistance are simultaneously located.'[29]

I poach the term 'third space' from Homi Bhabha (1990a, 1994) whose work, as geographer Gillian Rose (1995: 369) has recently noted, 'renders space a central problematic of (cultural) politics.'[30] Bhabha's sustained project, crucially, has been to understand how identity, hybridity, and the 'articulation of cultural differences' are produced through '"inbetween" spaces provid[ing] the terrain for elaborating strategies of selfhood – singular or communal – that initiate new signs of identity, and innovative sites of collaboration, and contestation, in the act of defining the idea of society itself' (1994: 1). His concerns span the 'emergence of the interstices' (1994: 2), border zones between margins, enunciative sites where abstract conceptualizations of 'the structure of symbolization' (1994: 36) explore the depths of psychoanalytic concepts, split subjectivities, and mobile subject-positions. Yet I wish to grapple with Bhabha's 'third space' of 'cultural practices and historical narratives' (1994: 217) insofar as it represents a point from which to engage contemporary accounts of power and resistance in cultural studies. Situated betwixt and between the structural certainties of any fixed binary, such a perspective explores the 'continuous "play" of history, culture and power' (Hall, 1990: 225).

As Steve Pile points out, 'radical geography needs to understand how the multiplicities of power operate with, off and against each other,' and he advocates a 'third space,' much like Bhabha, that 'refuses to settle down,' a 'process,' a sort of moving target, 'simultaneously structured by intersecting geometries of power, identity, and meaning' (1994: 265, 272, 273). Similarly, Soja and Hooper's (1993: 199) call for 'alternative geographies' advocates a 'cultural politics ... located and understood in [a] third space of political choice.'[31] And Radhakrishnan (1993: 755) advocates an 'authenticity' bearing the traces of the 'critical search for a third space that is complicitous neither with the deracinating imperatives of westernization nor with theories of a static, natural and singleminded autochthony,' 'an invention with enough room for multiple-rootedness.' Amidst the scramble for spatial metaphors, is it possible to appreciate how place, in the grounded livelihood struggles of Kaerezians, is always more than metaphor?

Problematizing sites of resistance and the politics of place may enable alternative cartographic maneuvers, but I remain troubled by my own discursive complicity, signalled by how smoothly I spin out spatial metaphors.[32] Do Angela's struggles and those of her neighbors become the ground on which 'theory' reflects, smugly inhabiting a complacent 'third space.' Like Benita Parry, I also worry over the 'inability of post-structuralism to relocate a space for transgressive discourse and self-affirmative resistance' (1991: 42). Soja (1995: 31) similarly wonders 'Can we create an effective postmodernism of resistance that involves more than bovine immobility or sitting on the fences like parodic Humpty Dumpties?' In turn, who are these solidarities, imagined or practiced, that constitute a 'we' of resistance? While I may have engaged in particular solidarities with those whose grounded struggles are historically inscribed in Kaerezi's landscape, those are not 'our' struggles, but rather 'theirs.' My mobility and ability to leave the Eastern Highlands engages global routes not available to Angela and most of her neighbors. Their mobility, as migrant laborers, war refugees living in Mozambique, and those internally displaced through post-colonial evictions, is hardly something to celebrate and romanticize.

Beyond the concern of turning Angela and her neighbors' livelihood struggles into fodder for conceptual refinement, I worry, too, that 'deterritorialization' and 'displacement' do not become simply trendy theoretical tropes, that the politics of place becomes subordinated to academic sites/cites of struggle, turf battles in what Donna Haraway (1991: 67) aptly terms 'theory wars': 'Are you for it or agin' it? The word "theory" has become a fetish: Either it's an enemy or something you embrace; you either are or are not a post-structuralist.' Angela lives in a world where the United Nations estimates that 1 out of every 130 people on the planet have forcibly taken flight, where an estimated 18 million have been forced out of their state of residence, and where an estimated 24 million people have been displaced within their own state's borders.[33] Southern Africa has been particularly hard hit. Estimates of forced removals in South Africa alone between 1960 and 1983 run to more than 3.5 million.[34] Political violence in Mozambique, where many Kaerezians lived during the 1970s when they were evicted from Rhodesian soil, has displaced nearly one-quarter of the country's 15 million people (Nordstrom, 1995: 133). In southern Africa, the ravages of war, the historical effects of colonialism, and a regional political economy heavily reliant on male wage labor – a process that has linked Angela's livelihood practices to place in a manner distinct from those of her husband and of Rekayi Tangwena, who worked in distant mines and cities while he maintained connective links to ancestral territory – have all shaped the politics of place and displacement.

Trying to steer clear of fetishizing or essentializing 'place,' I have sought to emphasize that *place matters*; it has a politics and is produced through myriad material and symbolic struggles. Places, and people's relationship to them, have histories woven into their very fabric. Resistance to resettlement policy in Kaerezi hinged, critically, on the social memory of past displacements, colonial evictions from Gaeresi Ranch. Refusal to enter the planned resettlement grids was shaped by attachments to the specific sites of ancestral homesteads and critical waterpoints. Angela and her neighbors 'located' the chief 'outside' their defiance of colonial evictions and hence their historical struggle for place and identity. This move, in turn, helped create room for Angela's maneuvering amidst the cross-cutting matrices of chiefly, patriarchal, and state power in Kaerezi. The complex texture of local politics was far from an 'autonomous' space of subaltern insurgency. Her limited mobility was constrained by those social relations and historical patterns that have long connected Angela's homeplace with distant mines and cities, the presence of state functionaries, and the flows, routes, and livelihood practices of her neighbors.

Bhabha's 'third space,' then, would do well to recall the cross-fire of power and history, political economy and grounded struggle pervading cultural practices situated in any landscape of 'resistance.' As Gonzalez and Habell-Pallan (1994: 82) underscore, 'identity is not simply a matter of choice or free will, but is rather a negotiation between what one has to work with, and where one takes it from there . . . it is often in relation to place that identity must be negotiated and transformed.' As Marx mused more than a century ago, people make history in conditions not entirely of their own choosing. In Zimbabwe's Eastern Highlands, identities and places, as well as histories, have been crafted from crucibles of contention where women and men have struggled, perhaps above all else, to procure a livelihood.

As Kaplan (1994: 146) notes, 'romanticizing nomad or guerrilla cultures is a frequent practice in contemporary poststructuralist theories,' while Young (1995: 173) stresses that 'if we recall the enforced dislocations of the peoples of the South ... nomadism is ... one brutal characteristic mode of capitalism itself.' Any attempt to remap resistance, to forge a provisional, mobile, and shifting 'third space,' would do well to meld theoretical humility with vigilant self-criticism, appreciating that 'there is all the difference in the world between understanding the politics of intellectual work and substituting intellectual work for politics' (Hall, 1992b: 286). I recall Angela's memories of living in mountain forests, told to me two decades after Rhodesian troops destroyed her hut in 1971: 'it's not easy sleeping in the forest. You have to be very brave.' Her politics of location, for me, demand that any brave new cultural studies must not lose sight of her grounded struggles.

ACKNOWLEDGEMENTS

I would like to thank, without implicating, Charles Hale, Gillian Hart, Steve Pile, Allan Pred, and especially Orin Starn for critical responses to earlier versions of this argument. Field research in Zimbabwe was supported by a Fulbright Fellowship, the Social Science Research Council, the US Department of Education, and the Institute of Intercultural Studies. The Centre for Applied Social Science hosted my research affiliation while Kaerezians, as well as Nyanga District government officials, kindly tolerated an anthropologist traversing a localized cultural politics of place, for which I remain grateful. A Ciriacy Wantrup Fellowship provided crucial funding for writing, and the Department of Geography and Institute of International Studies at the University of California, Berkeley, generously provided critical 'space' and provocation.

NOTES

1 'Identity' and 'The Cultural Politics of Difference' have been related growth industries in academic and activist scholarships. To cite only among the more prominent recent interventions, see Appignanesi (ed), 1987; Clifford, 1988, 1994; Rutherford, 1990; Ferguson *et al.*, 1990; Appiah, 1992; Appiah and Gates, 1992; Hall, 1992a, 1995a; Gilroy, 1993; Gunew and Yeatman, 1993; Keith and Pile, 1993a; Laclau, 1994a; Rajchman, 1995; and Socialist Review Collective, 1995). The launching of the transdisciplinary journals *Identities* (in 1994) and *Social Identities* (in 1995) further suggests the marking of a particular theoretical 'moment.'

2 In my own discipline of anthropology, Appadurai (1995) has recently criticized ethnography for eliding the 'multifarious modes for the production of locality.'

3 See Watts, 1991; Jackson, 1989; Soja and Hooper, 1993; Jacobs, 1994; Rose, 1994; Willems-Braun, 1994.

4 For sharp critiques of the uncritical use of spatial metaphors, see Nash, 1994; Sparke, 1994; and Willems-Braun, 1994.

5 For example, see Taussig, 1980; Willis, 1981; Comoroff, 1985; Fox, 1985; Scott, 1985;Ong, 1987.

6 I make no definitive claims here to either periodization or canonicity, and betray a North American perspective on the circuits through which 'theory' travelled. Earlier work by Birmingham's Centre for Contemporary Cultural Studies (especially Hall and Jefferson, 1976, and Hebdige, 1979) was extremely influential as was a deeper genealogy of anthropological perspectives. See Ortner, 1995, and Starn, 1995a, for insightful reflections on these trends.

7 For example, see Scott, 1985; Scott and Kerkvliet, 1986; Colburn, 1989; Kerkvliet, 1990. For a deeper genealogy of this perspective, see Bauer and Bauer, 1942.

8 The book was significantly subtitled 'Everyday Forms of Peasant Resistance.'

9 I use the past tense not to 'locate' Scott's analysis in a closed theoretical moment discontinuous with the present, but in the hopes of historicizing it.

10 See Moore, 1966; Wolf, 1969; Paige, 1975.

11 Despite covering related terrain, Scott's major interventions along these lines (1985, 1990, 1992) have not explicitly drawn on the work of de Certeau. Note, however, how closely de Certeau's reasoning, at points, echoes Scott's: 'Many everyday practices (talking, reading, moving about, shopping, cooking, etc.) are tactical in character. And so are, more generally, many "ways of operating": victories of the "weak" over the "strong" (whether the strength be that of powerful people or the violence of things or of an imposed order, etc.), clever tricks, knowing how to get away with things, "hunter's cunning," maneuvers, polymorphic simulations, joyful discoveries, poetic as well as warlike' (de Certeau 1984a: xix).

12 See Ong, 1990; Hart, 1991; Agarwal, 1994.

13 Verena Stolcke astutely criticizes this as the 'add gender and stir' approach (cited in Cooper and Stoler, 1989: 613).

14 In a similar vein, Haynes and Prakash (1991: 11) argue: 'Scott suggests that contestatory practice is to be found, above all, in an autonomous consciousness, thus resurrecting the concept of the self-determining subject, now in the arena of everyday life.' See O'Hanlon (1988) for a critical review of subalternity and subjectivity. For a thoughtful discussion of subject formation that locates 'popular culture as a site of *struggle* between dominant discourses and forces of resistance,' see Mankekar (1993: 557).

15 For a recent critique, see Gal, 1995.

16 Scott appears, at times, to appreciate struggles over the construction of boundaries between 'public' and 'hidden' transcripts: 'it is clear that the frontier between the public and the hidden transcript is a zone of constant struggle between dominant and subordinate – not a solid wall . . .The unremitting struggle over such boundaries is perhaps the most vital arena for ordinary conflict, for everyday forms of class struggle' (1992: 76). However, his binary opposition of onstage/offstage naturalizes separate spaces of domination and resistance, refusing to acknowledge the discursive boundaries themselves as effects of power.

17 See Rosaldo, 1980; Ortner and Whitehead, 1981; Collier and Yanagisako, 1987; Radcliffe, 1993; Grewal and Kaplan, 1994; and Yanagisako and Delaney, 1995.

18 As Orin Starn's (1992: 94) perceptive critique of Guha stresses, 'Peasant politics may be distinctive, but it is never autonomous.'

19 See Spivak, 1985; O'Hanlon, 1988; Prakash, 1994; Arnold, 1993; Chakrabarty, 1994. Subaltern Studies has encouraged scholars to explore a trope of 'traveling theory' (cf. Said 1983; Clifford and Dhareshwar, 1989; Kaplan, 1996) in contexts far beyond South Asia, reflecting on historical processes in China (Hershatter, 1993), Latin America (Mallon, 1994), and Africa (Cooper, 1994; Crush, 1994a). Cooper (1994: 1518) has gone as far as to suggest that the ossification of the currently fashionable concepts of 'resistance' and 'subaltern agency' 'risk flattening the complex lives of people living in colonies' such as those of India and Africa.

20 Note the similarity of the perspective offered by Homi Bhabha, who, while he does not explicitly situate his work within the project of Subaltern Studies, frequently grapples with the representation of colonial India: 'Resistance is not necessarily an oppositional act of political intention' (1985: 153).

21 I use the optical metaphor advisedly, with its implications of a gendered and racialized colonizing gaze (cf. Pratt, 1992). This is not to deny that the '"imperialist gaze" was not homogeneous but extraordinarily heterogeneous, and those gazed upon were by no means,

passive, engaged as they were in a variety of practices of resistance, complicity, subversion, and involvement' (D. Harvey, 1995: 162). Nor should one assume a monolithic theoretical perspective on 'the gaze'; as Deutsche (1995: 174), among others, points out, feminist work on vision remains deeply divided.

22 Rekayi Tangwena, quoted in International Defence and Aid Fund (1972: 31).

23 Rekayi Tangwena, quoted in 'New Tangwena eviction move expected soon,' *Rhodesian Herald*, May 28, 1970.

24 Rekayi Tangwena, quoted in '"They can kill me" Says Tangwena,' *Sunday Mail*, September 28, 1969.

25 FRELIMO, the Front for the Liberation of Mozambique, waged a guerrilla war against Portuguese colonial rule, attaining independence in 1974.

26 In both cases, '*nyika*' traverses the semantic fields of 'land,' 'territory,' 'country,' and 'chiefdom.'

27 I certainly do not mean to imply that Kaerezians shared a uniform historical consciousness woven into a seamless understanding of place. As Feierman (1990: 29) points out, there is always a 'coexistence of multiple historicities within . . . [a] particular locality.'

28 Similarly, Foucault (1983: 211) wrote of an 'antagonism of strategies' 'locate(d)' at the 'position' where power relations necessarily encountered resistance.

29 Bammiker's perspective bears more than a faint trace of Marx's (1963 [1869]: 15) famous dictum that while people make history, they do so 'not just as they please; they do not make it under circumstances given by themselves, but under circumstances directly encountered, given and transmitted from the past.'

30 I 'locate' my position in relation to Bhabha while appreciating alternative understandings of the term. At the time of writing, Soja has a (1996) book entitled *Thirdspace*. For a related formulation, see also Lavie and Swedenburg's (1996) notion of 'Third Timespaces.'

31 For this reason, Keith and Pile (1993a: 8) emphasize the 'host of competing spatialities' encountered by social actors traversing any landscape. In so doing, they stress how spatial practices and the production of locality are necessarily imbricated in struggles over identity.

32 To signal my citational complicity, noting only a few prominent *titles*, I remain struck by the diversity of approaches to the politics of identity within cultural studies gathered under the rubric of mapping, remapping, and cartography: see, in particular, Mohanty, 1991; Watts, 1991; Jackson, 1989; Boyarin 1994; Nash, 1994; Bell and Valentine, 1995; and Pile and Thrift, 1995b.

33 For the UNHCR figures, see O'Tuathail (1995: 260). Bammiker (1994: xi) puts the figure of worldwide refugees since 1945 at 60-100 million. For crucial anthropological citations on 'deterritorialization,' see Appadurai (1990) and Gupta and Ferguson (1992, 1997).

34 See Platzky and Walker (1985: 9).

5

DANCING ON THE BAR

sex, money and the uneasy politics of third space

Lisa Law

Stereotypical representations of sex tourism in the Philippines have abounded for the past three decades. Images of middle-aged Western men debauching adolescent Filipinas in the red light district in Manila, or of American servicemen 'letting loose' in the R&R districts at the American military bases in Angeles and Olongapo, have inspired extensive critiques of American colonialism, cultural imperialism and the commodification of Filipina female sexuality. It is difficult to imagine these commoditised sexual relations without at least contemplating the structural inequalities and patterns of globalisation which have enabled the development of sex industries catering to foreign men. Indeed, it has been through the important interpretive frames provided by feminist, nationalist and anti-colonial accounts of sex tourism that the stereotypical encounters between 'voyeuristic' Western men and 'submissive' Filipino women have gained a politically strategic coherence.

The popular representation of *a-go-go* bars as sites of foreign oppression underpins, and plays a major role in sustaining and authenticating, political mobilisation against sex tourism in the Philippines, as well as internationally. Within the language of advocacy, prostitute identity is metaphorically fixed within a rich-Western-male/poor-Filipina-female dichotomy, ultimately conveying a powerful subject/disempowered other. While this representation plays an important role in highlighting the economic, political and social bases of inequality, it simultaneously reinforces the hegemonically constructed identities of the 'oppressor' and the 'victim' through naturalising them as fixed identities and subject positions. In so doing, it offers little room to manoeuvre, to negotiate identity or to resist the complex power relations constructed at points where class, race and gender intersect.

The aim of this chapter is to destabilise the naturalness of the encounter between Western men and Filipino women through an analysis of space, and more specifically, to deconstruct and reconstitute the places of sex tourism as negotiated spaces of identity.[1] This is not to deny the relations of power in the sale of sex; it is a rather different conception of power which enables us to conceive more nuanced geographies of resistance. In this way, it can be suggested that the white, male gaze is not merely an autonomous voyeurism – it has its own difficulties and uneasiness due to the gaze from the supposedly powerless bar women, who have their own sights/sites of power, meaning and identity. It can also be conveyed that bar women are capable of positioning themselves in multiple and intersecting relations of power – which include, but are not exclusive to their encounters with men – and that it is in these spaces that subjectivities, capable of resistance, are forged. By analysing the space of the bar in this manner, the presumed identity of women *as prostitutes* and men *as customers* is called into question, as well as the moralising discourses upon which these social/political portraits of oppression depend.

To examine the places of sex tourism as negotiated spaces of identity, I situate this conception within a particular space: a typical bar in Cebu City.[2] The bar is typical in the sense that it is owned by an Australian, managed by a Filipina, employs approximately 20 Filipino women, and is primarily frequented by white, Western men. Despite its vernacular status as a 'foreigner bar', however, its space is far from being clearly indigenous or foreign; it is neither and both depending on how it is framed and experienced. Nor, as many pro- or anti-prostitution activists would argue, is the space of the bar clearly liberating or oppressive; it simultaneously offers the possibility of both liberation and oppression, together with a range of other experiences. While there is no 'true' reading of this landscape, the space itself is not entirely innocent. For the space of the bar is also situated within other spaces of political, religious and moral significance, and these conflicting and interpellative discourses play a role in how this real-and-imagined space is experienced and understood. Indeed, the bar is the place where the real and the imagined merge to mediate the performance of identity.

Because Cebu's bars, and the encounters occurring within them, are beyond dualistic economies of meaning and power, I begin by specifying an understanding of 'third space', as well as its implications for strategies of resistance. This is followed by a narrative of the everyday social relations in a typical bar in Cebu City, mapping a space for a Filipina female subject. The bar is a space where dominant images and stereotypes are contested, where people speak from spaces beyond conventional representations of sex tourism, and where resistance, in potentially subversive forms, is possible. At the same time, however, it is the uneasy politics of third space encounters which invite the performance/visibility of fixed identities – identities which are thoroughly intelligible though representations which simultaneously circumscribe the stereotypes of the 'oppressor' and the 'victim'. I conclude by asking questions about the constitution of subjectivity in space and the spatial constitution of subjectivity, and how an understanding of third space can aid in our conception of geographies of resistance.

The ontology of this approach is not apolitical; instead, elaborating the negotiation of power and identity – revealing ambiguity, displacement and disjunction – refuses to totalise experience and therefore offers the potential to locate where more subtle sites of resistance are enunciated. There have been many ways of articulating the 'problem' of sex tourism within various political discourses in the Philippines, but these discourses remain largely dominated by the interpretations of middle-class activists. This is at least partially due to the reality that comprehending the political, economic and colonial dimensions of sex tourism does not necessarily provide a basis for understanding the more personalised modes of identification which form around issues of race, gender and sexuality. In short, emphasising the determined character of prostitution ironically tends to alienate its subject. Elaborating a negotiated space in order to find contemporary sites of collaboration and contestation, and new perspectives on identity, power and resistance provides a means to imagine more ambivalent deployments of power, and how space is constitutive of this process. For what if meaning and power are not necessarily mobilised through stereotypes and their associated dichotomies? Culture generally, and strategies of power in particular, are not found in predetermined categories of identity and experience, and power has the potential to be mobilised through ambivalence.[3]

'THIRD SPACE', IDENTITY AND RESISTANCE

> Meaning is constructed across the *bar* of difference and separation between the signifier and the signified.
>
> (Bhabha, 1990a: 210, emphasis added)[4]

Drawing from recent writing in feminist and post-colonial studies, Pile (1994: 255) has argued that dualisms are 'intended to mark and help police supposedly fixed, natural divisions between the powerful and the disempowered', and that 'if we accept these dualisms then we collude in the reproduction of the power-ridden values they help to sustain'. As dualisms are more fluid than their architecture would suggest, Pile outlines an alternative to dualistic epistemologies which incorporates the notion of a new geometry of knowledge, or a 'third space' (see Bhabha, 1990a, 1994). This space is a location for knowledge which: (1) elaborates the 'grounds of dissimilarity' on which dualisms are based; (2) acknowledges that there are spaces beyond dualisms; and (3) accepts that this third space itself is 'continually fragmented, fractured, incomplete, uncertain, and the site of struggles for meaning and representation' (Pile, 1994: 273). An epistemology which uses the concept of third space therefore encourages a politics of location which recognises the 'social construction of dualisms as part of the problem', as well as 'places beyond the grounds of dissimilarity – collectively named the third space' (p. 264).[5]

This conception of third space is useful for examining the articulation, transgression and subversion of dualistic categories; indeed, in a metaphorical and material sense, this third space is the bar. The bar is the location of difference, particularly of the cultural, racial and sexual differences between white, Western men and Filipino women. It is, therefore, the location where various bar experiences are articulated and assigned meaning by bar employees, the men who frequent these establishments, and researchers such as myself. Yet this third space does not contain preconstituted identities which determine experience, nor does it possess an authentic character or identity. Instead, identities are continuously negotiated through this space of difference – apropos Bhabha's (1990a) opening quote to this section – and therefore constituted through encounters with otherness. To elaborate this concept of negotiation, Pile draws on Bhabha's notions of hybridity and identification. For Bhabha (1990a: 211):

> the importance of hybridity is not to be able to trace two original moments from which the third emerges, rather hybridity ... is the 'third space' which enables other positions to emerge. This third space displaces the histories that constitute it, and sets up new structures of authority, new political initiatives, which are inadequately understood through received wisdom.

Moreover, in enabling such other positions, this third space is therefore (Bhabha, 1990a: 211):

> not so much [about] identity as identification (in the psychoanalytic sense) ... identification is a process of identifying with and through another object, an object of otherness, at which point the agency of the identification – the subject – is itself always ambivalent, because of the intervention of that otherness. But the importance of hybridity is that it bears the traces of those feelings and practices which inform it, just like a translation, so that hybridity puts together the traces of certain other meanings or discourses. It does not give them the authority of being prior in the sense of being original: they are prior only in the sense of being anterior. The process of cultural hybridity gives rise to something different, something new and unrecognisable, a new area of negotiation of meaning and representation.

I quote Bhabha at length due to the complexity of his statements, drawn – as Pile (1994) observes – from the psychoanalytic and post-structuralist traditions. What is stressed here is the ambivalence, and not fixity, of the construction of identity. 'Prostitutes' and 'customers' are actively produced through inherently unstable social encounters, and it is through these intersections of power and difference that they locate their oppositional identities. Both bar women and customers, for example, are conscious of the moralising discourses of sex tourism, yet their constructions of prostitution draw from various and different cultural, historical and gendered positions. While some Western men might be inclined to construct the issue around the moral and secular perspectives of Western feminism (e.g. the subordination of women), Filipinas might be more inclined to situate their employment within the struggles of everyday life and Catholic beliefs (e.g. suffering, shame and martyrdom). Yet the result of negotiating these differences may be the articulation or subversion of prevailing notions of powerful/powerless in terms of how they experience their encounters: a 'submissive' Filipina may simultaneously be striving for self-actualisation, a 'voyeuristic' Western customer may be humbled by an awareness that his masculinity is yoked to his pocket book. The third space is therefore capable of disrupting the 'received wisdom' which portrays sex tourism as an uncomplicated relation of domination.

Homi Bhabha's 'interstitial perspective' has recently been criticised by feminist geographers as a disembodied – and gendered – perspective (Rose, 1995). While a feminist critique of Bhabha is beyond the scope of this chapter, it is important to acknowledge at least some of these criticisms. First, Bhabha's attempts to disrupt the dominant gaze in colonial encounters spoke to my understanding of the relationships between men and women in Cebu's bars (which were simultaneously about race, class and sexuality). Bhabha's perspective appeared to be a useful way beyond the stereotypical representations of the voyeur/victim, opening a space for alternative subjectivities capable of differentially engaging the relations of the bar. Second, and if our theories are to be embodied, then hybridity and third space reflect my position as a researcher, as a white, Western woman who simultaneously identified with both *Western* men and Filipino *women*. It was not possible to locate myself at either the margins or the centre in this space – such a move would be a dubious privilege. Third, while Bhabha's conception of third space stands accused of lacking critical possibilities or a radical potential, what I attempt here is to develop Bhabha's project in ways which take these concerns into account.

If the bar is a third space, an ambivalent space of negotiation, and a site of struggle for meaning and representation, then this approach to understanding power and identity poses interesting questions for geographies of resistance. In the context of Cebu's bars, these questions circulate around issues of how Filipino women find ways to engage Western men in ways which are not outside power, but are in interstitial spaces between power and identity. By dancing on the bar, and negotiating interstitial spaces through their own perspectives, women resist the power of the voyeuristic gaze through disruption rather than covert opposition. Yet this approach to resistance necessitates a different conception of power – one that is transient, flexible and ambivalent. Writing on resistance has traditionally relied on an assumption of power which was thought to oppress coherent human subjects. Resistance was therefore anchored to an essential measure of agency (usually class) as a point from which to undertake political projects.[6] The theoretical displacement of the human(ist) subject has most recently inspired critiques of these studies, which are said to neglect the complex layerings of meaning and subjection which inform social encounters, as well as how the deployment of power is frequently laced with contradiction, irony and compromise (Abu-Lughod, 1990; Kondo, 1990). Indeed, Kondo (1990: 219) has asked if 'articulating the problematic of power in terms of resistance may in fact be asking the wrong question'.

If everyday sites of struggle occur in places of overlapping and intersecting forms of subjection, and if domination is never achieved without ambivalence, then how are we to envision power and resistance? What is evident is that Foucault's (1978: 95) formulation 'where there is power, there is resistance' is no longer sufficient. One way forward is Abu-Lughod's (1990) reformulation 'where there is resistance, there is power', and resistance is understood as a 'diagnostic' of changing relations of power. Although Abu-Lughod's focus is on gender relations, by analysing resistance on its own terms, she displaces abstract theories of power with subjectivities which are capable of contingent and flexible modes of resistance. Through focusing on these resistances, we can begin to ask questions about precisely what relations of power are at play in different contexts. This approach therefore addresses the problem of attributing people with consciousness which is not a part of their experience, as well as opening the possibility for examining the more localised contexts of resistance. Within the internal politics of the bar, for example, bar women are not merely resisting white, Western men; they are resisting stereotypical encounters with men, their co-workers and researchers such as myself, where the categories of race, class, gender and sexuality place them in uneasy positions of coherence.

Now I introduce a new border – a shift to a narrative voice – to explore this conceptualisation further. Drawing on the critical style of Malcomson (1995), I strategically destabilise the boundaries between self and other by replacing a first person narrative ('I') with a second person narrative ('you'). Malcomson's essay on the multiple identities performed by customers in a bar in Bulgaria formed the inspiration for this narrative, particularly his emphasis on the fluidity of bar patron identity. The second person narrative is not meant to reject or dismiss my own gaze, but rather to decentre and destabilise it – allowing 'you' to imagine and experience presumably discrete identities in different contexts (see Murray, 1995). It is also a device of memory, where the academic 'I' remembers the fieldworker (in this case 'you'), and a modality which reflects my own resistance to the confines of an academic discourse which occludes the smoke, beer, laughs, intimate chatter and secrets which permeate the space of the bar.

THE DIALOGICS OF THE DANCE: AN EVENING AT THE BRUNSWICK

It's 5:30 pm and you're in the Brunswick waiting for Cora. She agreed to meet for a beer before work so you could talk about her life in Cebu. Last month she told you she was completing high school in Leyte a few months ago when her poverty-stricken parents forced her to migrate to Cebu to seek paid employment. Because of her lack of education, she resorted to working in a bar: 'I had no choice', she'd said, 'what else could I do?' You know this is a common story in and around the growing metropolis, but Edna – another employee of the Brunswick and your friend for almost a year now – said they worked together in a bar around the American military bases before the

Mount Pinatubo eruption. 'That's why she got those breast enlargements', Edna elaborated, 'and where did you think she learned to speak Pampangan? Of course she's been working in a bar before.' You're interested in hearing Cora's side of the story.

It's 6:30 pm and you sit at the bar watching women trickle into the Brunswick from the crowded streets of Fuente Osmeña. Anna, Edna, Maryann and some new women have arrived, but there's no sign of Cora. While you wait, you spend some time talking to Fely, the manager, who has proven to be an endless source of anecdotes about life in the Philippines. Most of the bar women believe Fely used to work as a prostitute in Manila, where she met her husband, the Australian owner of the bar. Fely disregards such tsismis *(gossip) as pure jealousy over her elevated social status, however, and tells you grand stories about working in various jobs around the Philippines.*

Fely spends time talking to you because she too is taking note of recent arrivals; indeed, she is waiting for them. Each night she sits at the bar writing each woman's name onto a list as they arrive. The list is very much like a roll call – Fely is very strict on attendance and punctuality – but will also serve as the dancing order and an inventory of the ladies' drinks each woman earns for the evening. Fely's air of formality and strictness contrasts with the chatting and laughter emanating from the back rooms where her employees are spending their time preparing for the night: dressing up, putting on make-up, eating dinner, chatting about last night's business or their boyfriends and wondering if the night's business will be good.

Cora shows up at almost 7:00 pm, and you know she doesn't have time to talk. She just strolls past by you, barely catching your eye and not saying a word. Later Edna will inform you that Cora cannot separate your identity from that of a religious or social worker, and after dealing with your frustration – you've emphasised your support and don't want people to hold that opinion of you – you accept that as a good strategy for women who have been previously condemned for their employment. 'Anyway', Edna will say, 'she'll never understand and you have lots of friends who want to talk to you in the bar already.'

This passage is simultaneously a story about identity, power and resistance. Cora's refusal to separate my identity (as a student/researcher) from a religious or social worker[7] can be read in many ways. First, it can be seen as resistance to confronting the power of moralising discourse; Cora resisted such power through refusing to talk. Second, Cora's identification of herself as a particular subject – an immoral Filipina – relates to prevailing attitudes about prostitution. Most bar women are aware that in many Western countries, as in the Philippines, prostitution is conceived of as an immoral form of employment, although Western men and women are more inclined to understand the context of poverty in the Philippines. While this identification may have been ambiguous for Cora, it nonetheless impeded our ability to communicate. Edna, on the other hand, was pleased to inform me that I had many friends in the bar already, and that my research did not require a formalised interview with Cora. This identification – where to Edna my interest developed from research to empathy and friendship – led to a new area of negotiation between friends like Edna and myself, where I learned that constructing her as a subject was only possible through my academic pursuits. To Edna I was merely a student at the local university who was interested in the lives of bar women and, while she patiently answered my many questions, I assumed part of that identity too.

Cora's breast enlargements are one example of how the bodies of Filipino bar women – or more appropriately, 'dancers', as they refer to themselves – become marked through the selective incorporation of features of an imagined Western sexuality. Other examples would be the use of cosmetics, and various drugs to control their weight. These practices are often interpreted by middle-class Filipinos as a purely Western influence – large breasts and skinny bodies are not highly valued in the Philippines, but are dominant in imported pornography and television programmes – but they could also be read as the psychological dynamics of resistance which allow women to retain control of their bodies. Dancers re-learn their bodies in new and often revolutionary ways during their employment in the sex industry, and this is often equated with 'becoming modern'. Yet while dancers' bodies express modernity they also bear the marks of being Filipino. Their dancing styles and entertaining costumes, for example, follow distinctively Filipino fashion trends. It is precisely this hybridity – exotic otherness and a sexualised modernity – that male customers are often attracted to (see Manderson, 1995).

Lastly, it is important to note Cora's assertion of having 'no choice' but to work in a bar, which bears resemblance to Foucault's (1978) notion of a 'reverse discourse'. In the Philippines, political mobilisation against sex tourism has primarily been based on analyses which situate prostitution within the political economy of colonialism, militarism and sex tourism.[8] By emphasising the structural determinants of why women enter the sex industry, however, Filipino dancers are cast as 'victims' who have no choice. While it has been recently argued that this representation victimises women, denies agency and distorts the complexity of experience, here it is important to note that the naturalness of the choice issue is appropriated by the women themselves, where having 'no choice' simultaneously becomes a source of agency, a resistance to moral judgements and justification for their employment.[9]

It's Maryann's night to do manicures so you go out to the back room to get one yourself. Maryann is very petite and shy, and is fussy about the customers she goes out with, so Fely has agreed she can earn extra money from this side-business one night a week. This used to be Maryann's regular employment in her barangay *(suburb) until her husband was sent to jail last year and she needed more money to support her two children. Everyone agrees Maryann's manicures are very good, and she needs the money, so you splurge and get your toes done with red polish too.*

While you get your manicure, you realise that your own placement in the bar is related to your relationships with the women – it took a long time until you were invited to this back room, for example, or until you were comfortable sitting out front by the dance floor. You also wonder about the recent government AIDS information campaign about sterilising manicuring equipment between customers, but know this is not the time nor place to mention it. Whenever you've tried to get to this aspect of your research – to discuss AIDS as an issue for sex workers – this space becomes awkward and silent. You've wondered if AIDS has been delimited as an impossible subject, and if there are too many borders for you, as an outsider, to cross. Of course they use condoms whenever they can, but by virtue of the fact that you are not a dancer, it is difficult for you to participate in discussion of the times when they don't. Dancers are obliging in terms of discussing their sexual lives with you, but condom use, if discussed in a rational or clinical way, is difficult. You end up feeling like a City Health Official and the answers you get are rehearsed.

The recent government AIDS education campaign did not go unnoticed by bar women in Cebu. This was partially due to the increased information drive at their weekly checkups at the City Health Department, but also because there had been several interested non-government organisations through the bar over the past year asking them to fill out KAPB (knowledge, attitude, practice and belief) questionnaires. The most profound effect, however, came from the release of a Filipino film, *The Dolzura Movie*, which was a more emotionally charged perspective on a woman who had contracted the virus. While bar women were happy to answer KAPB questionnaires and discuss the movie in terms of their fear, they located these discussions in two distinct discursive realms: KAPB questionnaires were part of an official education campaign they were required to participate in, while the film dealt with personal circumstance, emotions and love.

There are several academic and activist 'theories' regarding why women in the sex industry do not use condoms for every sexual encounter. In the Philippines, it is emphasised that women in the sex industry do not use condoms with their regular partners, and while at work, have less negotiating power due to the oppressive nature of the business. While it is true that bar women tend not to use condoms with their regular boyfriends in order to separate work from pleasure, to equate the non-use of condoms with general disempowerment distorts the issue. The non-use of condoms, in what would appear to be unreasonable circumstances according to middle-class sensibilities, has more to do with Filipino conceptions of *bahala na* – translated by Enriquez (1990: 302) as 'risk-taking in the face of the proverbial cloud of uncertainty and the possibility of failure' – than with disempowerment or a lack of negotiating skills. Many women employed in these establishments are looking for a future husband, and the use of condoms in these instances – that is, with a prospective husband – is seen to impede the achievement of intimacy, and ultimately, an opportunity to exit the industry.

Awkwardness, particularly in the form of silences and bodily discomfort, is a common phenomenon in the bar. Such awkwardness occurs in the space between attempting to comprehend a situation, and articulating a (suitable) response to it. In a sense this uneasiness may embody the third space concept, and is what makes the bar a 'special' space. If third spaces are merely spaces where dualisms are worked out, then all spaces in a sense are third spaces. But in the Brunswick, as in most bars, uneasy silences and bodily discomfort are better conceived as the result of uncertain identities in tension. In the back room of the Brunswick, for example, discussions of HIV/AIDS had previously provoked the performance of official, public health identities, but these identities were inappropriate/d for a space which was usually used for chatting and socialising. Rather than interrupting this space with what could be considered an interrogation, I opted instead to listen to stories about Maryann's children, her previous employment and the high quality of her beautician skills.

Although discussions on HIV/AIDS in the back room fell outside appropriate codes of conduct, the public area of the bar is a space where dominant codes can be transgressed. Dominant codes of morality, for example, prohibit commercial sex between Western men and Filipino women. It is the possibility of precisely this kind of sex, however, that creates a third space of bodily discomfort and a quest for an individual subject position. If the third space is a 'space' of negotiated identity, and more specifically, if it is a space beyond dualisms, then both the bar and the body can be conceived of as sites of ambivalence and negotiation. In this space dancers are awkward on their first nights in the business, and new customers tend to be rather gauche. Researchers are 'out of place' in this environment, so negotiating an appropriate subject position – whether it be in the back room discussing HIV/AIDS or sitting out front watching a dancer perform – is hardly surprising.[10]

A few customers arrive. You look at them and suspect they're probably tourists, and probably here for the first time. They look nervous and unsure of what to do or where to sit tonight, but by tomorrow you know they'll be acting like regulars. You've stopped being so self-conscious and aware of the white, male tourist gaze for some time now, and briefly wonder why they're staring at you. One of them looks as if he is about to engage you in conversation, but Edna comes over to remind you about the birthday party in her barangay *the next day. You're grateful you don't have to talk to this man after being berated by a drunken Australian the week prior. He had demanded to know why you were in the Brunswick, and when you told him, he accused you of wasting Australian tax money by carrying out research in bars, and particularly because you're Canadian.*

Edna's neighbour Lorna, who works at a nearby bar, comes in to see what's going on in the Brunswick. She's on a steady barfine this week; she's entertaining a young Australian 'boyfriend' who she's been travelling around Cebu with. Last night he took her to a beach resort where they went jet-skiing and skinny dipping, and today she's lost her voice so cannot work. She and the Australian are playing pool in a nearby pub, the Richmond, and she asks you if you'd like to join them. The Richmond is a staple hangout for foreigners, particularly for local expats and tourists who have come to Cebu to scubadive. Lorna knows all of the bartenders and waitresses – many of her customers bring her here to socialise – so you stop to chat and compare notes on Lorna's boyfriend. Fely has unofficially banned all Brunswick employees from entering the Richmond unescorted, however, because she says it makes them look cheap because they have no reason to be there apart from 'hunting'. Lorna points out the 'hunting girls' that are working the bar tonight. 'Hunting girls', she tells you, 'aren't dancers, and don't work in a bar. They're just working when they want to, and looking for a man who will give her all their money – no barfine – and maybe she'll tell them she's not working, its just for fun.'

You play a few games of pool and think the Australian is OK. You want to check him out because Lorna is pretty serious about him. This is his second trip to see her, and he sent her 12 red roses for her birthday last month. She confides that she's planning to have his baby, while you're happy she's happy, you wonder what she needs another baby for; she already has two Amerasian children from when she was working in Olongapo. But you say nothing of the sort because you're happy for her, you know how much Filipinas love children, and how dancers often see children as an opportunity to solidify a relationship.

The community where Lorna and Edna, and several other Fuente Osmeña bar women live has an extensive history of women migrating to work in the bars around the American military bases at Angeles and Olongapo. The first bar woman is rumoured to have migrated in the late 1960s, after which time a series of personal recruitments produced a steady flow of women to both cities. Many families rely on remittances from daughters elsewhere in the Philippines, or from daughters who have married American men and are now living in the United States. The Mount Pinatubo volcanic eruption in 1991 saw the unequivocal departure of American servicemen, however, and this event also displaced thousands of dancers in both cities, many of whom came from communities such as Lorna's. Both Lorna and Edna migrated back to Cebu in 1992, and obtained employment in the bars of Fuente Osmeña. In contrast to the perspectives of those who strive to abolish sex tourism, it is important to note that they do not author themselves as victims to American imperialism.

The phenomenon of 'hunting' should not pass unremarked, although my contact with these women was fleeting and occasional. The women I spoke to who were 'hunters' either currently, or had previously, worked in a bar in Cebu. They were frustrated with the bar owners' commissions on their 'barfines', which were perceived to be an unfair reduction to their salaries (primarily thought to be obtained from dancing and entertaining customers in the bar).[11] In this case hunting provides the opportunity to transgress the boundaries of a dancer identity which, as evidenced in Fely's strictness in attendance and punctuality, as well as her ban on employees in the Richmond, is also one of 'employee'. Hunting emphasises women's looking and agency, revealing their ambivalence in negotiating their identities with Western men in different spaces.

It's 10:00 pm and you walk back to the Brunswick. Some of the regulars are around and the space feels more comfortable: there are more people around, there is less overt staring and the women are happy because they're earning drinks. One of the regulars is May's German customer Hans, but that doesn't stop a group of women from coaxing him to buy them a drink: Alice is telling him about her sick mother, and Anna is telling him she needs a break from working but her family cannot afford it. Ladies' drinks cost the customer from P50 to P100, and the women receive a P25 to P50 commission for each drink. The women know Hans is sometimes generous and, anyway, May has a Filipino boyfriend at home. You look at May, she's acting so confident – she knows she'll be out making money on short time tonight – and you remember that she's almost 35 years old although she looks much younger. She's been working in the business off and on since she was 18, and has lived with two former husbands in Europe and the Middle East respectively. She always gets the German customers because she speaks their language.

You go and join the group and say hello to everyone. The women are chattering in Cebuano and you hear them discussing how cheap Hans is because he won't buy anyone but May a drink. Behind their backs such customers are 'cheap charlies' or 'kuripot' *(in Tagalog, the national language, this is the word for 'cheap'), but Hans has been in the Philippines long enough to speak Tagalog, so they're calling him* 'tahik', *which is the Cebuano translation. You ask him if he's buying rounds, but he tells you – for what is probably the fifth time – that he just runs a scubadiving shop and earns Filipino wages. You laugh but understand his point and so does Edna, who tells everyone to go look somewhere else for a drink. Hans buys the three of you a beer.*

You ask May about the women soliciting drinks and she reminisces about a time when things were different. 'These girls we have right now, they're only interested in money', she tells you, 'like if they can get a customer that night. You see how they react if a customer walks in, and I hate that. Our old group was not like that before. I mean, we wait, give him a chance. Not like now, one customer and five girls at the table, its supposed to be the customer wants her or likes her, you know, and then they cannot say anything because they are too embarrassed to say.' You agree but wonder if a round of drinks isn't really that expensive after all.

While Edna and Hans talk about scubadiving, you and May notice one of the new dancers on the stage. She looks extremely uncomfortable because tonight is her first night, and May informs you she's a virgin. She was recruited by a friend who also took a job in the Brunswick as a virgin but married her first customer, a Norwegian man, after only three days on the job. 'I talked to her tonight,' May tells you, 'and I said if you are dancing, dance with the music. Don't dance and look around because you don't know how you're reacting on the stage. You know, its boring. The customer, instead of saying hey she's nice, says hey she's boring. So I told her that if you dance, don't think about the girls around you. Listen to the music and dance with it, then you will feel good. That's what I did before.'

Calling Hans cheap to his face, but in another language, can be understood as a source of resistance; but it is a resistance directed not only at Hans, it is also directed at May. Hans is an expatriate with a Filipino wife, and only engages in 'short time' with May. Both May and Hans have Filipino partners they return to each night, and neither are interested in an emotional relationship. Because Hans is not a prospective husband for women in the bar, but also because they know his wife – she used to work at the Brunswick a few years ago – many dancers question both his and May's morality. It is frequently insinuated that Western men differ from Filipino men in that Filipino infidelity is more concealed, and therefore more respectable. Therefore, while Hans' ineligibility and infidelity should deter women from soliciting drinks so staunchly, their inclination to solicit May's customer in particular is derived from wanting Hans pay for his pleasure, on the one hand, while questioning May's morality on the other. Many women justify their employment in the sex industry through their desire for a better life, symbolised by their desire for relationships with customers. It is precisely this absence of desire in May that vexes them.

Although my purpose here is not to analyse 'romance', the centrality of love and emotional pursuits should not be denied. Indeed, May's attraction and 'short time' with Hans is criticised by many women because they are not interested in 'love'. It would be unfair to cast dancers as merely seeking the allure of wealth or an escape from poverty, because their relations with foreign men – even if they desert or disappoint them – are most often seen fairly favourably. Furthermore, Catholic ideas about victimhood and martyrdom play a constitutive role in terms of their emotional desire. It is difficult to place their desire outside the pursuit of happiness via foreign men, marriage and more children, and this desire also connects to the non-use of condoms to achieve intimacy with customers. Contra Hochschild's (1983) position on the professionalisation of emotions in the hospitality industry, however, there are 'real' relationships that form in the bar, happy marriages are consummated, and these relationships often provide both the dancers and customers opportunities for marriage. These contradictions, which are marginalised within current debates on sex tourism, are important in the constitution of dancer subjectivity.

Because of her age and long-term employment in the bar, May usually takes it upon herself to initiate new women through advice on interacting with customers, other bar women, as well as miscellaneous tips on personal hygiene (usually on condoms, personal and customer genital health and how to avoid being diagnosed positive for STDs at the City Health Department). May's advice on dancing in this passage is interesting because she identifies an important gaze as coming from the women themselves (i.e. not the customers). Most women identify their first few weeks of employment in the sex industry as tremendously difficult, and this is at least partially due to the look May specifies. Because the bar is a space which breaches dominant codes of morality, new dancers must negotiate the way this space is understood. As the opinion on May's involvement with Hans indicates, it is also a judgemental look which continues throughout the course of bar work.

It's almost midnight and a man you might call a holiday regular comes in. His name is Derek, he's a successful businessman in Australia, and he spends all of his holidays in Cebu. You've had the opportunity to talk to him before because his good friend is your current housemate's boyfriend; you've all been for dinner, dancing and hanging around to enjoy the relative luxuries of the Graduate Hotel. The bar is crowded and one of the few seats left is beside you. He comes over to join you, May, Hans, Edna, and what would appear to be Edna's impending barfine. Derek asks you how your research is going and you tell him you're doing fine. He always wants to talk about your research because he's certain that he's an expert whose opinion you should solicit. You do want to talk to him, since his presence plays an important role in dancers' lives, and you recognise that customers all have their own constructions of bar work. Tonight Derek wants to talk about how the bars in Cebu are no different from the pick-up bars in Australia, except that the women in Cebu get paid for their sexual liaisons. He dissociates prostitution in Cebu from prostitution in Australia because in Cebu the women are interested in more than sex, and are willing to spend more of their time with him. 'It's really a much better deal for your money,' he says, 'I've been coming here almost every month for two years.'

Derek's perceived parallel between Cebu's bars and pick-up bars in Australia is partly a denial that he is paying for sex, and therefore, by implication, that he is morally degenerate and sexually inadequate. At the same time, however, it must also be acknowledged that dancers in Cebu consider themselves more than prostitutes, they are also city guides, interpreters of local culture, prospective wives, and so on. Both dancers and customers maintain these impossible spaces for themselves – as pick-up artists, as prospective wives – spaces which subvert dominant codes of behaviour and morality in interesting ways. Resistance in these spaces is about negotiating uneasy and contradictory feelings and desires.

Furthermore, it is important to note Derek's dissociation of these bars, and his desire for relationships which are about, but are also more than, sex. As mentioned above, customers are often attracted to the hybridity of dancer sexuality, but in some emotional relationships the contradictions of these desires may not be resolvable (see Manderson, 1995). While some men desire the sexy selflessness of Asian women, for example, it is also true that employment in the sex industry is constitutive of dancers' self-actualisation.

In this sense the women's intentions contradict the assumptions of the 'submissive' stereotype, emphasising that this identity is not accepted without struggle. Yet the desire for a sincere relationship does exist on the part of some customers, a large proportion of whom are divorced and in their forties and fifties. While such desire meshes well with dancers' desires for marriage, it is also true that many of the women do not find many of the men physically attractive. These men are therefore in a position where their allure is their wealth. While this position is a form of power, it also underlines the contradictory relationships between sex, money and the uneasy politics of third space.

DANCING ON THE BAR: MAPPING GEOGRAPHIES OF RESISTANCE

> [Nietzsche] insisted on a new type of philosophy or knowledge, one which, instead of remaining sedentary, ponderous, stolid, was allied with the arts of movement: theatre, dance, and music. Philosophy itself was to be written walking – or preferably, *dancing*.
>
> (Grosz, 1994: 127, emphasis added)

Dancing on the bar, as a play on words, is one way to imagine the subjectivity of Filipino dancers. First, it bears in mind the importance of dancing to employment in Cebu's sex industry. Working in a bar is about, but also exceeds, the exchange of money for commercial sex. Second, dancing *on* the bar is an attempt to displace the dominant representations of sex tourism which, through emphasising the oppressive relationship between Western men and Filipino women, fails to recognise the production of identities through the negotiation of the sex tourism encounter. Lastly, it is a way to stress the importance of the kind of *dancing* to which Grosz refers; that is, a way to recognise that women in Cebu's sex tourism industry are capable of manoeuvring to position themselves within multiple and intersecting relations of power.

The relationship between resistance and subjectivity, however, is more problematic. It might be helpful first to differentiate between identity and subjectivity, demonstrating how the performance of particular identities can itself be understood as resistance, while at the same time recognising that the performance itself tells only part of the story. In their analysis of Rio Carnival, for example, Lewis and Pile (1996) have suggested that the performance and masquerade of identity (in this case, femininity) can be a form of resistance which renders that particular identity indeterminate and unknowable. Following Butler (1990a, 1993), they maintain that the most productive effect of power is to secure these identities as visible/intelligible. Within this framework, it could be argued that dancers in Cebu are aware of, and perform, their status as objects (of Western desire) and subjects (immoral Filipinas). Many incorporate features of an imagined Western identity (e.g. the use of cosmetics, tobacco, and the attainment of an ideal breast size or weight), for example, particularly as they 'become modern'. Furthermore, and within more traditional understandings of performance, women often perceive their sexual relations in emotional terms. Although these fantasies and incorporations only become visible through stereotypes which deny the lived experience of sex work, reading them in this way also has the potential to undermine resistance as either reproducing stereotypes or as participating in their own oppression. It is therefore important to acknowledge that as these women 'become modern' and experience emotional attachments, new images and desires are simultaneously being produced.

These new images and desires exist in a third space of possibilities which are not capable of being understood within current theorising on sex tourism in the Philippines. At the same time, however, it is precisely these possibilities which are constantly engaging meaning and power: the bar is one site; the body, however, is another. Yet by restricting an analysis of resistance to the performance of identity, demonstrating how dancers ritually escape dualistic economies of meaning and power, dancers are reduced to a series of 'subject effects' which have little internal coherence (see Ortner, 1995). Instead of focusing on these 'effects', it is useful to focus on what becomes defined as contested terrain, and how resistance can, and does, surface.

Cora's refusal to speak to me, reactions to Alma's engagement in 'short time' with Hans, and Derek's dissociation of prostitution in Cebu are good examples of how power can be produced through stereotypes (i.e. and not through a clear relation of domination), and how dancers can refuse to occupy particular subject positions which place them in uneasy positions of coherence. In a sense these resistances are, indeed, effects. An alternative approach would be to replace these effects with a subject capable of resistance, but this runs the risk of essentialising the subject and 'romanticising' the concept of resistance (see Abu-Lughod, 1990). To focus instead on how individual resistances reveal power in particular contexts and how these resistances, taken together, play a role in defining the contested terrain of politics in/at the bar may be another way to conceptualise resistance while simultaneously displacing dichotomised relations of power. Within these domains there are mobile and multiple points of resistance which surface in dancers' relationships with customers, researchers, management and between themselves, and at issue are more personalised questions of self-actualisation, personal ethics, friendship, morality and so on.

One evening in the Brunswick, and this approach to resistance, raises interesting questions about what assumptions of power, agency and encounters between people have been assumed in traditional analyses of resistance. In reconceptualising the space of the bar as a negotiated space, a third space where identities are negotiated and ambivalent, performed and not fixed, it becomes possible to question the positioning of dancers as 'victims'. While the 'reality' of sex tourism in the Philippines is a historically specific reality, the practice of sex tourism, and the various identities of the bar are certainly more ambiguous. Understanding the manner in which points of resistance surface in the contested terrain of the bar offers a way to conceptualise the politics of meaning within the sites of sex tourism, as well as our comprehension of geographies of resistance.

NOTES

1 This approach is gaining increasing recognition in geography. See, for example, Keith and Pile (1993a), Valentine (1993), Bell *et al.* (1994), Pile (1994) and Forest (1995).

2 Cebu City is the second largest city in the Philippines with a metropolitan population of approximately 1 million people. The bar is typical in the sense that the events described are not fictional; they occurred in many of the different bars in which my research was carried out (between November 1992 and March 1994). The names of all people and places have been replaced with pseudonyms.

3 See Bhabha (1994: 38, 66–7).

4 Bhabha's (1990a) 'bar' is derived from the space between signifier/signified, and he is primarily concerned with 'culture' and issues of meaning in translation. I have borrowed this quote and highlighted the *bar* to spatialise this concept within the entertainment establishments of Cebu City.

5 Many authors, particularly within feminist and cultural studies, have sought to theorise this space. See, for example, Irigaray, 1985; hooks, 1991; Grosz, 1994; Bhabha, 1990a, 1994; Keith and Pile, 1993a.

6 The work I refer to here is the well known literature on the English working class (Willis, 1977) and peasants in Southeast Asia (Scott, 1985). Ong's (1987, 1991) important work on factory women in Malaysia also focuses on class-based resistance, but she brings issues such as indigenous religious systems and Japanese management techniques to bear on the concept of resistance.

7 'Social worker' was the term Edna used to describe a range of government and non-government professionals, whether they be involved in health, employment or community development issues.

8 For a review of some of these perspectives, see Enloe (1989), Miralao *et al.* (1990) and Sturdevant and Stoltzfus (1992).

9 See Law (1997) for a discussion of the choice issue in the Philippines. While the Filipino feminist movement is somewhat polarised on the issue of free and forced prostitution, this paper argues that women in Cebu's sex industry perceive a negotiated tension between their free will to enter the sex industry, and the more structural determinants that make this type of employment a job opportunity for them.

10 Geographers have also discussed the phenomenon of personal comfort in terms of the production of private and public space (see Valentine, 1989), and gay and straight space (see Bell *et al.*, 1994).

11 A barfine is the amount a customer pays to take a woman outside the bar, although dancers often refer to their paying customers as barfines as well. Barfines in Cebu were approximately P300 during 1992-4, and were usually split 50/50 between the woman and the bar. A night's salary for dancing and entertaining customers ranged from P100 to P150. During this time the exchange rate was approximately US$1=P25.

6

THE STILL POINT

resistance, expressive embodiment and dance

Nigel Thrift

At the still point, of the turning world. Neither flesh nor fleshless;
Neither from nor towards; at the still point, there the dance is,
But neither arrest nor movement. And do not call it fixity,
Where past and future are gathered. Neither movement from nor towards,
Except for the point, the still point,
There would be no dance, and there is only the dance.

T. S. Eliot, *The Four Quartets*

INTRODUCTION

I have always had problems with the word 'resistance'. On the one hand, it all too quickly conjures up a history of a politics of 'overturning, questioning, revaluing of everything' (Hill, 1975: 14) which stretches all the way from the seventeenth-century 'world turned upside down' to the twentieth-century world transfigured by 'senseless acts of beauty' (McKay, 1996; Marcus, 1989). On the other hand it can, equally precipitately, conjure up a history of myriad fugitive acts all making mock of the system which 'slide from notions of individual and group "creativity" to cultural "production" to political "resistance" – which can lead to the kind of criticisms that a friend once parodied as the discovery that washing your car on Sunday is a revolutionary event' (Morris, 1988: 214).[1]

This kind of thinking about resistance with its implicit David versus Goliath romanticism has one very distinct disadvantage: everything has to be forced into the dichotomy of resistance or submission and all of the paradoxical effects which cannot be understood in this way remain hidden (Bourdieu, 1991).

What if, instead, we were to consider 'resistance' in other less fundamentalist ways?

In attempting to answer this question, I have been much influenced by the spirits of Walter Benjamin and Michel de Certeau. Their work forms a kind of background to the arguments I make in the rest of the chapter. Both Benjamin and de Certeau set much store by 'habitual' everyday practices. Benjamin was especially interested in how, at particular turning points in history, the new tasks facing perception are solved, slowly and gradually, by the reconfiguration of practices (e.g. Benjamin, 1969, 1973, 1979; Taussig, 1992). De Certeau (1984a, 1984b) was also interested in practices, and specifically those practices that challenge dominant orders. In particular, he was, of course, known for his appreciation of the swarm of 'tactics' which are constrained but not defined by dominant social orders. Yet, ultimately, both Benjamin's and de Certeau's accounts of practices which go, however briefly and in whatever location, against the grain both seem to me to suffer from at least two potentially crippling problems. The first of these is a kind of romanticism about resistant practices. Ahearne (1995), for example, notes the way that de Certeau turns to metaphors of the night and the sea to describe these practices in ways which are not always entirely helpful. Then, second, both authors seem loath to take their arguments very far, perhaps for fear of disabling what it is they write of. For example, Ahearne (1995: 155) comes to the conclusion that de Certeau gets caught up in an 'aesthetics of incomprehension' which involves 'covering over a disconcerting proliferation of practices with a series of potentially satisfying images'.[2]

Therefore, in this paper I seek to move one step on from the too general generalities of the accounts of writers like Benjamin and de Certeau. But to make this one step will require a considerable detour. It means, first of all, summarising a particular body of 'theory' (and straightaway it ought to be pointed out that, in a number of senses, 'theory' is what it is not since the goal is, in classic Wittgensteinian terms, not knowledge but understanding) which can come to terms with the proliferation of practices. Then, in the second part of the chapter, I am in a position to write at some length about issues of power, domination and resistance from the perspective of this 'theory', by commenting upon embodiment. In turn, this means that in the third part of the chapter, I can then be much more specific about resistance, that specificity being attained through dance as a 'concentrated' example of the expressive nature of embodiment.

Some readers will be disappointed by the account I offer. It is in the nature of the subject that 'resistance' is often thought of in heroic terms as heroic acts by heroic people or heroic organisations.[3] In this chapter, I am more interested in everyday practices and how they provide, especially through embodiment, alternative modes of being in the world which afford a continually evolving symbolic resource. I want, in other words, to touch the invisible in the visible.

NON-REPRESENTATIONAL THEORY

> We must finally face up to the lack of any pre-established orders in the world.
>
> (Shotter, 1993: 25)

Since the mid-1980s, something remarkable has happened; a major change has begun to take place in the way in which the social sciences and humanities are being thought and practised – but no one has really noticed. This change is the rise of what I call non-representational theory or the theory of practices. Such theory can now be found making its way into almost every branch of social sciences and humanities – for example, in philosophy (Searle, Dreyfus), in social theory (Foucault, de Certeau), in sociology (in the early work of Giddens,[4] and in the work of Game, Latour, Law, Rose and others), in anthropology (in the work of Bourdieu), in social psychology (Billig, Shotter), even in geography (Thrift).

This theory can now, I think, be seen to be attempting to provide a non-intentionalist account of the world. Such an account has three main goals. First, it is meant to provide a guide to a good part of the world which is currently all but invisible to workers in the social sciences and humanities, with their intellectualist bent, that part which is practical rather than cognitive. Second, in providing such a guide, it is attempting to reconfigure what counts as explanation and knowledge. Its concern 'is not primarily the discovery of how the world "really" is, its representation. Instead, it is active in making a sketch, a continuity of engagement that allows us to know how things are because of what we did to bring them about' (Radley, 1995: 5). Thus, it is not concerned with propositions and denials, but with insights. It wants to change our way of looking at things by 'moving' us towards

> a new way of 'looking over' the 'play' of appearance unfolding before us, such that, instead of seeing the events concerned in terms of theories as to what they supposedly 'represent', we see them 'relationally' – that is, we see them practically, as being embedded in networks of possible connections and relations with their surroundings, 'pointing toward' the (proper) roles they might actually play in our lives.
>
> (Shotter, 1996: 7)

Third, it is attempting to produce a new kind of politics, one built from the ground up, so to speak, which takes seriously E. P. Thompson's (1967: 13) famous call of many years back to protect ordinary people 'from the enormous condescension of posterity' by appreciating, and valorising, the skills and knowledges they get from being embodied beings, skills and knowledges which have been so consistently devalorised by contemplative forms of life, thus underlining that their stake in the world is just as great as the stake of those who are paid to comment on it. In turn, that means that people can only be memorialised to a limited degree in print since this medium renders legitimate only what is written or spoken, thus cutting away 'the content of experience from the form . . . in which its meaning originated' (Radley, 1995: 19).

What, then, is non-representational theory? We can summarise its main tenets as follows:

(1) Non-representational theory is about practices, mundane everyday

practices, that shape the conduct of human beings towards others and themselves in particular sites. The domain of investigation is the absorbed skilful coping of these practices and the concern is therefore

> not so much ... with us seeing the supposedly true nature of what something is contemplatively, as with attempting to articulate how, moment by moment, we in fact conduct our practical everyday affairs – something we usually leave unacknowledged in the background to our lives.
>
> (Shotter, 1996: 2)

This is not, then, a project concerned with representation and meaning, but with the performative 'presentations', 'showings' and 'manifestations' of everyday life (Thrift, 1996a).

(2) Non-representational theory is concerned with the practices of subjectification (note the crucial 'ion'), not with the subject. A number of consequences follow. First, the subject is radically decentred. Human being is thought of, as

> a kind of machination, a hybrid of flesh, knowledge, passion, and technique. One of the characteristics of our current regime of the self is a way of reflecting upon and acting up all the diverse domains, practices, and assemblages in terms of a unified 'personality' to be revealed, discovered or worked on in each: a machination of the self – that today forms the horizon of the thinkable. But this machination needs to be recognised as a specific regime of signification of recent origin.
>
> (Rose, 1996a: 144)

It follows that such an approach requires

> only a minimal, weak or thin conception of human material on which history writes ... the human being, here, is not an entity with a history, but the target of a multiplicity of types of work, more like a latitude or a longitude at which different vectors of different speeds intersect. The 'interiority' which so many feel compelled to diagnose is not that of a psychological system, but of a discontinuous surface, a kind of infolding of exteriority.
>
> (Rose, 1996a: 142)[5]

But, second, the subject is still embodied. This means much more than the actual shape and innate capacities of the human body. It also involves basic general skills and cultural skills. Merleau-Ponty (1962: 146) makes this distinction clear:

> The body is our general medium for having a world. Sometimes it is restricted to actions necessary for the conservation of life, and accordingly it posits around us a biological world; at other times,

> elaborating upon these primary actions and running from their literal to a figurative meaning, it manifests through them a core of new significance: this is true of motor habits [sic] such as dancing. Sometimes, finally, the meaning aimed at cannot be achieved by the body's natural means: it must then build itself an instrument, and it projects thereby around itself a cultural world.

After all, embodiment is tactile, it involves an active grip on the world. The body, in other words, is understood in terms of what it can do (Grosz, 1994).

> Merleau-Ponty takes the sensing to be active from the start; he conceives the receptivity for the sensuous element to be a prehension, a *prise*, a 'hold'. The concept is Heideggerian; Merleau-Ponty envisions looking – palpating with the eyes – tasting, smelling, and even hearing as variants of handling. The tactile datum is not given to a passive surface; the smooth and the rough, the sleek and the sticky, the hard and the vaporous are given to movements of the hand that applies that to them with a certain pressure, pacing, periodically, across a certain extension, and they are patterned ways in which movement is modulated.
>
> The hard and the soft, the grainy and the sleek, moonlight and sunlight in memory give themselves not as sensorial contents but as a certain type of symbiosis, a certain way the outside has of invading us, a certain way we have to welcome it.
>
> (Lingis, 1994: 7–8, citing Merleau-Ponty, 1962: 317)[6]

Thus, even an activity like studying, 'innocent in its unwinking ocularity, may itself be in for some rough handling too' (Taussig, 1992: 13; see also Game and Metcalfe, 1996). Third, the subject is affective. Thus desire is not something incidental or inimical to subjectification; it is the bedrock of becoming subject:

> It has been argued that while psychoanalysis relied on a notion of desire as a lack, an absence that strives to be filled through the attainment of an impossible object, desire can instead be seen as what produces, what connects, what makes machinic alliances. Instead of aligning desire with fantasy and opposing it to the real, instead of seeing it as a yearning, desire is an actualisation, a series of practices, bringing things together or separating them, making machines, making reality. Desire does not take for itself a particular object whose attainment it requires; rather it aims at nothing above its own proliferation or self-expansion. It assembles things out of singularities and breaks things, assemblages, down into their singularities. It moves; it does.
>
> (Grosz, 1994: 165)

Fourth, the subject is engaged in embodied affective *dialogical* practices, that is it is born into and out of *joint action*.

> as intellectuals, we pay most attention to those kinds of activity in

> which we suppose people to know what they are doing, in which they put their plans into action, or theories into practice. But ... there are many other human activities in which – though we may be loath to admit it – we all remain deeply ignorant as to what we are doing, or why we are doing it. Not because the 'ideas' or whatever, supposedly in us somewhere informing our actions, are too deeply buried to bring out into the light of day, but because the formative influences shaping our conduct are not wholly there, in our individual heads, to be brought out. Activity of this kind occurs in response to what others have already done, and we act just as much into the opportunities, or against the barriers and restrictions they offer or afford us, as 'out of' any plans our desires of our own. Thus, the stony looks, the nods of agreement, the failures of interest, the asking of questions, these all go towards what it is one feels one can, or cannot, do or say in such situations. This is joint action: it is a spontaneous, unselfconscious, unknowing (although not unknowledgeable) kind of activity.
>
> (Shotter, 1993: 47)

Joint action has a number of important features. To begin with, it gives rise to unintended consequences, that is it has consequences which are not intended by any of the participants in an interaction but are a *joint* outcome. Then, joint action has intentionality;

> that is, at any one moment, the outcomes people construct between them have a meaning or significance, such that only certain further activities will 'fit' and be appropriate, while others will be sensed as unfitting or inappropriate and will be ignored or even sanctioned.
>
> (Shotter, 1993: 47)

Joint action always has an active rhetorical-response form, as distinct from the passive, Saussurian referential-representational form. That is, it has a practical–moral character which draws on different knowledges (which do not sum to a system or framework but are dilemmatic in nature) to make sense of things *on the spot*, not to explain everything ahead of time.

(3) Non-representational theory is always and everywhere spatial and temporal. It is spatial–temporal in three different ways. First, because for non-representational theorists the world is always in process, becoming and thereby encountering. This is a prepositional world (Serres and Latour, 1995), 'nothing is more alien to our life-form than (a world) with no up/down, front/back orientation ... no preferred way of moving, such as moving forward more easily than backwards, and no tendency towards acquiring a maximum grip on the world' (Dreyfus, 1996: 10). Second, because its domain is the surface, taken as a process 'by which things become visible and are produced' (Probyn, 1996: 12) rather than as an object. 'Arranged on the surface, things take on their full relations of proximity' (Probyn, 1996: 35), intersecting, traversing, and disrupting each other

(Grossberg, 1992).[7] Third, because space therefore transmutes into a series of contexts in which encounters are produced:

> In other words, as an instance of the joint action between them, people find themselves 'in' a seemingly 'given' situation, an organised situation that has a 'horizon' to it and is 'open' to their actions. Indeed, its organisation is such that the constraints (and enablements) it makes available influence, that is to say, 'invite' or 'inhibit', people's next possible actions.
>
> (Shotter, 1993: 47)

As a result, geography 'wrests history from itself in order to discover becomings that do not belong to history even if they fall back into it' (Deleuze and Guattari, 1994: 96).[8]

(4) Non-representational theory is concerned with technologies of being, 'hybrid assemblages of knowledges, instruments, persons, systems of judgement, buildings and spaces, underpinned at the programmatic level by certain presuppositions about, and objectives for, human beings' (N. Rose, 1996a: 132). Such an approach has a complex genealogy, but it might be suggested that three major examples can currently be found.[9] The first of these is the genealogical approach of Foucault. Foucault attempted to provide a number of 'regional' examples of technologies which take modes of being human as their object. One, the disciplinary technology, operates in terms of a detailed structuring of space, time and relations amongst activities, through procedures of hierarchical observation and normalising judgement, and through attempts to fold these judgements into the procedures and judgements which each 'individual' uses to order their own conduct. Another, the 'pastoral' technology, attempts to build

> a relation of spiritual guidance between an authority and each member of their flock, embodying techniques such as confession and self-disclosure, exemplarity and discipleship, enfolded into the person through a variety of schemes of self-inspection, self-suspicion, self-disclosure, self-decipherment and self-nurturing. Like discipline, this pastoral technology is capable of articulation in a range of different forms, in the relation of priest and parishioner, the rapist and patient, social worker and client and in the relation of the 'educated' subject to itself.
>
> (N. Rose, 1996a: 132)

A second example is so-called actor-network theory (which, in some of its versions, owes a debt to Foucault). Actor-network theory is an attempt to produce an account of technologies of being which is more performative than that provided by Foucault. The world is told, performed, embedded and represented in materials that are only partly social. The ordering struggles that take place using these materials produce more or less fragile networks that are more or less successful according to their ability to 'translate' the heterogeneous set of materials to hand to their ends. Agents are the *effects* of these endless struggles.[10] In particular, actor-network theory provides three differences from the Foucauldian model. First, it gives a much greater role to non-human agency. Human and non-human agents are associated with one another in networks and evolve together with these networks. Actor-network theory is therefore symmetrical with respect to human and non-human agency. Second, it gives a greater emphasis to contingency,

> a mode of understanding is always limited. It sometimes generates precious pools of apparent order. Certainly it doesn't hold the world in the grip of a totalising hegemony. But if this is right, then there are questions that Foucault tends to refuse. Though there are exceptions in his work – as for instance in the opening chapters of *The Order of Things* – most often he avoids explaining the ways in which discourses or modes of ordering interact as they are usually told, performed and embodied in the networks of the social. But my argument is that questions of changes in the mode of ordering, on the one hand, and their interaction on the other, are closely related. Agents, decisions, machines, organisations and their environments, speech, action, texts – I want to say that all of these change because they are recursive interording or interdiscursive effects. They all, that is, tell, embody or perform a network of multiply-ordering relations.
>
> (Law, 1994: 22)

In other words, as Law (1994: 22) puts it, 'poststructuralism meets symbolic interaction'. Third, actor-network theory provides a much more central role for space than normally found in Foucault's accounts (Law 1994: 102). Space is not just a set of different compatible or incompatible orders. In actor-network theory, space is constructed by networks; space is constantly made mobile, durable and productive.

> At root, the argument is simple. It is that some materials last better than others. And some travel better than others. Voices don't last for long, and they don't travel very far. If social ordering depended on voices alone, it would be a very local affair. Bodies travel better than voices and they tend to last longer. But they can only reach so far and once they are out of your sight you can't be sure that they will do what you have told them. So social ordering that rests upon the somatic is liable to be small in scope and limited in success. Texts also have their drawbacks. They can be burned, lost, or misinterpreted. On the other hand, they tend to travel well and they last well if they are properly looked after. So texts may have ordering effects that spread across the time and space. And other materials may have similar effects. The Palais de Versailles has lasted well, and machines, though they vary, may be mobile and last for longer than people.
>
> This, then, is the simple way of putting it. The argument is that large-scale attempts at ordering or distanciation depend on the creation of

> what Bruno Latour calls 'immutable mobiles', materials that can easily be carried about and tend to retain their shape. But to put it in this way is too simple. It sounds as if I am saying that mobility and durability – materiality – are themselves relational effects. Concrete walls are solid while they are maintained and patrolled. Texts order only if they are not destroyed *en route* and there is someone at the other end who will read them and order their conduct accordingly. Buildings may be adopted for other uses – for instance, as objects of the tourist gaze. So a material is an effect. And it is durable or otherwise as a function of its *location* in the network of the social.
>
> (Law, 1994: 102)

A third example is provided by the work of Deleuze and Guattari. For Deleuze and Guattari, the world is made up of 'assemblages' or 'machines'. These assemblages are provisional linkages of elements, fragments, flows, all of heterogeneous substance and status, which are never circumscribed by any fixed exterior coordinates and which can therefore, at any moment, extend beyond themselves, proliferate or be abolished (Guattari, 1996). Ideas, things, animals, humans; all of these bear the same ontological status. There is, in other words,

> no hierarchy of being, no preordained order to the collection and configuration of the fragments, no central organisation or plan to which they must conform. Their 'law' is rather the imperative of endless experimentation, metamorphosis or transmutation, alignment and realignment. It is not that the world is without strata, totally flattened; rather the hierachies are not the result of substances and their nature and value but of mode of organisation of disparate substances. They are composed of lines, of meetings, of speeds, and intensities, rather than of things and their relations. Assemblages or multipicities, then, because they are essentially in movement, in action, are always made, not found. They are the consequences of a practice, whether it be that of the bee in relation to the flower and the hive or of a subject making something using tools or implements. They are necessarily finite in space and time, provisional and temporary in status, they always have an outside, they do not or need not belong to a higher-order machine.
>
> (Grosz, 1994: 167–168)

Deleuze and Guattari's 'mechanics of existence' is therefore somewhat akin to actor-network theory in its attention to non-human agency, contingency, and spatiality. If anything, these attributes, are powered up. Non-human agency is emphasised by the degree to which the body is no longer seen as a unified or unifying entity; 'an organism centred either biologically or psychically, organised in terms of an overarching consciousness or unconscious, cohesive through its intentionality or its capacity for reflection and self-reflection' (Grosz, 1994: 169). The body moves promiscuously with all manner of assemblages, it is a teeming mass of multiplicities (Deleuze and

Guattari, 1987).[11] Contingency is emphasised through the vocabulary of connections, intensities, flows, speeds, lines of flight. In Deleuze and Guattari's world there is only direction and movement, never any fixed stations or final places. Spatiality is also given an extra turn of the screw: space becomes continual encounter, and thought is a consequence of the provocation of the encounter (and not vice versa). Geography therefore becomes crucial as a means of thinking through. Thus,

> Geography is not confined to providing historical form with a substance and variable places. It is not merely physical and human but mental, like the landscape. Geography wrests history from the cult of necessity in order to stress the irreducibility of contingency. It wrests it from the cult of origins in order to affirm the power of a milieu ... It wrests it from structures in order to trace the lines of flight that pass through the Greek world across the Mediterranean. Finally, it wrests history from itself in order to discover becomings that do not belong to history even if they fall back into it.
>
> (Deleuze and Guattari, 1994: 96)

What is common to each of these three approaches to technologies of being is that, in contradiction to the grand *narrativisations of being* by theorists of modernity or postmodernity, which, in truth, are not much more than a rediscovery of modernisation theory (Alexander, 1995), they set in their stead more modest *spatialisations* of being which

> render being intelligible in terms of the localisation of routes, habits and techniques within specific domains of action and value: libraries and studies; bedrooms and bathhouses; courtrooms and schoolrooms, consulting rooms and museum galleries; markets and department stores ... To the apparent linearity, unidirectionality and irreversibility of time, we can counterpoise the multiplicity of places, planes and practices. And in each of these spaces, repertoires of conduct are activated that are not bound by the enclosure formed by the human skin or carried in a stable form in the interior of an individual: they are rather webs of tension across space that accord human beings capacities to the extent that they catch them up in hybrid assemblages of knowledges, instruments, vocabularies, systems of judgement and technical artefacts.
>
> (N. Rose, 1996a: 143–144)

POWER, DOMINATION AND RESISTANCE?

If these are some of the basic tenets of non-representational theory, what is their relation to notions of power, domination and resistance? They suggest, I think, that such notions need to be reformulated in a number of different ways.

First, and following Foucault, these tenets form a positive conception of power. Thus for example;

> the individual is not to be conceived as a sort of elementary nucleus, a primitive atom, a multiple and inert material on which power comes to fasten or against which it happens to strike, and in so doing subdues or crushes individuals. In fact, it is already one of the pure effects of power that certain bodies, certain gestures, certain discourses, certain desires come to be identified and constituted as individuals.
>
> (Foucault, 1986: 234)

Second, these tenets indicate a contingent (but not idiosyncratic) conception of power. Technologies of being impute 'quite general patterning strategies to the materially heterogeneous networks of the social' (Law, 1994: 95). Thus, they tend

> *not* to want to say that God, or the scientific method, or human nature, or the functional needs of society, or the economic relations of production, determine how things turn out generally. [But] interpreted narrowly [this] could (and indeed *is* sometimes) treated as an invitation to celebrate idiosyncrasy. No doubt such a celebration has its place. But as a mode of social inquiry, this doesn't follow at all. For to talk of contingency is not to give up the search for pattern, but to assume that patterns only go so far.
>
> (Law, 1994: 96–97)

Third, it follows that this notion of power is 'local', in line with Foucault's 'capillary' notion of power.

However, technologies of being are about how 'large actors' (whether these be class or Louis Pasteur) are constituted from these local situations as 'an effect, a product of a set of alliances, of heterogeneous materials' (Law, 1994: 12). Thus, a large actor, is

> a spokesperson, a figure head, or a more or less 'black box' which stands for, conceals, defines, holds in place, mobilises and draws on a set of juxtaposed bits and pieces. So symmetrically, power or size are network effects. There is no a priori difference between people and organisations: both are contingent achievements. And if some things are bigger than others (and to be sure, some are), then this is a contingent matter. 'Macro-social' things don't exist in and of themselves. Neither are they different in kind. It is just that, in their propensity for deletion, they tend to *look* different.
>
> (Law, 1994: 101)

And deletion will depend upon the ability of networks to juxtapose bits and pieces, human and non-human, needed to build a coherent 'large actor' sometimes extending its scale, and its 'lines of force' (Latour, 1993), across the whole world (but only along the narrow corridors of the network). This is, therefore, an intensely *practical* view of power, which is implicitly spatial and temporal since it points to a ceaseless task of spatial and temporal *extension*. It deals in verbs, not in nouns. But it does not follow that, because of this, the sense of oppression is drained away:

> at their best ... modest sociologies ... are all about distribution, unfairness, pain. And, most importantly, they are all about how these are done in practice. So when the sociology of order complains that inequality is absent I now hear a different kind of complaint: an objection to the fact that the sociologies of ordering do not buy into a reductionist commitment to some final version of order: that they are not, for instance, committed to a particular theory of class or gender exploitation; that they refuse to adopt what some feminists call 'a stand point epistemology'; that their materialism is relational rather than dualist; that there is no a priori distinction between the macro-social and the micro-social. These complaints are right, but I don't believe that they are justified. For ordering sociologies, whether legislative or interpretative, prefer to explain *how* hierarchies come to be told, embodied, performed and resisted. But to choose to look at hierarchy in this way is neither to ignore it, nor to deny it. Rather it is to tell stories about its mechanisms, about its instances, about *how we all do it day by day*.
>
> (Law, 1994: 134, emphasis added)

However, in making these points, it is also important to make clear the position I want to take. Putting it crudely, there are two lines of approach to power in non-representational theory, which differ in important ways.

The first of these might be called hard-line Foucauldian. The second might be called, rather clumsily, de Certeauian. Let me start by pointing out where these two approaches agree when they write on power. First, they agree that all subjectivities are, to use Bakhtin's term, 'polyphonic', plural, working in many discursive registers, many spaces, many times. Thus,

> people inhabit many different domains at once ... and the negotiations of identities, within and across groups, is an extraordinarily complex and delicate task. It is important not to preserve either unity or single membership, either in the mingling of humans or nonhumans. We are all marginal in some regard, as members of more than one community of practice (social world).
>
> (Star, 1991: 52)

Second, they agree that this variation in regimes of subjectification is, in many ways, sufficient to account for resistance since 'resistance' is inherent in the constitution of power;

> it is no longer surprising that human beings often find themselves resisting the forms of personhood that they are enjoined to adopt. 'Resistance' – if by that one means opposition to a particular regime for the conduct of one's conduct – requires no theory of agency. It needs

> no account of the inherent forces within each human being that have liberty, seek to enhance their own powers or capacities, or strive for emancipation, that are prior to and in conflict with the demands of civilization and discipline. One no more needs a theory of agency to account for resistance than one needs an epistemology to account for the production of truth effects. Human beings are not the unified subjects of some coherent regime of domination that produces persons in the form in which it dreams. On the contrary, they live their lives in a constant movement across different practices that address them in different ways. Within these different practices, persons are addressed as different sorts of human beings, acted upon as if they were different sorts of human beings. Techniques of relating to oneself as a subject of unique capacities worthy of respect run up against practices of relating to oneself as the target of discipline, duty and docility. The humanist demand that one decipher oneself in terms of the authenticity of one's actions run up against the political demand that one abides by the collective responsibility of organisational decision-making even when one is personally opposed to it. The ethical demand to suffer one's sorrows in silence and find a way of 'going on' is deemed problematic from the perspective of a personal ethic that obliges the persons to disclose themselves in terms of a particular vocabulary of emotions and feelings.
>
> (N. Rose, 1996b: 140–141)

Third, they also agree that practices can always be 'turned':

> the existence of contestation, conflict, and opposition in practices which conduct the conduct of persons is no surprise and requires no approach to the particular qualities of human agency – except in the minimal sense that human being, like all else, exceeds all attempts to think it, simply because, whilst it is necessarily thought it does not exist in the form of thought. Thus, in any one site or locale, humans turn programmes intended for one end to the service of others. One way of relating to oneself comes into conflict with others.
>
> (N. Rose, 1996b: 141)

But the common account now breaks down, as the Foucaldian approach and the de Certeauian approach begin to diverge. These two approaches divide in three ways.

They diverge first of all in the matter of the status given to different actors in technologies of being. In the Foucauldian approach, as in the actor-network approach, there is a 'tendency to assume a sort of uniform agential status across all actors' (Michael, 1996: 61). This arises from the post-humanist perspective of non-representational theory where 'human actors are still there but inextricably entangled with the nonhuman, no longer at the centre of the action and calling the shots. The world makes us in one and the same process as we make the world' (Pickering, 1995: 26). As a result, the body, like any other object, is a construct, an inscribed surface. I believe that such an approach needs to be tempered, at the very least, by a recognition of the special status of embodiment of which de Certeau writes in his description of;

> the ordinary practitioners of the city ... [who] live 'down below', below the thresholds at which visibility begins. They walk an elementary form of this experience of the city; they are walkers, *wandersmanner*, where bodies follow the ups and downs of an 'urban text' they write without being able to read it ... The paths which respond to each other in this intertwinement, unrecognised poems of which each body is an element signed by many others, elude legibility.
>
> (de Certeau, 1984a: 93)

In other words, a Foucauldian approach cannot capture the special qualities of the body which flow from its tacit nature (Polanyi, 1967): it simply demonstrates that the body's functions and movements are shaped by discourse without considering the other, *expressive side* of its existence. Thus, through representation as a text the body as flesh is marginalised.

> The consequence of this situation is that, within discourse, the lived body is rendered knowable only through the constructions that are its multiple realities but its existence as a lived entity is effectively denied. Nevertheless, this absence continues to command the attention of theorists who repeatedly seek to represent that which is simultaneously being disavowed. From the perspective of discourse, the body is a subordinated term which, banished to the conceptual shadows, continues to exist as a nameless mass, charged with significance. That significance, arising from a materiality and a source of power beyond words, ensures that it is the object of repeated attempts to banish it on the one hand and to include and contain it on the other.
>
> (Radley, 1995: 7)

This tension becomes particularly acute when the distinction between the bodies of men and women are minimised, usually by decorporealising the figure of the woman. For example, in the radical antihumanism of Deleuze and Guattari women's bodies and subjectivities are deterritorialised only to be reterritorialised as part of a more universalist movement of becoming (Grosz, 1994). Then, there is the radical downgrading of the conversations of everyday life: a whole set of 'basic' communicative activities are elided. They simply do not figure in the monological Foucauldian model in which it often seems that the kind of 'ongoing argumentative conduct of criticism and justification, where every argumentative "move" is formulated as a response to previous moves' (Shotter, 1993: 14) is regarded as a secondary phenomenon. But the dilemmas of discourse are resolved *in practice*, not in theory. Something else then follows. The attempt to make humans and non-humans into simply different locations in networks or assemblages, with humans becoming effects performed by them is not as simple (to carry through) as it may at first seem. It is difficult to deny that human bodies have certain powers within the plane of practice which have real consequences for cultural extension (Pickering, 1995).

A second divergence concerns 'unsuccessful' or 'marginal' networks; networks with very little power. How is it possible, in the language of actor-network theory to take into account the fact that 'every enrolment [in a network] entails both a failure to enrol and a destruction of the world of the non-enrolled' (Star, 1991: 45). As Star notes, other, marginal and weaker networks are rarely completely destroyed. They work in the background. Of course, they can still inform the actors enrolled in larger networks. But the fact is that it is difficult not to see this as a significant lacuna with which the Foucauldian account only deals tangentially. In contrast, the Certeauian approach offers at least the beginning of an account through its emphasis on tactics and 'la perruque'.

In turn, this approach might be widened out to a more general focus on 'weapons of the weak' (Scott, 1985) which provide a whole set of affordances which can be deployed in marginal networks. Such weapons can exist even in the most heavily monitored societies. For example, Jowitt (1992) has identified how in Russia 'dissimulation', the withdrawal from an official sphere which can only mean trouble, was a marginal practice in Russian peasant society. Then it became a central element of everyday life in Soviet-style societies. In turn, it has now become a major block to the development of post-Soviet societies since the dissimulating subject refuses all public commitment;

> The identification of a dissimulative posture as the means of integrating public and private is Jowitt's genuine theoretical contribution, which singles him out amongst Western students of Russia who perceive that there is something specific about the way public and private hang together in communist societies but fail to capture it in a concept. Jowitt managed to pinpoint the central social practice familiar to every Soviet citizen in the 1970s and 1980s; saying something while believing the opposite to be true; participating in social rituals just for the sake of participation; even bringing up children by telling them a schism exists between what is and what ought to be, a schism not to be mentioned in public statements that should describe the world as if the ideal were real.
>
> (Kharkhordin, 1995: 209)

A third divergence concerns the means by which 'turning' is accounted for. In the Foucauldian account the constant mutation which everyday life injects into discourses is a given rather than an object of investigation. But, the elements of another account of a world of permanent dialogue, and dilemmas resolved *in practice* can be assembled from the work of de Certeau and others.[12] This account would rest on five closely related feet. One would be the importance of the Wittgensteinian insight into rule-following: to follow a rule often means not following it, all kinds of creative improvisations are often necessary to keep the rule in play, which in turn, suggests new ways of doing things. Another foot would be the unintended consequences stemming from the to and fro of joint action, from the response to what others have already done in particular concrete situations. These consequences can often not be controlled but have to be creatively worked through, so throwing up new actions and interpretations. Another foot would be provided by mistakes. The mistakes which are normal in practical everyday conduct are not only negative: they can have positive consequences by forcing participants to think in new ways of how to cope with a new situation. Yet another foot is the way in which stories about the world are changed as they are passed through concrete situations which will normally involve *many* evaluative positions: thus discourses are constantly mutating as new elements are fed into them as a creative response to new circumstances. Then, one more foot is the body which is a key expressive mechanism which is able to be used ambiguously and ambivalently, in what Radley (1995) calls an *elusory* way, to configure realms of experience: 'significant meaning would not be better achieved were we able to make everything visible, nameable; gesture depends upon its embodiment for communicating that which is elusory' (Radley, 1995: 17). Or as Merleau-Ponty (1968: 150) puts it, '[many] ideas would not be better known to us if we had no body and no sensibility ... they owe their authority to the fact that they are in a transparency behind the sensible, or in its heart'.

EMBODIMENT WITH ATTITUDE: THE CASE OF DANCE

> If I could tell you what it meant, there would be no point in dancing it.
>
> (Isadora Duncan, cited in Bateson, 1977)

I now want to try to add to these thoughts on power, domination and resistance by considering some of the points I have made about embodiment in more detail. Let me start by making some general points about embodiment. I am assuming that 'embodiment' (note the use of a process word rather than a noun) has the following characteristics. First, it has to be seen through Merleau-Ponty's notion of '*the flesh*', as a reversible and reflexive fold between subject and object.

> The flesh is that elementary, precommunicative domain out of which both subject and object, in their mutual interactions, develop. The subject can no longer be conceived as an enclosed identity or as an empty reciprocity ready to take in the contents provided by objects. And objects can no longer be viewed as a pure positivity or simply as an aggregate of sensations. Subject and object, mind and body, the visible and the invisible, are intercalated; the 'rays', the lines of force, indelibly etch the one into the other. The flesh is composed of the 'leaves' of the body interspersed with the 'leaves' of the world: it is the chiasm linking and separating the one from the other, the 'pure

difference' whose play generates persons, things, and their separations and unions.

(Grosz, 1994: 102–103)

Second, embodiment is *practical.* That is, it is *involved* in a relation with the world, which is always ongoing and joint with others and thereby creates 'an extensive background context of living and lived (sensuously structured) relations, within which [people] are sustained as the kind of human beings they are' (Shotter, 1993: 12)

Third, embodiment always involves *other objects.* The object world is not 'out there' but is folded into embodiment as simple extensions of the one into the other, or through more complex effects. Let me amplify this latter point briefly by pointing to the literature on new electronic technologies. For what is striking about this literature is its lack of a sense of embodiment. In most cases, the human body appears but only as a surface to be read for the signs of the effects of new electronic technologies, 'the body, as an experiential field, disappears from ... consideration' (Bukatman, 1995: 260). Yet what is also striking is the embodied nature of the new electronic technologies, the way in which they underline how being means being in touch with the object world. Thus Star (1995: 2) argues that even at academic conferences on the new electronic technology and embodiment academics still often seem unable to grasp the world as 'maximum grip':

> our business at hand is to discuss the relationships between RW (real world) bodies and VR (virtual reality) bodies and to think about issues like gesture, presence, and the divisibility of bodies in communication. I watch a demo tape with little cartoon figures lifting blocks and carrying them across a room, and another with photographs of people's faces pasted over the blocky bodies, sitting round a 'table' and 'conversing' with each other through the VR device across geographical distance. When someone in Sweden moves her hand, the block body on the screen moves; when someone talks from Georgia the words are broadcast to all the participants. I find my thoughts drifting during the discussion to the idea of embodiment and how I have felt my body around computers. Right now, typing this, my neck aches and I am curled in an uncomfortable position. I try to think about my fingertips and the chips inside this Macintosh as a seamless web of computing ... But chips made me think of the eyesight of women in Singapore and Korea, going blind during the process of crafting the fiddly little wires; of clean rooms I have visited in Silicon Valley and the Netherlands, where people dressed like astronauts etch bits of silicon and fabricate complex sandwiches of information and logic ... I think, I want my body to include these experiences. And if we are to have ubiquitous, wireless computing in the future, perhaps it is time to have a less boring idea of the body right now – a body positive, not just the substitute for meetings or toys.

Hayles (1996: 1) puts it equally well:

> Cyberspace, we are often told, is a *disembodied* medium. Testimonies to this effect are everywhere, from William Gibson's functional representation of the 'bodiless exultation of cyberspace' to John Perry Barlow's description of his virtual reality (VR) experience as 'my everything has been amputated'. In a sense these testimonies are correct; the body remains in front of the screen rather than within it. In another sense, however, they are deeply misleading, for they obscure the crucial role that the body plays in constructing cyberspace. In fact, we are never disembodied. As anyone who designs VR simulations knows, the specifications of our embodiments matter in all kinds of ways, from determining the precise configuration of a VR interface, to influencing the speed with which we can read a CRT screen. Far from being left behind when we enter cyberspace, our bodies are no less actively involved in the construction of virtuality than in the construction of real life.

Then, fourth, embodiment is *expressive*. That is, the body is not just a passive surface on which society is inscribed. As I hope I have made clear previously, embodiment can be seen as a positive force with its own resources. These resources are of a number of inter-related kinds but especially the symbolic, the sensual and the affective. Drawing on Merleau Ponty's previously cited phrase, 'the body is our general medium for living a world', we can move away from an essentially negative position in which the body only has the capacity to be elusive, that is, it can avoid compliance with social controls, towards a more positive position where the body becomes a 'body-subject' which jointly configures a number of different realms of experience.[13] In other words, embodiment becomes what Radley (1995: 5) calls *elusory*:

> What this elusory nature of the body-subject makes possible is not primarily the discovery of how the world 'really is', or its representation. Instead, it is active in making a sketch, a continuity of engagement that allows us to know how things are because of what we did to bring them about . . .
>
> The body-subject configures the physical body, *with which it must not be confused*, either in the latter's internal functions or in its external appearances. Unfortunately, social theory has too often divided the body along this line, being concerned with explaining the control of its functions on the one hand, or its appearances on the other, as if these constituted it entirely.

In turn, the elusory expressive power of the body does provide a resource for 'resistance', although this is not the resistance of the oppression versus resistance model.

> The expressive form of display in general, as with metaphor in particular, works not only to carry the meaning of the whole but, in

> being reflected through the features that it takes up, inflects these with a new significance.
>
> (Radley, 1995: 11)

This exaggeration of ordinary, everyday actions is a commonplace but it can be used by fringe social groups in explicitly political ways:

> the members of fringe groups inflect (ordinary, everyday actions) in ways that provide a picture of the everyday reality of the majority. It is not just the spelled-out value differences between groups (denotations) which annoy the establishment; for more threatening are the distortions and caricatures of the dominant order that are refracted through deviant actions. Displayed in analogical form, through gestures that hint, invite or mock, these meanings are not specifiable within analytic discourse. They are at once a comment on society and yet elusive to its ministerings (such refractions can, of course, be expressed in a poetic language form; but that is neither the language of authority nor of social theory).
>
> (Radley, 1995: 11)

To summarise these four characteristics, my sense of embodiment is about the body-subject, not the body, engaged in joint body-*practices* of becoming.

I now want to turn back to the problem of power by considering in more detail how embodiment is expressive. I will illustrate my argument through the example of *dance*. I could have chosen other body-practices, for example swimming, running, horse-riding, or even pet-owning (Freadman, 1988, Probyn, 1996) but I think that my reason for choosing this body-practice will become clear.[14]

A DANCE THROUGH HISTORY AND GEOGRAPHY

Let me begin by sketching a short history of dance. Dance has almost certainly been a constant in recent human history. Records of dancing date from the late Palaeolithic era but it seems likely that dance was already part of the repertoire of human behaviour long before this. One of the reasons why it is so difficult to tell is that dancing has routinely gone unrecorded. For example,

> For most of recorded history, portrayal of ordinary village dances in which rude, unskilled men, women and children took part, did not attract the attention of artists working in durable, expensive materials. Those who danced could not pay artists' fees, and those who could thought such festivals unworthy of attention. Only when urbans became so far removed from village life that prosperous city dwellers could afford to yearn for the moral simplicity they had left behind, did skilled painters take up such themes. A half-page painting of a peasant dance in a prayer book from the fifteenth century offers the earliest example I have seen of this sort of idyllic visualisation of rural reality.

> In the next century, Peter Breughel's (d. 1569) paintings of peasant concerns supplied the same market. As far as I am aware, it is the first time that a famous artist bothered to record such thoroughly vulgar, utterly commonplace scenes for the delectation of rich and noble patrons.
>
> (McNeill, 1995: 40)[15]

Another reason why the history of dance is difficult to trace is because it is stylistically diverse. Multiplicity reigns and different authors classify what they regard as dance in different ways.

The cautions concerning the paucity of the historical record and cultural understandings of what counts as dance notwithstanding, dance seems to centre around a number of overlapping expressive body-practices which it both expresses and is expressed by. Four of them seem particularly important. The first of these is ritual, the form seen most obviously in religious practices.

> Dance and dance-like behaviour continue to play a very prominent role in religion, whether or not trance states are involved. The emotional import of keeping together in time at religious ceremonies was, indeed, second in general importance only to community-wide dancing at festivals and, of course, the two regularly merged. Religious meanings often suffused festival celebrations, and religious ceremonies (with attendant dancing) frequently involved the whole community. The enrichment of meanings and elaboration of practices which such blending allowed became very prominent in most human lives, from the time the idea of a spirit world was first adumbrated until our own time.
>
> (McNeill, 1995: 47)

The second body-practice is associated with work and, specifically, relieving tedious or repetitive tasks by emphasising their rhythmic character.

> How unusual such behaviour may have been is impossible to say, but ethnographers of the nineteenth century observed rhythmic field work in many parts of the earth. In Burma, for instance, three to six men joined together to thrash grain, and village tradition held that six was the proper number because then the rhythm for swinging the flails attained an optimum pace that could be maintained all day long. In Polynesia women sung and moved in unison while shredding and pressing cassava roots to extrude their natural poisons. And in Madagascar, a Frenchman reported seeing women planting upland rice by forming in line across the prepared field, bending to place a single seed in the ground and then stepping it into the earth, thus moving forward rhymically 'like a troop of dancers'.
>
> (McNeil, 1995: 49)

The third body-practice is the construction of community. Dance has played an important part in the construction of many different types of community. There are, first of all, established religious communities, ranging from *formalised* communities like the Dervishes in the Muslim world (McNeill, 1995), and the Shakers in the Christian world (Stein, 1992) through to *spontaneous* movements such as the dance manias that broke out in medieval Europe (Hecker, 1970; Blackman, 1952) and the Ghost Dance religion which rose amongst Native Americans in the later nineteenth century (Mooney, 1965). These are all extreme examples. More commonly, dance has simply been one element of community practice in the complex history of European folk dance, or the religious and secular dance groups of East Africa (Ranger, 1975).

The fourth body-practice is leisure. Dance has, over a period of time, simply been a means of relaxation and/or celebration. It has, of course, become increasingly interwoven with commercial opportunities, as modern club cultures attest (Thornton, 1995). But what is most striking is the sheer diversity of opportunities for participating in dance in modern urban cultures (Thrift, 1996b). Here is 'a bodily activity in which many people participate at various stages of their lives, sometimes by themselves, in couples, or in groups, in a range of social settings, from street dancing to dance halls, discos and raves, to parties, dinner dances, weddings and church socials' (Thomas, 1995: 3).

Then, there is also, dance as formal entertainment. This can be divided into dance with artistic pretensions, which includes ballet (which dates from the French court dancing of the sixteenth and seventeenth century and which attained its classical form in the romantic ballets of the nineteenth century) and, now modern dance (which dates especially from the work of early dancers like Isadora Duncan, Ruth St Denis and Martha Graham in the middle twentieth century). [Think only of the sheer diversity of contemporary dance styles, each with their own sites and complex mixtures of class, gender and ethnicity, which currently circulate, from children's ballet classes to sequence dancing to morris dancing, from leroc to line dancing, from step dancing to, most recently, the macarena. And this is to ignore other forms of dancing like ice-dance.]

Modern dance attempted to overturn the pictorial logic of the ballet in favour of the affective.

> For the modern dancer, dance is an expression of interiority – interior feeling guiding the movement of the body into external forms. Doris Humphrey described her dance as 'moving from the inside out', for Graham it was a process of 'making visible the interior landscape'. This articulation of interior (material) spaces creates forms which are not, however, ideal or perfected ones. The modern dancer's body registers the play of opposing forces, falling and recovering, contracting and releasing. It is a body defined through a series of dynamic alternations, subject both to movements of surrender and movements of resistance . . .

> The modern body and the dance which shapes it are a site of struggle where social and psychological, spatial and rhythmic conflicts are played out and sometimes reconciled. This body – and it is specifically a female body – is not persuasive but dynamic, even convulsive.
>
> (Dempster, 1988: 42)

But note must also be made on the large number of other forms of formal dance entertainment from music hall to Riverdance. These dances, too, have an expressive potential.

DANCE, EMBODIMENT AND RESISTANCE

Why is it that such a rich, diverse and remarkable history has been all but ignored in the social sciences and humanities and how does it speak to the theme of power, domination and resistance? The peculiar invisibility of dance relates, I think, to the fact that it lives beyond the rational auspices of Western societies (Ward, 1993). It does this *in* two ways, both of them to do with the difficulty we face in grasping expressive embodiment in societies based on texts.

First of all, dance is not self-evidently about discourses of power and control. It is about *play*.[16] But there is a problem here. In Western societies, the notion of play has been emptied of all content (Game and Metcalfe, 1996). It is regarded as peripheral to the real business of life, at best adding a little oil to the wheels of social structure, at worst a trivial distraction. 'In the West, play is a rotten category, an activity tainted by unreality, inauthenticity, duplicity, make-believe, looseness, fooling around and inconsequentiality' (Shechner, 1993: 27). Certainly, it seems, 'play does not fit in anywhere in particular; it is a transient and is recalcitrant to localisation, to placement, to fixation – a joker in the neuro-anthropological act [sic]' (Turner, 1983: 233). What, then is play? Classically, play is defined as 'as-ifness': it is 'not-for-real' but it is enacted as if it were:

> this 'as-ifness' quality being precisely what sets it apart from real reality. One plays when knowing that the assumptions are what they are: assumptions which have been freely accepted and may be freely dropped. We speak of reality when we do not have such knowledge, or do not have to believe it, or suspect it to be untrue.
>
> (Bauman, 1993: 170)

Play is, in other words, a process of performative *experiment*: 'the ongoing, underlying process of off-balancing, loosening, bending, twisting, reconfiguring, and transforming the permeating, eruptive/disruptive energy and mood below, behind and to the side of focused attention' (Schechner, 1993: 43) which is brought into focus by body-practices like dance and which 'encourages the discovery of new configurations and twists of ideas and experience' (Schechner, 1993: 42).

Thus, play has six chief characteristics. First, it is *gratuitous*. It serves no useful purpose. 'When called to justify itself in terms of the function it serves, play reveals its utter and irremediable redundancy' (Bauman, 1993: 170). Second, play is *free*. Therefore;

> it vanishes together with freedom. There is no such thing as obligatory play, play on command. One can be coerced to obey the rules of the game, but not to play ... This is why play remains so stubbornly non-functional. Were it to serve a purpose, were I to play in order to bring about or protect certain things I or the others like or want me to like, there would be little freedom left in my act of playing. The act is truly and finally free only when gratuitous.
>
> (Bauman, 1993: 170)

Third, play is *not cumulative*. It does not add up. It speaks the moment.

> Play may be restarted and repeated; even its end is 'as if', not really real. No defeat (no victory either) is formal and irrevocable. The chance of revenge sweetens the most bitter of failures. One can always try again, and the rules may still be reversed. Because it can be repeated, played again, because its end just clears the site for another beginning, makes new beginnings possible – to play is to rehearse eternity: in play, time runs to its appointed era only to start running again. Time has a 'direction' only inside the play, but the responsibility of playing cancels that direction, indeed the flow of time itself. Nothing accrues ... nothing 'builds up', each new play is an absolute beginning.
>
> (Bauman, 1993: 171)

Fourth, play is *rule-bound* but the rules are internal to the game. Thus:

> each play suits its rules. Play is the rules: play has no other existence but a number of players observing the rules. 'One cannot be sceptical regarding the rules of the game' says Huizinga ... If you do not observe the rules, you do not play.
>
> (Bauman, 1993: 171)

Fifth, play is *difficult to command and control*, precisely because, unlike, say, work, it is not constituted out of fixed means-end relationships. But, at the same time:

> Work and other daily activities continuously feed on the underlying ground of playing, using the play mood for refreshment, energy, unusual ways of turning things around, insights, breaks, openings and especially looseness. This looseness (pliability, bending, lability, unfocused attention, the long way around) is implied in such phrases as 'play it out' or 'there's some play in the rope' or 'play around with that idea'.
>
> (Schechner, 1993: 42)

Finally, play is *located*. Play can have its place – 'the race course, tennis court, dance hall, sports hall, sports stadium, discotheque, church, chess board'

(Bauman, 1993: 171). But equally, play can and does take place outside such places.

Then, second, and clearly related to these points about play, dance is therefore about using the body to conjure up 'virtual', 'as-if' worlds by configuring alternative ways of being through play, ways of being which can become claims to 'something more' (Pini, 1996).

> dance can . . . be considered as the fabrication of a 'different world' of meaning, made with the body. It is perhaps the most direct way in which the body-subject sketches out an imaginary sphere. The word imaginary is used here in the sense 'as-if', suggesting a field or potential space. The dance is not aimed at describing events (that is, it is not representational) but at evolving a semblance of a world within which specific questions take their meaning.
>
> (Radley, 1995: 12)

This point deserves elaboration. The 'point' of dance, and of other forms of presentational communication, lies in their being able to 'articulate complexes of thought – with – feeling that *words cannot name, let alone set forth*. It is a way of accessing the world, not just a means of achieving ends that cannot be named' (Radley, 1995: 13, emphasis added).

This kind of accessing of the world relies on four main ways in which the body-subject is able to work to create non-denotative meaning. To begin with, the body can symbolise through *semblances*:

> In the world of everyday relationships, the creation of semblances involves the attribution of figural operators to persons and settings. The gambler's 'dangerousness' or the hostess's 'coldness' are examples of social qualities which are not objectifiable precisely because they are not literal. I may feel cold and put on a coat, and that coldness is applied to my body, just as it might be to the inside of my refrigerator. I can denote this feature by objective means, using a thermometer. However, if I actually move towards someone this can only be appreciated through the way that I appear. Being a virtual quality it cannot be denoted in a form sufficient to capture its full meaning. It can only be shown forth or displayed, and the medium for such display is the body. The whole panoply of writing in sociology about 'playing a role' is based upon this distinction. Rarely, however, is the role of the body-subject acknowledged in this.
>
> (Radley, 1995: 13–14)

Another way in which the body-subject can create non-denotative meaning is through the *senses*.

> The body's material aspect allows it to penetrate and to partake of the mundane world while remaining within it; it does not create an aethereal sphere of make-believe. The body's potential to configure the world uses its character of being material, of being able to 'dwell in' the

particulars of things, of being able to press, caress or resist other bodies.

(Radley, 1995: 15)

In other words, people and things can be lent significance through the way that they are approached/touched/gripped.

The body's imaginary grasp involves a transmutation of sensory distinctions that anticipates its meaning. In the sphere of objects, this might be the careful application of wax to a wooden table with a woollen cloth that lends warmth to everything it touches. In the sphere of social relationships, the way that a dancer touches her partner with a lightness that signifies (that is opens or invites) gentleness rather than distance.

(Radley, 1995: 15)

Yet another way way in which the body-subject can symbolise is in the *testing of its physical boundaries*. This is a particularly important aspect of certain physical activities like dance, certain sports and even certain kinds of entertainment like amusement parks.[17] What these activities all have in common is the production of experience which 'has no meaning outside of a world of sensations, of movement, of the loss and recovery of physical control' (Radley, 1995: 4). They are, in other words, a physically sensed way of being. Then, there is one more way in which the body-subject can symbolise. That is through the *space–time configuration of bodies and objects* which it can achieve or adjust to in order to symbolise. But, it is also about configurations that are, in a sense, only inferred from bodily movements: it can, in other words, create an appearance of influence and agency .[18]

To return to dance, dance shows these four attributes of embodiment in action. Leisure dancing is 'invitational' in that 'the movements of dancers together create a *potential space* in which individuals can evolve imaginary powers of feeling' (Radley, 1995: 14, emphasis added). But formal dance can go further since it is there to be enjoyed by an audience for whom the dancers' bodies are a spectacle:

what collective performances make possible is not just the maintenance of imaginary worlds, as if these stood apart from everyday 'reality'. This sets apart the virtual from the real, when the whole point of social activities is that they can be expressed as more real, more vital than the mundane sphere. In effect, such liminoid activities, play or ritual, have their significance because of the way they mirror the remainder of life (more precisely, because of how the remainder of life is refracted through them). The semblance of meaning that such bodily performances create are repeatedly accessed in the case of everyday, ordering experience. This accessing is often done by means of expressive devices that the body makes possible; the most familiar is the use of music, sung or even hummed to oneself, to reactivate the virtual sphere that can transpose the mundane present (Dennis Potter's television dramas

> *Pennies from Heaven* and *The Singing Detective* are elaborated illustrations of this point).
>
> This potential has long been recognised to attach to rituals, particularly of a religious kind, in which the transposition of bodily feeling draws upon occult powers that, at the same time, it seems to strengthen. Ritual acts display and re-present (that is refract) aspects of society in a way that renders its everyday experiences more comprehensible and more significant to those involved. They share with all forms of art a depiction not of how things inherently are, but of how they imaginatively might be. Geertz (1972) makes it clear that such episodes are 'not merely reflections of a pre-existing sensibility analogically represented' but that they create these sensibilities (semblances). It is not the material effects of play and art that matter, but their capacity to 'colour experience with a light they cast it in'.
>
> (Radley, 1995: 14)

Of course, dance can be used to build marginal networks which can combat powerful networks. For example, Tomko (1996) documents the way in which the founders of the New York Girls' Branch folk-dance clubs in the early twentieth century attempted to use dance both as a physical discourse through which to argue gender and public space and as a quite conscious attempt to distance themselves from rational models of gymnastics and other physical training imported from Europe.[19]

Equally, powerful networks are consistently able to subvert dance for their ends. Dance has been subverted in all manner of ways. By the dictates of organised religion. By the dictates of the commercial sphere (Cowan, 1990). And, most particularly, of late, by the project of the nation-state. For example, the French revolutionaries experimented with liberty dances and civic festivals as a means of giving expression to the principles of liberty, equality and fraternity (Schama, 1989; Ozouf, 1988).[20] Gymnastics, which had a part-ancestry in dance, was a major element of German and Scandinavian state building in the nineteenth century (McNeill, 1995). And in the twentieth century Wollen (1995) shows how the dance projects of Rudolf von Laban and Mary Wigan, with a catholic range of influences like the Dervishes, Cabaret Voltaire, nature worship, a kind of immediate phenomenology of the intense and primal body, and an explicit opposition to a mechanistic 'robotic' German culture, became, for a time at least, a part of the Nazi project.

But, in part, to state that dance can be used to subvert power or to combat it is to sorely miss the point. Play eludes power, rather than confronts it and for two reasons. First, because, as a world of virtual forms, it cannot be commanded in the way that is true of work, since it is not made up of fixed means–ends relationships. Second, because, as a world of virtual forms, it can be described by words but ultimately it cannot be written or spoken.

To summarise, dance can be seen as an example of play: a kind of exaggeration of everyday embodied joint action which contains within it the capacity to hint at different experiential frames, 'elsewheres' which are here (Pini, 1996). Equally play demonstrates, that 'the body is not merely a vehicle for departing from social norms, for escaping from the strictures of moral codes. It is, in its positive aspect, the grounds for configuring an alternative way of being that eludes the grasp of power' (Radley, 1995: 9). Thus, for example, dance 'can create a fantasy of change; escape and of achievement for girls and young women who are otherwise surrounded by much more mundane and limited leisure opportunities' (McRobbie, 1991: 192).

CONCLUSIONS

Recently, Grossberg (1996: 88) has made a call for cultural studies

> to move beyond models of oppression, both the 'colonial model' of the oppressor and the oppressed, and the 'transgression model' of oppression and resistance. Cultural studies needs to move towards a model of articulation as 'transformative practice', as a singular becoming of a community. Both models of oppression are not only inappropriate to contemporary relations of power, they are also incapable of creating alliances; they cannot tell us how to interpellate various fractions of the population in different relations to power into the struggle for change. For example, how can we involve fractions of the empowered in something other than a guilt-ridden way? My feeling is that an answer depends upon rearticulating the question of identity into a question about the possibility of constructing historical agency, and giving up notions of resistance that assume a subject standing entirely outside of and against a well-established structure of power.

This chapter has been an attempt both to answer this call and to at least begin to supply an answer to the question of the possibility of recognising an embodied historical agency which is based upon 'an unstoppable uncontainable speaking as we cast our bodies into space' (Dempster, 1988: 52). In particular, I have tried to show that such a possibility is intimately bound up with the project of non-representational 'theory' which, in its de Certeauian form at least, allows much greater attention to be paid to the geography of embodiment and play than has heretofore been thought necessary, possible or even desirable by most commentators.

NOTES

1 Thompson (1995) provides a useful account of the romance of resistance associated with subcultural theory.
2 The area where de Certeau really fills out his account is in the matter of space and spatial stories.
3 For an early statement of the problems revisited in this paper see Thrift and Forbes (1983).
4 Before Giddens became another modernisation theorist (see Alexander, 1995).
5 This vision follows from the increasingly popular but also rather nebulous Deleuzian metaphor of the fold (Deleuze, 1988). But perhaps it is better put by those working in contemporary psychoanalysis who have been influenced by Castoriadis:

> psychosocial life is portrayed as a stream of fantasies, representational wrappings, bodily sensations, idioms, envelopes, containers, introjects and memories. Yet although the internal world is understood as comprising such multivalent psychical forms, it is also found to be patterned through an interpersonal field of interactions with significant other persons. The individual, as representational flux, is embedded in both intrasubjective and intersubjective relations.
>
> (Elliott, 1996: 104)

6 To an extent, it is this sense of tactility that I think Benjamin was trying to simulate in his writings (see Taussig, 1992).

7 This is akin to Serres' 'geographies of the skein': 'what I seek to form, to comprise, to promote, I can't quite find the right word – is a syrrhese, a confluence not a system, a mobile confluence of fluxes. Turbulences, overlapping cyclones and anticyclones, like on the weather map. Wisps of hay tied in knots' (Serres and Latour, 1995: 122).

8 It is clear by now, I hope, that non-representational theory is highly contingent.

9 There are other examples. For instance, there is Haraway's (1992) notion of the actant, taken from Greimas, which allows her to bind humans and non-humans together in a 'functional collective'. Then there is Gumbrecht and Pfeiffer's (1994) approach to 'materialities of communication' which attempts something similar. Each of the approaches that I consider below is also highly correlated. Thus, for example, in some incarnations (such as that of Law) actor-network theory is highly influenced by Foucault. Similarly, both Deleuze and Guattari wrote at length of their debt to Foucault (see Deleuze, 1988; Guattari, 1996).

10 Thus, to quote Law (1994: 33),

> people are networks. We are *all* artful arrangements of bits and pieces. If we can't be organisms at all, this is because we are networks of skin, bones, enzymes, cells a lot of bits and pieces that we don't have much control over (though if things go wrong then we are in dire trouble). And if we count as people rather than organisms this is because of a lot of other bits and pieces – spectacles, clothes, motor cars and a history of social relations – which we *may* have some control over. But we are equally dependent on these. We are composed of, or *constituted*, by our props, visible and invisible, present and past. This is one of the things that we may learn from reading the symbolic interactionists and Erving Goffman. Each one of us is an arrangement. That arrangement is more or less fragile. There are ordering processes which keep (or fail to keep) the arrangement on the road. And some of these processes, though precious few, are partially under our control some of the time.

11 As in the examples of the automatic door closer provided by Latour (1992) and the stirrup provided by Deleuze and Guattari (1987).

12 De Certeau's debt to Wittgenstein usually goes unacknowledged, but it is there (see de Certeau, 1984a).

13 In other words, I am here concerned not with the individual body but with embodiment as expressed in joint action.

14 See Foster, 1986, 1996; Lange, 1975; McRobbie, 1984, 1991; Spencer, 1985; Thomas, 1993, 1995.

15 I am ambiguous about this assertion. For example, Hutton (1996) provides examples of stained glass windows and wooden panels depicting dancing which date from the sixteenth century.

16 See Huizinga, 1949; Bateson, 1977; Geertz, 1972; Bauman, 1993; Winnicott, 1971.

17 And in its extreme form, can result in a trance state, as in Dervish religion, raves, and so on (see Katz, 1982; McNeill, 1995).

18 I think this is what Lefebvre (1995) was, in particular, trying to get at with his notion of rhymanalysis.

19 One of the most interesting things about dance is that, partly because it is an activity which has often been regarded as 'feminine', its possibilities as a form of 'resistance' have more often been explored by women than men. The gendered nature of dance also points to the way in which women are probably more skilled at creating virtual worlds than men, because in a man's world, these are the worlds they can inhabit. It is significant that nearly all the recent writing on dance has been by women.

20 As in, as well, dances like the Carmagnole.

7

RADICAL POLITICS OUT OF PLACE?

the curious case of ACT UP Vancouver

Michael Brown

> Articulatory practices take place not only *within* given social and political spaces, but *between* them.
>
> (Laclau and Mouffe, 1985: 140, emphases original)

The AIDS Coalition To Unleash Power (ACT UP) has been touted as the most radical AIDS organization because of its penchant for disrupting public spaces of civil society (Crimp and Rolston, 1990; Kramer, 1989). So when I began my research on AIDS politics around Vancouver, Canada in January of 1992, the first organization I contacted was its local chapter. I was vaguely aware of the movement's actions in other cities, and I had heard about a few of its demonstrations in the local media. ACT UP seemed to be precisely the form and location of resistance that radical democratic theory (e.g. Mouffe, 1993) had been discussing all too abstractly (Massey, 1995).

After 120 interviews and a year and a half ethnography, however, my first interviewee remained the only ACT UP member on hand. Others had either died, left the city, or were no longer politically active. There were no demonstrations during my two years of fieldwork, and the lone member himself hinted at the group's dissolution during our interview:

> I don't know what the problem is these days. ACT UP is going through a difficult period where we aren't getting the people turning up. I think actually that ACT UP could be folding on a certain level. It just might not be necessary right now.

His suspicions were reiterated across the city. The emerging theme was that ACT UP Vancouver had been only modestly successful at resisting British Columbia's ultra-conservative, Christian-fundamentalist, Social Credit (Socred) government;[1] and the AIDS voluntary sector organizations were much more successful and popular.

In political theory 'civil society' classically defines public voluntary social relations that are not the state.[2] And theorists of radical democracy have been prodding us to look towards civil society for new forms of politics or citizenship (Offe, 1985; Magnusson, 1992; Mouffe, 1993), where ACT UP and grassroots AIDS service organizations (ASOs) originated (Gamson, 1991). But these theorists tend to ignore the inherent spatiality of social relations like the 'state' or 'civil society'. When we begin to think radical democracy spatially the limitations of this advice become all too clear.

ACT UP Vancouver provides just such a lesson for both scholars and activists interested in resistance. Precisely where we should expect a politics of resistance, transgressive civil disobedience and spectacle proved ineffective. Why? The simple answer is that ACT UP failed in Vancouver because Canada lacks a radical tradition more salient in America, and this argument has some purchase. But I think a spatial reading of ACT UP's politics in civil society sensitizes us to an all-too-static dichotomy between state and civil society befalling political theory and activism. Like many political theorists, the group treated the state and civil society as discrete social relations. This fallacy can be traced out quite spatially in Vancouver. ACT UP failed there (at least partially) by misunderstanding and misrepresenting the amalgam of state and civil society in Vancouver. So this chapter warns how fixed and static notions of space in radical democratic theory may fail to capture the ongoing shifts between actual spaces of state and civil society. I sketch out this dualistic geography implicit in ACT UP's philosophy and actions, showing how both its successes and failures drew on (and tried to structure) a hard and fast separation between state and civil society in the city.

ACT UP BACKGROUND

Radical democratic citizenship

ACT UP seemed to exemplify Mouffe's notion of new, radical democratic citizenship in at least four ways. Foremost, it is well known for its alternative, and transgressive approach to the political (Berlant and Freeman, 1993; Crimp, 1993. ACT UP's premiere chapter in New York began on March 10, 1987, out of a speech given by gay activist Larry Kramer.[3] It was a stinging critique of the political quiescence in the AIDS community. He noted the lack of progress in drug availability, the profit-mongering of drug companies, and the weakness and bureaucratization of local AIDS organizations. Two days after the meeting, ACT UP New York was formed to accelerate drug release in the US. Its functions quickly multiplied, as did its chapters.[4] Its official definition is, 'a diverse, nonpartisan group united in anger and committed to direct action to end the AIDS crisis' (Crimp and Rolston, 1990: 13). ACT UP provided a radicalizing opportunity for many who joined. Kramer's invectives were often mirrored back at the apathy of the gay community. Rather than ignoring the mounting deaths or viewing AIDS as a punishment, Kramer argued they must see the epidemic as an attack on the very existence of gay men. Crimp and Rolston (1990: 22) specifically state that it is committed to radical democratic change. ACT UP does not seek to change the system from within, but rather attacks the very assumptions and premises that underlie the practices of actually existing democracy and capitalism (Callen, Grover and Maggenti, 1991; Olander, 1991; Saalfield and Navarro, 1991).

Secondly, like Mouffe's notion of citizenship, it challenges existing hegemonies around AIDS. For instance, one ACT UP New York member described how this oppositional discourse inspired her towards civil disobedience against the Catholic Church because of its clear logic:

> And Cardinal O'Connor, especially at that time, was telling the general public that monogamy would protect them from HIV infection and that condoms didn't work. As far as I was concerned, those were both major lies. And while people were certainly entitled to make their own decisions about their lives – and far be it from me to tell them what to do – I would not sit by silently while they were being lied to. So when the group decided to target St. Patrick's, it just made absolute perfect sense to because it is an extremely important target in this epidemic.
> (Northrop, 1992: 484–485)

Similarly, Gamson (1991) has centred the coalition's purpose on the creation of an alternate way of seeing AIDS: one that uses spectacle to resist media, bureaucratic, and scientific tropes that normalize AIDS. On these grounds theatrical and cultural activism is used fervently by ACT UP, in order to displace meanings.

Thirdly, the organization itself is committed to radical democratic principles (Gamson, 1991; Northrop, 1992). There tends to be no hierarchy in decision-making. Chapters are notoriously lacking any formal organizational structure. The coalition in Vancouver, for instance, operated primarily through a telephone tree of 80 people, any of whom could call a meeting. There are no formal membership requirements; anyone may attend ACT UP meetings, or take part in their rallies. People participate to varying extents. No one is obliged to perform acts of civil disobedience at any given demonstration. Modes of actions are decided upon by consensus. Nevertheless, fervent discussion and debate are the rules at ACT UP meetings, signalling the antagonistic quality of modern democracy Mouffe (1993) pinpoints. ACT UP also tries to link disparate struggles around AIDS, a key component for the radical democratic project. The ignorance around women and AIDS, for instance, has been a key target for ACT UP campaigns (Banzhat, 1990). As Russo (1992: 415) puts it:

> This is a new kind of activism. It's a coalition that we were never able to achieve in the 1970's. Back then the ideal and the dream were that gay people would come together with other oppressed groups like blacks and Asians and women to form a coalition. That didn't happen because we had too many differences. Lesbians were fighting with gay men, the black community didn't want to admit that there was a gay community in their midst and blah, blah, blah. Now AIDS has brought

> us together in ways that we could not have foreseen. ACT UP is composed of gay people and straight people, women and men, black and white. And all these people have one thing in common: They want to put an end to the AIDS crisis by any means possible.

Finally, ACT UP works at these aims decidedly in the venue of civil society. Typically, actions are highly theatrical disruptions of public spaces. As Berlant and Freeman (1993: 224) praise, it takes on heretofore hegemonic spaces of meaning. The will to disrupt hegemony spatially draws a strong parallel with the Situationist International Movement of the 1960s (Ball, 1987). ACT UP hijacks cultural codes (around AIDS, disease, welfare, and sexuality) with irony and places them into heavily coded, unfamiliar contexts.[5] So, for example, dozens of activists kissing in public becomes transgressive when it is done by same-sex couples insisting on accurate safe-sex education. Like the situationists, ACT UP relies on what Kaplan and Ross (1987: 24) call 'acts of cultural sabotage' that critique mainstream society.[6] It has taken symbols used to oppress and subverted their meanings. In addition to the graphics, acts of civil disobedience and spectacle confront heterosexed space with gay and lesbian realities: alternate sexualities exist, and they exist in a context of a rather invisible epidemic that has been killing gay men.

ACT UP demonstrations appropriate locations in civil society – confronting strangers deliberately – in order to make the AIDS crisis visible to them, to press rights claims and to demand some forms of political obligation on the part of fellow citizens.

ACTING UP IN VANCOUVER

On July 21, 1990, ACT UP Vancouver officially formed (Shariff, 1990; Wilson, 1992: A-8).[7] The group's first meeting attracted 35 people, most between the ages of 18 and 25, and was spearheaded by a gay lawyer, Kevin Robb, and David Lewis, an outspoken, former leader of the Vancouver Persons With AIDS (PWA) Society.[8] At the meeting, the chapter was officially designated, and a mission statement was drafted. In part, it read:

> We demonstrate and protest; we challenge and demand governments and health institutions take positive action; we research, act on, and make available the latest medical information.
>
> (quoted in Buttle, 1990: A-17)

Apparently there were people there from across the different AIDS groups in the city. One former AIDS Vancouver employee recalls the meeting:

> I was at the meeting that David Lewis called. He called a meeting of people that would be interested in the formation of ACT UP. That would have been four or five [weeks] before he died. That would have been the summer of 1990. And we had it in the back yard at his place ... I was working at AIDS Vancouver at the time.

> So a co-worker and I went to this thing. We were curious and had heard about ACT UP and read about ACT UP and seen things in the media and so on. And thought, well maybe this is something we can be involved with. But we were sort of walking a fine line because we were also employees of AIDS Vancouver. And so we have to be careful about – you know we were really caught between a rock and a hard place at that point because some things that would reflect negatively on the organization.

ACT UP Vancouver's first official action attracted media attention across Canada. Some 150 people held a die-in during rush hour at Robson Square, a prominent public space in downtown, less than a month later (Wilson, 1990: A-1). Outside a Socred fundraiser (ironically a performance of *Les Miserables*) at the Queen Elizabeth Theater on August 24, 1990, Premier Vander Zalm was spat on, and his wife was knocked to the ground by ACT UP protesters ('AIDS protesters ... ', *Winnipeg Free Press*, 1990). Five protesters were arrested and later released.

After that confrontational summer of 1990, however, ACT UP ebbed from Vancouver's political scene. The only recorded action that year took place in December, when ACT UP occupied the then Provincial Health Minister John Jansen's Vancouver office to demand British Columbia provide better AIDS services and funding ('Seven AIDS Protesters Arrested', *Vancouver Sun*, 1990: A-2). By 1991, the group had only three press-recorded actions. In the most notable one, on January 29, an ACT UP member was arrested for allegedly spilling ketchup (fake blood) on and denting Premier Vander Zalm's limousine in front of the U-TV television studio, where he was scheduled to give the State-of-the-Province address. A moving die-in was held along Vancouver's upscale-retail Robson Street during AIDS Awareness week in October. Another '0' was added by ACT UP to the city's centennial monument, a cement '100' at the south foot of the Cambie Street Bridge, denoting the 1,000th case of AIDS in British Columbia. It then took its demands to Progressive Conservative MP Kim Campbell's local office (Buttle, 1990). By early 1992, however, its meetings were drawing no one. ACT UP never held another event.

ACT UP's political geography also fuelled my expectation that it was the best local expression of radical citizenship. Seven out of nine recorded demonstrations in Vancouver took place in public spaces of civil society (see Figure 7.1). While all the actions were directed against the provincial government, only two took place at explicitly state-centered sites. Moreover, none of these spaces lay inside the gay neighborhoods; instead, they were in visible yet neutral spaces, especially concentrated in the downtown core. Thus when ACT UP directed action against the state (i.e. the Socreds), it did so in spaces of civil society, for example in front of the Queen Elizabeth Theater, or during lunchtime at Robson Square. If the state was ACT UP Vancouver's target, civil society was its arena.

What inferences might we take from this geography of resistance? The

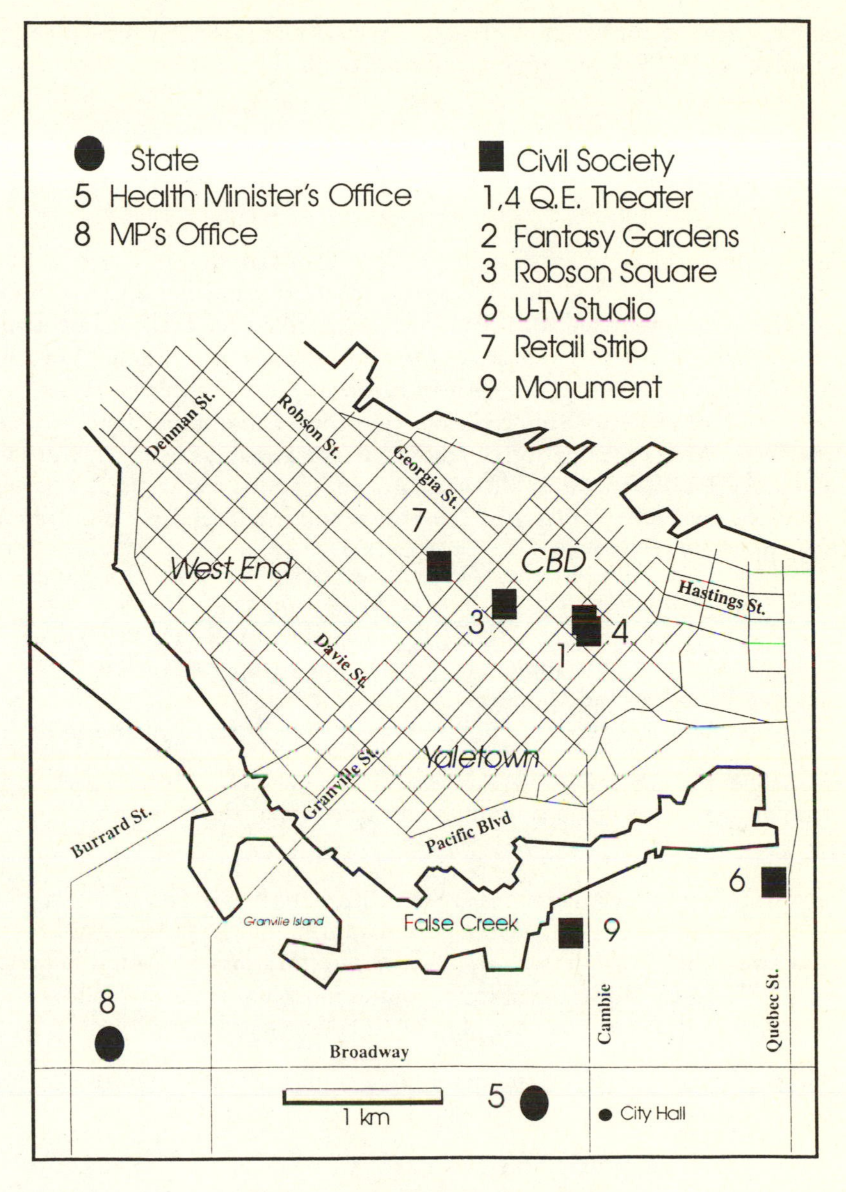

Figure 7.1 ACT UP Vancouver protests

theoretical point I would draw is that ACT UP's strategy was premised on a hard-and-fast dichotomy between civil society and the state, which it expressed and structured geographically through its tactics. The state was rarely engaged on its own turf; rather it was challenged in public spaces of civil society. Civil society, operationalized in public space, was a weapon of resistance against the state. To paraphrase the ACT UP member, 'I don't believe the streets belong to the state. The streets belong to the people!' I want to examine the wisdom of such a spatial distinction in resistance by outlining ACT UP's (limited) successes and its (broader) failures below.

THE SUCCESSES OF ACT UP VANCOUVER

Anger

Perhaps the most effective function ACT UP served was as a very public outlet for people's anger, directed against the virus and the Socreds, both enemies, to be sure. Civil society (as a space) enabled ACT UP members to vent publicly their enormous anger over the amount of death and government inaction *vis-à-vis* AIDS. That function of public spaces helps explain the location of protest within civil society, rather than in more state-centred sites. By drawing the attention of fellow citizens in public spaces of Robson Square, ACT UP Vancouver allowed for a very public, cathartic exercise of anger. One local writer's description of the group in a local paper defined the group precisely by its anger:

> To join ACT UP, you don't have to be gay or have HIV (the virus widely believed to cause AIDS). You just have to be angry – full of bilious rage in fact – about the government's response to the AIDS epidemic, and be ready for countless bouts of civil disobedience .
>
> (Shariff, 1990: A-11)

Several people, in fact, were quite critical of ACT UP's tactics, but acknowledged its need as an emotional outlet given the context of Socred indifference:

> And I've never been interested in the politics. It's just hot air . . . I mean, knocking down the Premier's wife and spitting on the Premier and so on. Well, they both deserve to be strung up. My fantasy would be to see Bill and Lillian strapped to the windmill, spinning around. But we can't all have *our* fantasy.[9] But, I mean, that doesn't play well.

One man living with AIDS, however, tried to describe the context of anger in his life at the fate of having a terminal illness:

> Sometimes you can get very frustrated and very angry. You've got to be able to let that out. It's always been very difficult to explain to people that this is a very high-stress lifestyle. You cannot walk away from it. If [an HIV-negative person] starts to get burned out, you go take a vacation and sit on a beach. I can't do that! No matter where I go, or whatever, I *have* this thing and, you know, it's the first thing I think of in the morning and the last thing I think of at night. And I'm immersed in it. Constantly.[10]

Later, he was more specific about how ACT UP serves as a tactic to vent the incredible anger people feel around AIDS:

> It also serves as an outlet for those people who have gone beyond anger – who are not just frustrated, who are not just irritated. These people are *damn Goddamn angry*!

It is important to recall that by the summer of 1990, when ACT UP Vancouver exploded onto the local scene, Bill 34 (a rather draconian quarantine measure) had been passed, and British Columbia remained the only province in Canada not to fully reimburse people for AZT. These policies, combined with the callous public remarks made by Social Credit ministers a year before, and the government's ongoing neglect of AIDS explain the anger some activists felt.

If ACT UP did vent anger we begin to see some usefulness for the state–civil society distinction that it presumed so spatially by its resistance. For instance, many members of ACT UP were also involved in the Vancouver Persons With AIDS Society or AIDS Vancouver. Yet because of these organizations' emerging financial and policy linkages to the state, they could not 'be political' as shadow state organizations. Once links to the state were made the society had to remain careful not to engage in political activities. The importance of this seemingly apolitical stance cannot be overemphasised. It is a persistent concern at the PWA Society.[11] As ACT UP was distinctly located within civil society individual PWA members (and their allies) under the guise of 'ACT UP', however, could angrily resist the state without jeopardizing their positions as members of the state apparatus. In this way, the group broadened the tactics of resistance in Vancouver. An executive director at AIDS Vancouver explained:

> I think it's important that there be that, sort of, radical–activist element out there to rabble-rouse. I mean, I think it's very important because it makes it easier for us (as a mainstream group) to then develop our programs because people are concerned. People are scared of ACT UP. It's quite ironic. I'm not scared by them And I really understand why that organization needs to exist. And, I mean, we all know each other. It's not that big a group of people that you don't know each other.

In the press, another former director at AIDS Vancouver was very careful to distance his organization from ACT UP:

> AIDS Vancouver completely understands the frustrations and the anger that was expressed (at the Socred protest). We welcome an organization that gives vent to those feelings of frustration and anger. Beyond that, we'll take a look at each political action they take.
>
> (Buttle, 1990: A-17)

The member of ACT UP, (who was also a member of the PWA Society) reiterated the theme that ACT UP could take the radical position that incorporated local ASO's could not:

> We already had groups like AIDS Vancouver and PWA Society. At that stage, PWA could not take drastic actions because of their funding.

> They would feel repercussions. Their aims are to meet the needs of the infected. And that's what they have been doing. I mean they've been working very hard. They can go to the proper agency or department and request that needs be met, although they can't demand them. ACT UP can demand them. And it's obvious that if PWA isn't going to be listened to, the next stop will be an AIDS activist organization taking that responsibility on. That means that there is a position for ACT UP to take a more radical stand. I mean, most people you talk to individually at AV or PWA will agree that there's a place for us: when, you know, polite gladhanding doesn't work any more then there is a need to take action.[12]

ACT UP VANCOUVER AS A DISCIPLINARY FORCE

ACT UP could also be a potential, imminent threat against the state or medical authority from the gay community – even if the local chapter had no such kinetic energy! During my interviews, for instance, it was credited for its ability to bring important ethical and moral issues to civil society in a very public and immediate way. That threat, it seemed, could be enough to challenge the medical authority around HIV and AIDS. The man quoted above also explained the need for ACT UP by the utility of its potential threat:

> It's like having the IRA on your doorstep. If somebody gives us a hard time and we're not getting through by talking, then we ACT UP on them.

Here, it seems, ACT UP Vancouver's cultural capital in civil society is emphasized. Located within civil society, with strong grassroots in the gay community, ACT UP could be used to threaten the state apparatus without the fear of hurting existing shadow-state funding or service delivery. ACT UP signalled that it could strike at any time, should some future injustice warrant retaliation. One HIV positive person recalled an instance where the threat of ACT UP was enough to change a hospital policy[13] almost immediately:

> And they have to listen to us. An example: the hospital board of directors was being a bunch of jerks, so ACT UP had said that they were considering picketing the individual doctors – not only at the hospital but at their homes. Well, they freaked out all over the place. The knee-jerk response reaction was to say that if that happens they wouldn't treat any more AIDS patients. And I went and saw them and said, 'Guys, you are living on the edge of a gay ghetto! There are gay people stacked up one on the other in this neighbourhood. You want to do a scene from *Frankenstein* where you're in the castle and the mob's out front? You're going to tell these people that you're not going

to deal with their brothers and sisters? Don't be a Goddamn bunch of assholes!'

This role as an imminent threat of escalating political force allowed ACT UP Vancouver to discipline institutions like the state and medical authority. Here ACT UP's assumption of a clearcut distinction between state and civil society provided some utility. It directed anger and grief outwardly, and enabled ACT UP to work as a disciplinary force in the city.

THE FAILURES OF ACT UP VANCOUVER

Misunderstanding political culture

If ACT UP Vancouver's successes pivoted on an assumed state–civil society distinction, its failures rested there too. Locals argued that ACT UP's demise could be traced to two broad and related factors: it did not fit the local political culture, and the Vancouver Persons With AIDS Society did a better job at advocating for AIDS issues. Even the Vancouver media did not take to ACT UP as in other cities. For instance, at least two of the demonstrations held in Vancouver were not reported locally, but instead in Toronto and Winnipeg papers (Wilson, 1992; 'AIDS protesters ...' *Winnipeg Free Press*, 1990). Coverage of ACT UP demonstrations remained sparse when written up in the local press. The city's tabloid, known for its attraction to sensational media events, even chastised the group's tactics as unproductive:

> The violence at the theatre was not justified and did nothing to elevate the AIDS cause the way a more reasoned and intelligent approach would. ACT UP should smarten up before it acts again.
>
> ('Acting stupid', *The Province*, 1990)

During my interviews the issue of political culture was described in a number of ways, on a variety of geographic scales, from the nation to western Canada to the city. Some people drew a strong national distinction, with Canada being described as a much more quiescent place than the US, where ACT UP originated.[14] This alleged culture of complacency was often tied to the different structures of rights available in Canada, universal health care being paramount. The greater accessibility of drugs, for instance was noted even by ACT UP's lone member, who offered the example of cross-border smuggling to illustrate plainly the political–cultural differences between the US and Canada:

> There's a glut of AZT in Vancouver. There's a huge surplus of it. And there are a lot of individuals in the United States who are forced to pay for their AZT. So what's happening is that it's being shipped across the border from Canada illegally and being given to people who have made the decision to take it.

A more generous welfare state combined with a political culture that allegedly lacks an activist flavour, meant that ACT UP was awkwardly transplanted from a US model into a Canadian urban context. As one woman half-jokingly asserted:

> I mean, the relatively small role ACT UP played here: this is Canada! There's your answer! There's nothing else to say. [Laughs] I'm being a little sarcastic . . . [But] I mean, think of the Lion's Gate Bridge.[15] If that were New York, do you think people would politely braid one car at a time? They do that here! [In the States its] 'shoot out the tires! Me first!'

Later, she acknowledged the limited, state-centered targets the Vancouver group could aim for, noting the availability of universal health care in Canada left the Socreds as a lone target:

> Other than getting rid of Vander Zalm, what would ACT UP here do? That was basically its moment. Basically here it was just people who wanted to be radical and copy what was happening in the States.

A member of the Advocacy committee at PWA also drew on the distinction in political cultures between the US and Canada in his explanation for why ACT UP seemed so out of place in a Canadian context. He identified the better protected private sphere for gays and lesbians in Canada.[16] Being from Toronto originally, he also emphasized that ACT UP was especially out of place in western Canada:

> There's always been a sort of Canadian complacency, I think . . . [T]he problem with ACT UP here was that they tried to use the same strategies that had been used in other places, and while it's appropriate in New York for everyone to chain themselves to some building, or in Washington, or go to the Center for Disease Control in Atlanta and pour red paint over everything, it's appropriate there, but it isn't here, in Canada, especially in the West it's not. That's not how you do things.

Another member of the PWA Advocacy committee concurred:

> ACT UP, well it doesn't exist. There is an ACT UP Vancouver, but there is a completely different political dynamic in Canada than in the United States. I don't believe ACT UP ever had been or ever will be successful in Canada because Canadians find that type of political activism to be 'not appropriate'. And this sort of extends to how you lobby in Canada, which is entirely different than in the United States. Canadians do not generally like public outbursts or messy-type confrontations. And if they're presented with that they tend to tune out the issue and the people involved. That's why by and far environmental activism has not had the same effect in Canada as it has elsewhere. The same with ACT UP: it has not had the same effect. We have had demonstrations. We have had die-ins. We have had the appropriate

> stopping of political figures. And it has had little or no effect on the political dynamic of this country.

Seeing the cultural politics in ACT UP, one reflective AIDS Vancouver volunteer stressed that AIDS events need to be safe and inclusive to affect significant change. He cited other local AIDS campaigns as examples of cultural politics that had been successful and less confrontational.[17] Hence he claimed that ACT UP Vancouver only served to alienate the public:[18]

> Things like the Walk for AIDS, AIDS Awareness Week, Joe Average: they're a lot softer, basically [than ACT UP] . . . And I think that the Walk for AIDS is very effective, but it's also very generic. And they make a big deal of inviting the whole political spectrum, which I think is wise politically. But for more radical cultural institutions like ACT UP to be effective there has to be – you can't just impose ACT UP in Vancouver. For those kinds of organizations to be effective they have to grow *out* of a community that's already quite sophisticated and quite diverse, and quite politicised. I just don't think there's a whole lot of interest in radical analysis, or more fundamental analysis of power structures around gay issues in Vancouver.

As a tactic of resistance, ACT UP did not work because of its disjuncture with the local political context of Vancouver, and Canada more generally.

Misunderstanding political structure

If ACT UP Vancouver failed to grasp the cultural sensibilities of civil society, it also misinterpreted and *misrepresented* the structuration between civil society and the state. For example, key state bureaucrats were secretly funding AIDS service organizations, much to the Socred's chagrin. In other words, by attacking the state *from* civil society, ACT UP failed to acknowledge the ironic linkages between the two spheres, which were common knowledge in local AIDS politics.

Instead, its vehement attacks on the state from public spaces in civil society erroneously assumed a clear, structural distinction that misunderstood the overlap and webbing of the two spheres in city politics, exemplified in places like AIDS organizations.

As well, ACT UP seemed blind to the valuable advocacy work done by the Vancouver Persons With AIDS Society. By 1990, PWA had a large membership and had developed successful programs and advocacy work with the government and medical system, which had been effective. Several members of the society stressed the overlap of function between the two groups, all the while favouring PWA:

> ACT UP has not been the force of change in Vancouver. I mean, we used those kinds of tactics for a while. We used to do demonstrations at Fantasy Gardens, and we'd generate a lot of public opinion through the media. Certain people in our organization have always been

> eloquent spokespeople, so we've always given those people that rein for as long as the group was satisfied. But then you have the issue of accountable representation. In some sense, PWA *was* ACT UP at that time, but we were everything to everybody. We've been quite militant in the past.

A PWA affiliate member drew the contrast between PWA and ACT UP as follows:

> To get back to the issue of ACT UP and the PWA Society, the activism is different here between the two. I don't think that that's universally true. I don't think ACT UP here is a movement. I think ACT UP came out of a number of very well intentioned people who wanted to hit the streets more than it was prudent for the PWA Society to do at that point. I think that's really laudable, and I'm not sure it won't resurface at some point. But I think the particular group that got together ... didn't know their stuff ... And whereas I think ACT UP in the states there's a very strong advocacy arm that stayed with the Society in Vancouver, and it has never left the Society here. So without that strong, fundamental base, ACT UP Vancouver just ended up being kind of demonstrations. I mean, if you look at AIDS Action Now! in Toronto, for example, where AAN is a strong force, AIDS Action Now is very much along the lines of advocacy that the Vancouver PWA Society does – with more of an activist overtone. The Toronto PWA Foundation is like the service arm of the PWA Society here. So the organisational structures are different here.

ACT UP failed to grasp the shifting relations between state and civil society in Vancouver, evinced by shadow–state organizations like PWA.[19]

One might argue that ACT UP Vancouver was at least radical in that it refused to be incorporated. Unlike the city's AIDS organization, it stayed squarely within civil society. Admittedly many members of the shadow state lament the increasing ties with the state in local AIDS organizations (Brown, 1994). Overall, though, I am not sympathetic to this critique because ACT UP failed to achieve any of the cultural or material gains made by the shadow state. That criticism also smacks of exactly the error that Laclau and Mouffe try to resist: insisting on an a priori form or location for what counts as 'radical citizenship'. As Laclau and Mouffe argue:

> This has led to a failure to understand the constant displacement of the nodal points structuring a social formation, and to an organization of discourse in terms of a logic of 'a priori privileged points' which seriously limits the Left's capacity for action and political analysis. This logic of privileged points has operated in a variety of different directions. From the point of view of the determining of fundamental antagonisms, the basic obstacle, as we have seen, has been *classism* ... From the point of view of *social levels* at which the possibility of implementing changes is concentrated, the fundamental obstacles have

> been *statism* – the idea that the expansion of the role of the state is the panacea for all problems.
>
> (Laclau and Mouffe, 1985: 177)

AIDS politics in Vancouver cannot be judged solely or primarily as exercises in theoretical consistency; they were about educating the public and meeting people's needs in very immediate, tangible ways during a crisis.

We might further question ACT UP's radical potential by its misplaced and exclusive focus on the state, even though it claimed civil society as its venue. Mohr (1993), for instance, has argued that ACT UP does not challenge the modern state, but actually *worships* its power and reproduces it, by demanding the state extend its responsibility through rights claims. ACT UP Vancouver's failure compared with the PWA Society's success substantiates this point. By focusing so much attention on what the state ought to do, one might argue that it drew attention away from the effective and sensitive workings of organizations that began in the gay community, in civil society. Clearly, the state could have (and eventually did) do more around AIDS issues. However, because the provincial government was ACT UP's *only* target, it failed in the sense of increasing strains between the state and the local ASOs. A deputy minister of health under the Socreds recalled that ACT UP only served to alienate the Socreds further into indifference. Many ASO employees and volunteers acknowledged ACT UP's utility as an outlet for anger, but wondered about the cost of its venting for already tenuous links between state and shadow state. Here again, ACT UP Vancouver's misreading of the state–civil society distinction is significant.

CONCLUSIONS

ACT UP Vancouver clearly had a unique origin and trajectory, and imposing the radical successes of the chapters in other cities onto this local context is fraught with difficulties. But I think the more interesting point to take from this paper is an appreciation of how a spatial reading of radical democracy demonstrates the limited successes and broader failures of a particular form of resistance. Imagining a clearcut separation between the state and civil society allowed ACT UP to vent mounting anger and grief, and act as a possible disciplinary force in city politics. In these ways the group commandeered civil society to attack the state as its other. But there were significant errors in this map of city politics. ACT UP's political geography neglected the contemporary webbing of these two spheres of life, evinced by the local voluntary sector. In Vancouver, the group's erroneous cartography limited its ability to survive and get needs met, an important strategic lesson as state apparatuses continue to diffuse (e.g. Boston, 1995). There are also theoretical lessons at stake, however, for scholars interested in new locations of resistance. The case study warns against merely dichotomizing 'old' state-centered locations of politics and 'new' ones in civil society when we look for radical democracy. Canel (1992: 37) has recently levied just such a criticism:

> Laclau and Mouffe ... insist that every social conflict is political, as politics expand to civil society, but they fail to discuss the institutional aspects of politics, the relationship between new social movements and political parties through which the democratization of the state can be achieved. Such a separation between social movements and the political system can potentially contribute to a depoliticization of social movements. This is most ironic given that the purpose of the argument is to demonstrate the expansion of the political.[20]

ACT UP Vancouver was out of place because the division between state and civil society that it assumed had been remapped – both culturally and structurally – by a different political culture and an effective shadow state. Scholars interested in new sites of resistance may risk the same mistake. Radical democracy's quest for new locations of political engagement should not blind us to changes in the way that 'old' and 'new' locations amalgamate, *because their spatial interaction is itself a new site of politics.* Consequently, the geography of these sites becomes crucial to political theory's often all too abstract tenor.

ACKNOWLEDGEMENTS

I want to thank several commentators on previous drafts of this paper, including, Nick Blomley, Gordon Clark, Kevin Cox, Robyn Dowling, Derek Gregory, Michael Keith, Bob Lake, David Ley, Pete Mayell, Steve Pile, Gerry Pratt, and Bruce Willems-Braun.

NOTES

1 The Socreds not only neglected AIDS issues; when they did pay attention it was through invectives that gays in BC could 'take care of their own kind' or that AIDS was 'a self-inflicted wound' (see Brown, 1994).

2 The state–civil society dichotomy can be found in a wide variety of historical and contemporary political philosophy, including Mouffe's work. See for instance Hegel, 1991; Kymlicka, 1990; Keane, 1988; Cohen and Arato, 1992.

3 Atranscript of the speech can be found in (Kramer, 1989: 127–139).

4 Chapters have been formed in (*inter alia*): Atlanta, Boston, Chicago, Cleveland, Dallas, Denver, Halifax, Kansas City, London, Los Angeles, Milwaukee, Montreal, New Orleans, Paris, Philidelphia, Portland ME, San Francisco (also Golden Gate), Seattle, St Louis, Toronto, and Washington. See Sword, 1991; Crimp and Rolston, 1990; Rayside and Lindquist, 1992; Chew, 1993.

5 The situationists referred to this process as *detournement*.

6 ACT UP's most well-known actions illustrate the use of civil society to disrupt hegemony radically. A banner reading 'Sell Welcome' was unfurled by activists posing as bond traders, actually interrupting the global shifts of capital on the floor of the New York Stock Exchange for a full five minutes on September 14, 1989, to protest the profits made by the drug company on AZT. Several months earlier, on April 25, 1989, Burroughs Wellcome's headquarters in North Carolina were infiltrated by four ACT UP members, dressed as businessmen, who talked their way past security points and sealed themselves in the building, demanding the reduction of AZT's price, which they eventually won.

7 Though there had been an earlier demonstration. In early September of 1989, AIDS activists targeted BC Premier Vander Zalm, holding an elaborate protest at Fantasy Gardens theme park in Richmond (an inner suburb), which the Premier and his wife

owned. There were die-ins held, mock graves and coffins with tombstones reading, 'He Died On A Placebo Study' and 'I Died In Poverty Paying For AZT'. Protesters picketed the park, which was at the time sponsoring a fundraiser for the Lions Club Timmy Telethon. 'HIV Is NOT a Gay Disease' and 'Homophobia is a Sickness', read their placards. Not surprisingly, there were many members of the Persons With AIDS Society on hand (Rebalski, 1989: D 12).

8 The Vancouver Persons With AIDS Society is part of the local shadow state. It is a non-profit, self-help organization dedicated to empowering people with HIV/AIDS. It provides education, support, and advocacy functions.

9 Fantasy Gardens, the Premier's religious theme park, houses a replica of a windmill, evoking Vander Zalm's Dutch heritage.

10 This person died in 1992.

11 It was especially pertinent in the summer that ACT UP Vancouver formed, since one of the founding members of ACT UP (and a former director of the PWA Society) announced that he would kill himself in late August. His death coincided with an ACT UP demonstration, in fact (see Kines, 1990). His suicide created an acute and very political problem for PWA, since it was feared that his action would be interpreted as the society's condoning of euthanasia. In other words, fear hinged on the confusion over whether he was acting as an individual or as a member of PWA. I explore the overlap between state and civil society more fully in Brown, 1994.

12 This person died in August of 1994.

13 The interviewee did not specify the policy in question. He merely indicated that the issue was resolved with the threat of demonstration.

14 On measures of difference between the US and Canada see Goldberg and Mercer, 1986.

15 The Lion's Gate Bridge is an often-congested transit across the Burrard Inlet that links downtown Vancouver with its northern suburbs.

16 We should not assume too much on this point. Canada Customs has often arbitrarily banned gay and lesbian material specifically from entering the country, causing at least one Vancouver bookstore to take the government to court recently. See Fuller and Blackley, 1995.

17 Dr Peter Jepson-Young, (who died in November 1992 at 35) held a weekly series of public education 'diaries' broadcast on the CBC evening news in British Columbia from 1990 to 1992 about his life with HIV and AIDS (see Gawthrop, 1994). Through his intensely personal and riveting segments, viewers could watch the progression of illnesses and his ability to cope with them. He received far more press than ACT UP ever did locally. See, for instance, Wigwood, 1992; Fraser, 1992; Easton, 1992. Joe Average is a popular Vancouver artist living with HIV, whose colorful paintings are locally associated with AIDS issues and charities. See Smith, 1993.

18 In fact, later on this respondent compared the cultural politics of the Quilt with ACT UP explicitly by saying, 'You know what I think? I think the Quilt is the smartest piece of art that's ever happened.'

19 Here again we see the importance in stressing Mouffe's decentered subjectivity for citizens. Contradictory processes were going on that allowed some PWA members to see themselves as opposed to the state (especially drawing on PWA's grassroots, gay-community heritage), while organizationally the society was moving towards a closer relation to the state. Members could thus position themselves in a number of ways. As well, bear in mind that individual PWA members would have varying levels of commitment and allegiance to the organization at any given time.

20 Mouffe does not necessarily exclude the state from radical democracy: 'it is clearly impossible to identify either the state or civil society as the surface of emergence of democratic antagonisms' (Laclau and Mouffe, 1985: 180). Nonetheless, because the state has dominated the geography of political theory, it – or its amalgam with civil society or the family – tends to be neglected by those interested in radical democracy.

8

RINGS, CIRCLES AND PERVERTED JUSTICE

gay judges and moral panic in contemporary Scotland

Lawrence Knopp

INTRODUCTION

In the early morning hours of July 19, 1992, someone broke into the Fettes Avenue headquarters of the Lothian and Borders Police Department in Edinburgh, Scotland, stealing a number of files and spray-painting slogans suggesting that a group called the Animal Liberation Front was responsible. Shortly thereafter it was reported that in fact one or more 'gay criminals' had staged the raid, stolen files on themselves and others and tried to cut deals with the police for their return.[1] Indeed, a deal apparently was struck and the individual responsible for the break-in never went to prison for his part in the burglary (although he did later on separate charges of robbery and assault).[2]

Then, in September 1992, a copy of an internal police memorandum, prepared in December 1991, was leaked to the press. The memorandum, which had already been dismissed as unsubstantiated and ordered shredded by the department's Chief Constable, alleged 'the existence of a well-established circle of homosexual persons in Edinburgh with influence in the judiciary ... who have formed associations which in themselves lay them open to threats or blackmail ... [H]omosexuality may well have been used as a means to seriously interfere with the administration of justice' (Campbell, 1993a: 9). This appeared to confirm rumors that had been circulating for at least three years in Scotland of a so-called 'magic circle' of gay lawyers and judges involved in the 'perversion' of justice.[3]

The leak of the 1991 memorandum precipitated an investigation into both the memorandum's claims and the leak's source. On January 26, 1993 a 101-page report was released which concluded that there was no evidence to substantiate claims of a 'gay threat to justice' (see Figure 8.1) and that certain frustrated officers in the fraud squad, motivated by anti-gay and anti-lawyer bias, were responsible for leaking the 1991 memorandum.[4] In the process, however, it confirmed rumors that one Scottish High Court judge, who had

Secret police report claims
senior lawyers could be
vulnerable to blackmail

GAY THREAT TO JUSTICE

by David Forsyth
NEWS REPORTER

Figure 8.1 Gay threat to justice
Source: Edinburgh Evening News

resigned in late 1989, had done so because of a fear that his involvement in a gay relationship would be exposed in the tabloid press.[5] In the tabloid *Sun* newspaper, this, not the other conclusions of the report, is what was emphasized (see Figure 8.2).

The discourse surrounding this set of events (which came collectively to be known as 'Fettesgate') is the subject of this chapter. In particular, I focus on certain of the power relations which were reflected and reproduced in popular representations and understandings of these events (especially in the mainstream and tabloid print media). I am inspired by the very productive efforts of many contemporary geographers (and others) who have brought a critical spatial perspective to the interpretation of cultural forms and identity politics.[6] These writers have argued persuasively that: cultural forms are both products and shapers of identity; that they embody, contradictorily, both domination and resistance; that their constitution, expression and reproduction are inherently spatial; and that their spatiality operates in such a way as to call into question seriously the usefulness of a distinction between the material and the metaphorical (see also Rose, 1993b). Most have been

THE Sun

PRINTED IN SCOTLAND

Wednesday, January 27, 1993 25p FIGHTING FOR INDEPENDENCE

DERVAIRD: I DID HAVE GAY FLING

Fettes probe bombshell

DISGRACED Scots judge Lord Dervaird quit the bench after confessing to a homosexual fling, it has been revealed.

He admitted the affair after he was spotted with a male friend during a secret meeting in London.

The judge — real name John Murray — caused a sensation when he resigned in December 1989.

But his real reasons for stepping down from the £72,000-a-year post were not made public despite strong rumours.

By ALAN MUIR

Dad-of-three Dervaird, 57, is the only prominent legal figure revealed as having a homosexual link in a report published yesterday, which cleared the Scots legal system of gay corruption.

JUSTICE

The Fettes probe — by QC William Nimmo Smith and fiscal James Friel — said there was no evidence of a gay conspiracy to pervert justice.

But for the first time it detailed why Dervaird — now a law professor at Edinburgh University — bared his soul to Lord President Lord Hope in secret talks after claims a newspaper was about to print allegations of romps at a country cottage.

The respected lawman denied the story but admitted he had "homosexual relations" while he was a judge.

His bid to keep it a

Continued on Page Four

Lord Dervaird quit after homosexual affair

The Report on an Inquiry into an Allegation of a Conspiracy to Pervert the Course of Justice in Scotland

Revealed . . . the report published yesterday

SAS men tortured me says riot con

HOODED SAS men tortured a prisoner and threw him down two flights of stairs as they broke up a jail riot, a court heard yesterday.

John Devine told how the crack troops pounced as he and two other inmates held a prison officer hostage.

He said they grabbed him after throwing a CS gas canister into the attic of D Hall at tough Peterhead Prison.

Three men wearing boiler-type jumpsuits with gas masks and respirators threw him 12ft from the loft to the gallery.

Bundled

He claimed one of them told him as they hurled him to the ground: "You are going for a spin, pal."

Then they beat him with batons and bundled him down the stairs before handing him over to prison officers who stripped him and gave him another beating.

Devine, 33, is suing Scots Secretary Ian Lang for £36,000 over incidents at the jail more than five years ago.

A jury at the Court of Session heard that Devine and two other inmates Sammy Ralston and Malcolm Leggat were the last to surrender after the five-day siege.

He told Alistair Dunlop

Continued on Page Seven

6% INTEREST RATES

Figure 8.2 I *did* have gay fling
Source: The (Scottish) Sun

influenced by one or more of postmodern, post-structuralist, feminist, anti-racist, post-colonialist and queer theories. Judith Butler's (1990a) work on the *performativity* of identities has been particularly influential (see Bell, Binnie *et al.*, 1994; Kirby, 1995; Knopp, 1995; Probyn, 1995; Walker, 1995).

The reason for engaging in this analysis is to demonstrate empirically just how complex and contradictory the relationships between identities, political action, space and place can be. In the events discussed, a highly spatialized, sexualized, racialized, gendered, and classed set of material and discursive practices constituted an extraordinarily powerful force in the reproduction of a *very* conservative social order. The determined agency of individuals representing dominat*ed* as well as dominat*ing* cultural groups played important parts in this reproduction (as well as in more predictable forms of resistance). Many of the problems and contradictions inherent in attempts to fix certain spaces, places and identities (e.g. 'gay') as inherently counterhegemonic, as well as in attempting to adhere rigidly to an analytical distinction between notions of material and metaphorical space, are thus laid bare by this analysis.

RINGS, CIRCLES AND PERVERTED JUSTICE

The first allegations of a 'magic circle' emerged in the context of a gay solicitor's trial for mortgage fraud in 1989 (see Table 8.1). This individual (Colin Tucker) and his attorney were said to possess a list of other gay lawyers, as well as gay judges and police officers, who were supposedly involved in illegal activities of various kinds (many of them sexual offenses). Furthermore, there was the strong implication from Tucker's advocate that this list might appear at trial. As it turned out, the list was little more than an unsubstantiated statement by Tucker in which exactly one person, the recently appointed High Court judge John Murray (Lord Dervaird), was named as having had an affair with a male former client of Tucker, but not of having perverted justice.[7] Nonetheless

Table 8.1 Chronology of events

1980s: Deregulation of financial institutions; cutbacks in social services; increases in mortgage, DHSS frauds; periodic sex scandals

May 1988: Sensational trial and conviction of gay DHSS landlord on charges of sexual assault

December 1989: Trial and acquittal of solicitor on charges of mortgage fraud; first allegations of 'Magic Circle'; High Court judge resigns

January 1990: Police 'Operation Planet' investigation begins

February 1991: Planet case ends with few convictions

December 1991: Under pressure from Tam Dalyell, MP, internal memorandum is prepared

19 July 1992: Break-in at Fettes Headquarters of Lothian and Borders Police

August 1992: Gay link to break-in reported

September 1992: December 1991 memorandum leaked to press; Nimmo-Smith investigation ordered

September 1992–January 1993: Height of 'Magic Circle' moral panic

26 January 1993: Nimmo-Smith report concludes no 'Magic Circle'; blames disloyal detectives in fraud squad and certain others; confirms High Court judge's resignation in 1989 was because of gay affair

Tucker's advocate and certain police with whom he shared this statement talked this statement up as a 'list' and suggested that something was terribly rotten in the Scottish legal establishment. Indeed, it was argued in both the mainstream and tabloid presses that the information might, if made public, 'blow the lid off' the Scottish legal establishment (Campbell, 1993a). In December of 1989, Tucker was easily acquitted on the basis of a defense which claimed he had been blackmailed into participating in the frauds by his deceased partner, who had allegedly threatened to expose him as gay. A few days later, under intense pressure from the tabloid *Sun* newspaper, Lord Dervaird resigned from the bench.

Almost immediately (in January 1990), Lothian and Borders Police began an investigation (first called 'Operation Uranus', but later changed to the more discreet 'Operation Planet') into an alleged 'ring' of men involved in conspiracies to commit sexual offenses with other men and boys (Campbell, 1993a). A link was thus forged early on (and was repeated in the press) between gay sexuality and a wide range of illegal activities, including non-sexual violations such as mortgage fraud (more on this below). One informant in the operation was a young man named Stephen Conroy, who had worked at Colin Tucker's law firm. He claimed an affair with a former officer and now sheriff in the Edinburgh area, and added the names of other judges and lawyers to the previously established rumors of corruption and perversion in the Scottish legal world. His claims, repeated in both the tabloid and mainstream presses, continued to fuel speculation and rumors about a 'gay threat to justice' in Scotland.[8] When all but 10 of 57 counts against 10 men in the Operation Planet case were dropped, some police officers became angry and frustrated. They voiced their concerns (mostly based on Conroy's claims) to a Scottish Member of Parliament named Tam Dalyell, who was known for his willingness to demand inquests and investigations into all manner of state misconduct and/or ineptitude.[9] Dalyell then raised the issue with the Lothian and Borders Chief Constable, who ordered preparation of the December 1991 police internal memorandum (Forsyth, 1992).

All of this came on the heels of over two years of other gay-related scandals, including ones in which certain gay figures had been arrested for Department of Housing and Social Services (DHSS) landlord fraud, for sexually assaulting young male tenants, and for telephone fraud involving gay sex-lines.[10] It was in the wake of the failure of Operation Planet, the embarassment of the Fettes break-in, persistant rumors in the press, and difficulties prosecuting mortgage and other forms of fraud, however, that frustrated officers, acting without authorization, leaked the December 1991 memorandum (which was supposed to have been destroyed) to the press (Kinsey, 1992).

A number of things are interesting about the public discourse surrounding all of this. First, it featured gay men as instigators and perpetuators of rumors about gay criminality as well as police, politicians, and the press. Second, it was extremely spatialized. Both spatial metaphors and references to specific sites were constantly invoked to give allegations a sense of groundedness and drama. Third, most elements of the discourse were quite hegemonic and only weakly contested. Responses from established gay organizations and spokespeople were generally limited to defensive comments about the homophobic nature of the reporting; few if any alternative readings of the events in question were offered. The only comprehensive one was developed by the journalist Duncan Campbell (well-known throughout the UK for his work in the print and electronic media) in a Channel 4 television documentary and a two-part feature in *Gay Scotland* magazine during the summer of 1993. This focused, however, primarily on disentangling the homophobic actions of police and popular press from the activities of 'criminals' who, according to Campbell, simply happened to be gay. Thus the legitimacy of dominant notions of criminality was never questioned. Finally, virtually nowhere in the public discussion of these events were the politics of market deregulation, social service cutbacks, and criminalized consensual sex explored as a context in which mortgage and other forms of fraud, as well as the exploitation of homeless young men, became possible. From an exhaustive search of the local print media around the time of the Fettesgate affair, I found exactly one article, out of several hundred, which even raised this issue. But this article, by a criminologist at Edinburgh University, went on to examine the failings of the police bureaucracy rather than to develop this idea (Kinsey, 1992).

Gay male complicity

In the 1989 mortgage fraud trial the defendant, Colin Tucker, was a gay man. He and his advocate clearly exploited recently reinforced myths and stereotypes about gay sexuality and vulnerability to blackmail to obtain his acquittal. The implicit *necessary* link between gay male sexuality and mortgage fraud, however, was more the invention of police and press than of these individuals. There was, after all, a great deal of mortgage fraud going on throughout Britain at the time (largely as a result of the deregulation of financial institutions), and there is no evidence that it was the particular province of gay men, nor that it was linked inherently to gay sexuality. Another gay man, former rent-boy and later alleged pimp Neil Duncan, also named names (and sex acts) in the 1990 Operation Planet vice operation, but again none of his claims posited a necessary link between gay sexuality and mortgage fraud (or other non-sexual 'perversions of justice'). Then, in 1991, Stephen Conroy, the former associate of Colin Tucker, tried to get acquitted using the same strategy that Tucker had. He failed, and has sinced recanted many of his accusations from prison (Campbell, 1993a: 20). The Fettesgate thief himself (apparently a man named Derek Donaldson, who was never named publicly in the press until Duncan Campbell's piece in *Gay Scotland* magazine) also seemed quite happy to play the homophobic press, the police and certain elements of the gay community off against each other for personal and political reasons.

But whether one sees these people, as Campbell apparently does, as petty criminals who bear responsibility for the link between gay sexuality and criminality being made (i.e. who were quite willing to do damage to the gay community and gay political and social movements for their own narrow aims), or as simple survivors of and/or resistors to a virulently homophobic and heterosexist culture, depends upon one's frame of reference. The police and tabloid press, obviously, saw them as *criminally perverse*. That is, they saw the alleged fact of their criminality as unproblematic and the link between this and their homosexuality as confirming the sociopathic and therefore criminal nature of homosexuality itself. This is the reading that overwhelmingly dominated public discourse, notwithstanding Campbell's 1993 attempt to rework this slightly. But it appears that these 'criminals' (and others besmirched by their associations with them) saw themselves quite differently: the Fettesgate thief, for example, saw himself, in a letter to *Gay Scotland* magazine in late 1992, as a hero, someone who had exposed homophobia in the police and press as well as the timidity and quisling-like posturing of the institutionalized gay movement in Scotland. He claimed that 'Fettesgate and the explosion that quickly followed will act as a catalyst for gay law reform that some of the chumps in the SHRG [Scottish Homosexual Rights Group] have been only bleating about for years' (*Gay Scotland*, 1992: 5). Another figure, implicated by association and innuendo in various of these and other dealings (but never convicted of anything, and who apparently is in all likelihood innocent of most if not all of the charges levelled against him), sees people like Donaldson, who have long histories of trouble with the law, as very dodgy but chooses to emphasize the hypocrisy of the legal and journalistic establishments. He maintains that the Scottish legal system is rife with institutionalized illegalities (such as plea bargaining and other forms of deals, many of which are driven as very hard bargains, with implied threats of exposure and intimidation being used all the time), and that some of his own detractors who accuse him of DHSS landlord fraud have profited legally from practices whose ethics and morality (if not their legality) are at least as questionable as, and certainly more exploitative than, the technical fraud of splitting DHSS reimbursements with young, poor tenants (which he also denies doing):

> the whole legal profession is based on crime ... the legal system has nothing to do with justice, you expect people to do you favours and some solicitors will go so far, others will go further ... Plea bargaining [is] informal and nobody admits that it happens, but it does. And it gets tied up. You then phone up the judge and say as part of the plea bargaining 'What's the tariff going to be?' That's legal. So there are the law enforcers breaking the law and they can't admit it ... Fixing goes on all the time.
>
> When I moved to Edinburgh I was on the dole and realised there was a business opportunity to be had by getting large flats and renting them out to kids on the dole. Money was being paid to you direct

> by the state, no skin off the kids' noses. Better than living in the streets or slums. Other people were in the same business and were doing it more or less ethically as I was. Quite frankly, some of the things that went on in DHSS properties were appalling, raping of males and females. I bumped into [name deleted] in London whilst I was doing this ... and he said to me 'Why don't you split the money with the kids?' I said, 'Well ... that would be a criminal offence, it would be a fraud, charging more rent and giving it back to them.' I must admit, when I started in the business I thought this would be a reasonable thing to do. I soon realised no, because it's something that could be held over you.
>
> (Personal interview)

So while some gay men were consciously instrumental in the exploitation and perpetuation of cultural fears surrounding homosexuality, their intentions are much more complex. To dub all their practices, and those of others who share their views, as 'criminal' is very problematic. For these individuals were engaged in a quite damning critique of dominant constructions of justice and criminality in what they were doing, albeit with self-serving motives as well.

A spatialized discourse

The spatialized nature of the discourse was quite vivid. Among other things, it relied heavily on the use of spatial metaphors to imbue the alleged conspiracy with the actual power to pervert justice. The conspiracy was called the 'magic circle' but was also referred to by use of such terms as 'ring', 'secret network', 'underground community' and 'dark underworld'. Military and epidemiological metaphors of movement were also used to suggest the active subversion of justice: 'infiltration', 'penetration', 'infection', 'disease', etc., all of which were seen as threats to an unproblematized notion of something called 'justice'.[11]

But the discourse also made reference to a huge number of specific sites in Edinburgh and elsewhere to legitimate the existence of this 'underworld'. There was, for example, a house in Edinburgh's West End (owned by former business manager of the Bay City Rollers, Tam Paton) where the Operation Planet investigation began with a raid and where numerous acts of rent-boy procurement, bondage, group sex, and sadomasochism were alleged to have taken place.[12] There was a flat in central London (dubbed a 'palace') where Edinburgh judges and solicitors allegedly had orgies with young boys to avoid being seen in Scotland (Rimmer, 1992b). There was a network of gay bars and discos, including one from which one judge supposedly 'stormed out in disgust' and another where a supposedly gay sheriff had had 'dates'.[13] There was Lord Dervaird's family cottage in southwest Scotland which was supposedly also a site for depraved sex acts (*The Sun*, 1990b). There was Regent Road outside the Scottish Office, and other cruising grounds, where gay judges were supposedly picking up rent-boys (Burrell and Forsyth, 1991b). And of course there were DHSS-subsidized hostels and B&Bs where exploitation of young men was alleged to be taking place and which served as fronts for telephone and mortgage fraud operations.[14] This referencing of sites, along with the use of spatial metaphors, keyed specifically to (but suggesting a much larger and nefarious geography than that constituted by) those sites, had the effect of grounding (literally) and fixing certain of the activities, identities, spaces and places associated with the 'gay threat to justice'. Gay male identities were constructed, quite stereotypically, as being all about exploitative, abusive, painful, narrowly self-serving, and ultimately socially (as well as personally) *dangerous* activities. Gay male *spaces* and *places*, meanwhile, were represented as the at once visible and invisible, ordinary and extraordinary facilitators of these activities. When viewed under 'normal' circumstances, and from the 'outside', they appeared unproblematic and unthreatening. When 'exposed', they became active agents in the construction, through various hidden (and thus protected) sexual and social practices, of gay men. They were thus presented quite literally as in need of reappropriation by the dominant culture – in its own defense!

This discursive representation of gay male spaces and places as two-faced and active – with the particular face which is seen depending entirely upon the social and spatial location of the observer – was in a very real, though narrow and distorted sense, accurate. It referenced both a metaphorical world of gay subversion *and* aspects of an actual geography of survival (or resistance) on the ground in Scotland (bars, discos, cruising grounds, social networks, etc.). These were then tied together discursively into a single package, linked ideologically to other aspects of gay as well as non-gay life (mortgage fraud, sadomasochism), and constructed as a criminal gay conspiracy to pervert justice.

The metaphorical/material conflation which took place, therefore, reflects not so much an analytical *error* as a discursive strategy. By associating gay male spaces (through metaphorical language) with diseased bodies, secret agendas, darkness, and danger, consumers of the discourse were presented with what was in effect a variation on the Jews-poisoning-the-well myth (Campbell, 1993a). Gay spaces became protectors of a deeply subversive threat to the social and moral order. They not only contained but *empowered* sinister forces of *dis*order. Moreover, these forces were, by virtue of their ability to move seamlessly and invisibly from protective gay spaces to spaces of great power (e.g., the judiciary), potentially able to strike deeply and effectively into the heart of 'enemy' territory (Scottish society). These constructions thus both created and exploited deeply held human fears of closed, dark, unknown spaces – made all the more frightening by their association with closed, secret circuits of power linked to institutions of the dominant culture.

Indeed, the 'magic circle' myth deployed, from the start, a particular ideologically constructed geography to perpetuate particular constructions not only of homosexuality but of criminality, justice, age, gender, class, markets and Scottishness as well. These constructions are detailed in the next section, in which I consider the weakly contested nature of the public discourse surrounding the 'magic circle'.

Dominant constructions and the lack of effective resistance

In virtually all of the discourse surrounding Fettesgate and the 'magic circle', dominant notions of justice, criminality, age, gender, class, markets, Scottishness (and, by extension, Englishness), and sexuality were rarely if ever questioned. The only significant challenges mounted were to the homophobic nature of the reporting and certain other activities surrounding the affair. These came largely from representatives of the Scottish Homosexual Rights Group (SHRG) and Duncan Campbell, both of whom emphasized the distance between most gay people and the criminal activities of people like the Fettesgate thief. Said Tim Hopkins, Secretary of SHRG:

> Saying he [Donaldson] is responsible for any good things over the past few weeks is like saying Hitler was responsible for setting up the United Nations after the Second World War.
>
> (*Edinburgh Evening News*, 1992)

Justice, therefore, was implicitly constructed as synonymous with legality (*except* in a few areas, such as Britain's infamous Clause 28, which prohibits government 'promotion' of homosexuality, and the unequal ages of consent for gays and straights). Criminality, then, was associated with anyone who broke the law *except* in these areas. Two areas of the law are particularly pertinent here: one has to do with age, and the other has to do with land and housing markets.

In the area of age, one of the offenses with which the generally unnamed gay lawyers and judges were continually accused was having sex with 'underaged' males, meaning males under 21 (until recently the legal age of consent for gay sex in Scotland).[15] This was always presented unproblematically as exploitative, and usually in the context of rent-boy procurement. The geography of these allegedly exploitative practices was, as discussed above, central to 'proving' their existence. But while issues of power and consent in prostitution and intergenerational relationships are extremely problematic, some of the relationships alleged were clearly between consenting older and younger men and did not involve exchanges of money. Yet this issue was virtually never raised in public discussions of the 'magic circle' and related events. Nor were the geographies of homelessness and other forms of victimization experienced by young gay men expelled from their families, and alienated from their peers because of their sexualities, discussed. Instead, youth was universally constructed as meaning that one was under 21 (or at best under 18), powerless, capable of consenting only to certain kinds of sex acts (if any), and incipiently heterosexual (as evidenced by the higher age of consent for homosexual than heterosexual relations).

This is problematic on a number of counts. First, notions of the ability to 'consent' to sexual relations, and the relationship of this to age, have always been contested in modern Western societies. Generally, the ways in which this has been resolved have reflected the interests of those constructed as heterosexual, white and male. Thus in many jurisdictions females have been deemed capable of consent at younger chronological ages than males (allegedly because of an earlier onset of puberty, but most likely to facilitate access of older males to younger females), and relations between members of the same (socially constructed) anatomical sex, or social categories constructed as 'different races', have required higher ages before one is deemed capable of consent (if such relations are permitted at all). This suggests a shifting and strategic construction of personal empowerment which uses the category 'age' to advance particular interests in very different sets of power *relations*.

Second, and related to this, is the damage that such constructions can actually do to the individuals they purport to protect (children and young adults). By denying the existence of homosexual youths, of homophobia and heterosexism, and of any capacity young people might have to make wise choices (and by legitimating disciplinary intervention into the sexual development of young people by a heterosexist state), young gay people are made to pay an extraordinarily high price for being different. The result is often homelessness, poverty, illness, and alienation. Yet, again, these issues were totally absent from the discourse surrounding the 'magic circle' and Fettesgate affairs. With respect to land and housing markets, all of the public discussions interpreted the illegalities of mortgage fraud and DHSS landlord fraud as unproblematically criminal and unjust. While much of it certainly was used to line the pockets of property developers and landlords (and to empower them to exploit tenants, sexually and otherwise), and while it was all certainly illegal, questions of the motives of the perpetrators, of the selectivity of the prosecutions, and of the justice of the market environment in which all this took place, were never raised. For example, some of the DHSS fraud apparently involved giving tenants a kickback on their payments and may have been motivated by unexploitative gay landlords trying to help young gay people survive in a dangerous and hostile environment (personal interviews). And some of the mortgage fraud involved financing these and other developments that had positive as well as negative consequences for many gay men, and were not profitable for the individuals involved (personal interviews). The same can be said of some of the for-pay telephone sex lines. So the markets for housing and property were constructed implicitly as just, and *all* activities which appeared to threaten them (and even some which probably didn't, in the sense that the deregulated environment of the 1980s was *encouraging* highly speculative deal-making), were dubbed criminal and unjust. They were also associated with particular constructions of gay male sexuality, which was seen as equally criminal and unjust, because of its violation of all sorts of rules governing markets, ages of consent and 'adult' sexual practice.

Particular constructions of gender and of Scottishness were also advanced in the discourse surrounding Fettesgate and the 'magic circle'. With respect to gender, representations of both women and female sexuality (especially lesbian sexuality) were noticeably absent from discussions. A link between male sexuality and social power, however, clearly was not. This reinforced the idea that history and geography (including sexual histories and geographies) are made by men, and in particular that sexuality is a predominantly male prerogative, intimately linked to power, subject to abuse, and rife, therefore, with potential social consequences (both positive and negative). Women and female sexuality, by contrast, were presented, by omission, as irrelevant to the issues at hand.

With respect to Scottishness, the fact that contemporary Scottish identity has much to do with the preservation and practice of a distinct form of law means that the sexualized 'assault' on the Scottish judiciary was much more than a potential perversion of an abstract 'justice'. The Scottish legal system has long been represented in nationalist media (and popular discourses generally in Scotland) as more rational, efficient and 'just' than its English counterpart (McCrone, 1992).[16] To the extent, then, that the discourse surrounding the 'magic circle' was aimed at protecting this system from an alleged 'gay threat', it was about protecting Scottish identity as well. This was reinforced by a frequent focus in the Scottish editions of the tabloid *Sun* and *Daily Record*[17] newspapers on an English connection to Fettesgate and the 'magic circle'. Many of the most heinous sexual offenses, these papers insinuated, took place in dens of decadence and iniquity south of the border (Rimmer, 1992b).

Related to this are the ways in which class and the occupational culture of the judiciary were represented. The anti-lawyer bias of the police and popular culture were clearly reflected in language such as 'pin-striped pervert',[18] which was common in the tabloid press. Even the mainstream press insinuated that there was a link between Fettesgate, the 'magic circle' and upper-class decadence (Douglas and Dinwoodie, 1992). In a slightly different way, this was the case as well in the 1993 Nimmo-Smith report (which dismissed allegations of a 'magic circle' but confirmed Lord Dervaird's departure from the bench due to a perfectly legal, but homosexual, love affair) and the Fettesgate raider's statement in *Gay Scotland* magazine:

> As a participant in the Fettes break-in, I want to express my disgust at so-called 'respectable' members of the gay rights organisations ... Let me remind establishment lovers ... of some hard facts.
>
> (*Gay Scotland*, 1992: 5)

All of these constructions, of course, informed each other, and served to reinforce a social order which was at once sexist, heterosexist, classist, ageist and nationalist (i.e., racist). They must be seen, therefore, as mutually reinforcing and intimately intertwined. The lack of any discussion of the two affairs' economic, social, and political contexts makes this particularly clear.

The absence of context

Discussions of Fettesgate and the 'magic circle' virtually never addressed the social and political contexts within which the affairs took place. Yet the mortgage and other forms of fraud that were so much a part of 1980s Scotland were not only made possible but in many ways encouraged by Thatcherite economic policies at the time, most notably the deregulation of financial institutions. This precipitated a wave of reorganization for which large quantities of capital were necessary. Building societies in particular (the equivalent of Savings and Loan Associations in the US) were keen to encourage short-term, high-return (but, because of their speculative nature, high-risk) projects. Until a 1988 reform in DHSS housing allowances, this gave would-be DHSS landlords (among others) a real opportunity to make quick profits at low risk to their financiers; but lack of close oversight also encouraged potential fraudsters to partially represent or misrepresent their projects. Arguably this was common practice, with the boundary between legal and illegal misrepresentation being quite fuzzy.

Similarly, the DHSS reform of 1988, which tied housing allowances to property assessments, created an opportunity for landlords to demand that the amounts they lost be made up out of their tenants' pockets or, possibly, be taken out 'in trade'. Either way this reform, this cutback, can be seen as empowering ruthless DHSS landlords and disempowering their already vulnerable tenants. Finally, the absence in the discourse of any concept of oppression (except in a very narrow sense on those occasions when established groups like the SHRG talked about homophobic reporting and police practices), much less of any non-essentialized notions of homosexuality (or sexuality in general), contributed to the power of a homosexual–heterosexual dualism to shape and control people's lives. It allowed the association of all kinds of evil (criminality, immorality, injustice, etc.) with one side of the divide and the association of virtue (legality, justice, morality) with the other.

CONCLUSION

The moral order which was at stake in the discourses and other practices surrounding Fettesgate and the 'magic circle' (and which emerged from these largely unscathed) was a fundamentally spatialized one. By disciplining sexual practices and relations, not only were traditional sexist and heterosexist constructions of sexualities affirmed, so were the traditional workings and dominance of various other practices and institutions (e.g., land and housing markets, the Scottish legal system). So were dominant constructions of youth and childhood, justice, criminality, gender, and Scottishness.

Gay people on both sides of the 'criminal' divide participated in the reproduction of these social relations, though in different ways. People like the Fettes thief entered into an unholy alliance with homophobic police and members of the tabloid press to exploit and perpetuate certain cultural myths surrounding homosexuality. But they did so, they argue, with the intent of exposing both the hypocrisy of the Scottish legal establishment and other insitutions which oppress gay men – forces which they saw as ultimately much more powerful than the attitudes, perceptions, and opinions of a largely apathetic mass 'public'. They also sought to undermine the position of gay 'mainstreamers', whom they saw as quislings in the struggle against this institutionalized hypocrisy. Other gays, those within SHRG for example, participated in and reproduced dominant constructions of criminality, justice, age (and by extension gender, Scottishness, and the homosexual–heterosexual dualism) as they focused their contributions narrowly on homophobia and heterosexism. In this way it can be argued that the spatialized moral order which was at stake found many aspects of itself reproduced by the determined agency of parties who were at the same time victims of it.

The significance of this contradiction is not limited to the role it played in reproducing a particular spatialized moral order, however. The social and spatial locations of the various actors also shed light on the relationships more generally between identities, political action and moral orders. That's because the locations had everything to do with the particular identity-performances and political actions which were possible for these actors. Those who positioned themselves 'outside' both the law and the insitutionalized gay political movement, for example, were able to challenge dominant constructions of justice, age, markets, etc. in ways that gay 'insiders' (not to mention people located relatively unproblematically within the dominant culture) could never – almost by definition – have done. By performing their identities in ways (and in places) which might be described as stereotypical, they were also able to claim many of these places (e.g. gay-oriented B&Bs, hostels, and even citadels of power such as the judiciary) as spaces of empowerment. They were able by this strategy to selectively appropriate and reinforce cultural myths and stereotypes which, while generally negative, were correctly grounded in an understanding of space as an active and contradictory force in human social and political life, rather than just a container of events. More 'mainstream' gay activists, meanwhile, were in an excellent position to expose those myths and stereotypes as the ideological distortions that they were (and are), but only at the cost of reproducing other institutions of gay oppression (including the stigmatization of gay spaces and places, markets, national economic policies, and the legal system). Thus the staking-out of social and spatial 'turf' by these various actors was both a strategy and a determinant of their particular identity-performances and political actions.

The capacity of some actors (e.g. gay lawyers, judges, journalists, rent-boys, and the Fettes thief himself) to travel *between* locations, affecting different identity-performances and political strategies in different sites, also reveals the tenuousness and instability of any non-contingent interpretation of the relationships between identities, political actions, space and place. While all of these, as well as the moral orders from which they spring and against which they sometimes act, may *seem* fixed, they clearly are not. They are always in the process of *becoming*, since they are always being negotiated and contested by actors who are at once struggling to change the world and to survive in it as it is. The question of power, then, becomes one not primarily of identifying oppressors and oppressed, or even oppressive and liberatory institutions and practices (though it may well involve these). Rather, it becomes one of identifying how highly contingent (and, usually, contradictory) practices, effectuated by particular actors and institutions in particular places at particular times, are, to varying degrees, oppressive and/ or liberatory *in those situations*. Oppressive institutions and practices surely do exist – at all spatial scales – as do the injustices which they perpetrate. But attempts to *fix* these, whether in the form of particular identities, political subjectivities, social movements, or places, ultimately deny the contradictory dimension of power relations. The political consequences of this can be serious indeed.

Finally (and in a related vein), the events of Fettesgate and the 'magic circle' illustrate quite concretely the problematic nature of maintaining a distinction between metaphorical and material space in the realm of social relations. The 'material' geography of gay life on the ground in Edinburgh (and, indeed, elsewhere in Scotland and the world) was at once a reflection of, a response to, and a creator of the imagined (or metaphorical) geography which was so effectively used against gay men in this affair. Put differently, the imagined geography of a gay underworld was at once an ideological reading of, an assault on, and a re-creator of the geography of everyday gay life in Edinburgh. Attempts to disentangle the two miss this crucial point. The material shaped the metaphorical and vice versa, so much so that they are most properly viewed as different faces of a single whole – one space, one geography, not two (or three or four).

ACKNOWLEDGEMENTS

Thanks to Jacquie Burgess, Alan Gilbert, Peter Jackson, Michael Keith, and Steve Pile for their useful comments on an earlier version of this chapter, and to the staff of the Scottish Room at the Edinburgh Public Library for their help locating source materials. A special thanks to Loretta Lees for some late 'emergency' photocopy work, which was above and beyond the call of duty.

NOTES

1 Hutchison, 1992a, 1992b, 1992c, 1992d; Forsyth and Douglas, 1992; *Scotsman*, 1992.
2 Campbell, 1993b: 25; Hutchison, 1993a.
3 Campbell, 1993a; Douglas, 1992a; Douglas and Dinwoodie, 1992; Farquharson, 1992; Forsyth, 1992; *Scotland on Sunday*, 1992.
4 Douglas, 1993; *Edinburgh Evening News*, 1993; Hutchison, 1993b.
5 In the end, this happened anyway (*The Sun*, 1990a; 1990b).
6 See Bell *et al.*, 1994; Keith and Pile, 1993a; Jackson, 1989; Rose, 1993a; Smith, 1993.
7 Campbell, 1993a; Douglas, 1993; Hutchison, 1993b.

8 *Daily Record*, 1990; Douglas, 1992a; Forsyth, 1992; McCartney, 1992; Rimmer, 1992a; *The Sun*, 1990c, 1992.
9 For example, Dalyell demanded an investigation into the British sinking of the Argentine war ship *Belgrano* after the Falklands War.
10 Campbell, 1993a; Hutchison, 1993b.
11 Muir, 1992; *The Sun*, 1992.
12 Burrell and Forsyth, 1991a; Crow and McIlwraith, 1992.
13 *Daily Record*, 1990; *The Sun*, 1992.
14 Burrell and Forsyth, 1991a; 1991b; Douglas, 1992b; *Scotland on Sunday*, 1992.
15 In the Operation Planet case, however, charges involving 18–21 year olds were dropped by the Crown.
16 In a parallel argument, Robert Miles and Anne Dunlop (1987) contend that racism has been represented (wrongly) in the print media as a distinctly English problem, thus setting Scotland apart from and 'above', in a moral sense, its southern neighbor.
17 The *Daily Record* is the Scottish edition of the tabloid known south of the border as the *Daily Mirror*.
18 Muir, 1992; Rimmer, 1992b.

9

PERFORMING INOPERATIVE COMMUNITY

the space and the resistance of some community arts projects

Gillian Rose

FOREWORD

For about a year now, I've been interviewing people involved with community arts projects in Edinburgh, trying to get to grips with how they understand their work as one kind of radical, local, cultural politics. One of the many things that's surprised me about this project is just how much paper it's generated – or, rather, how much writing it's produced. All the interviews have been transcribed. I read lots of books on analysing qualitative data and made copious notes on them. I settled on a method for interpreting the transcripts and typed it up neatly, loading it with footnotes to all those books. I experimented with coding categories, making a list of several possibles before writing one probable – which was later revised and re-written. I coded the transcripts using a computer package and then printed out all the bits from the transcripts that I'd allocated to each category – thousands of quotes, piles of paper. I pored over these, making more notes, and then, on large pieces of paper, I drew boxes, circles and lines around and between what seemed to me to be the structuring themes of the interviews. I started to write papers about these structures, planning, drafting, redrafting. All the time too I've been jotting down ideas in a ring binder. And I recorded all of this in a research diary.

But more recently I've been wondering (or maybe I just got tired of all the writing) – what if the most challenging, the most subversive aspect of the politics articulated in these interviews isn't the words at all? What if it wasn't the meaning I thought the interviewees were making that was important, but the silences in a narrative or an explanation, or the eruption of a brute facticity into discourse – something in their voices beyond signification? Some things beyond discourse, representation, interpretation, translation? And if this was possibly the case, what were the implications for my interpretive project, with its transcripts, codes, categories, records and papers (Rose, 1996, 1997)?

THE POLITICS OF COMMUNITY: IDENTITY AND EXCESS

'Community' has long been one of the most powerful terms through which collective identity can be named and collective action legitimated (Williams, 1976: 66). Perhaps, however, now, the hegemony of community is nearing its end. Many critics – Stuart Hall (1995b), David Harvey (1989), Doreen Massey (1994b) and Iris Marion Young (1990) among them – are suggesting that it may no longer be possible to use the term in a radical political project. While the specifics of their arguments vary, all these writers are agreed that it is the ways in which 'community' constructs the distinction between its members and non-members that are intolerable for a radical politics. These processes of inclusion and exclusion are based on the assumption that members know themselves and each other absolutely (Young, 1990), and this depends on a vision of the subject as 'a rational transparent entity which could convey a homogeneous meaning on the total field of her conduct by being the source of her actions' (Mouffe, 1995: 260). The subject of community thinks they know their self through the same transparency with which they imagine they see their others. Such subjects constitute a 'community' which considers itself as made uniform within by sharing something all members can recognize: an identity. Difference from this self-same disturbs and threatens its transparency, and produces both a denial of difference and a desire for it. The different other, placed beyond the bounds of community, becomes a source of both fear and fascination: condemned and idealized, needed and negated, always exoticized, it is only ever represented through the fantasies of those inside the borders of the same. The marginalized other is denied its own difference by this construction of community.

Much of the recent critical discussion has focused on the oppressive uniformity of 'community' membership. Young, for example, argues that:

> Community represents an ideal of shared public life, of mutual recognition and identification [...] the ideal of community also suppresses difference among subjects and groups. The impulse to community often coincides with a desire to preserve identity and in practice excludes others who threaten that identity.
>
> (Young, 1990: 12)

For Young, as for many other writers, it is crucial to move away from this purity of communal identity, with its territorialized and terrorizing boundaries between same and other. And, again like many other writers, Young chooses to do this by insisting on the extraordinary diversity of social identities and of their overdetermined, contingent intermediations. This produces a spatiality of identity which, as Massey (1995) and Natter (1995) among others have argued, is not a territory but a multi-dimensional matrix of mobile, fusing axes of identity within which individuals are complexly, contingently, multiply and contradictorily positioned.

This reworking of 'community' is not without its difficulties, however, and Young's discussion of social difference in terms of a 'group' rather than a 'community' exemplifies one of them:

> A social group is a collective of persons differentiated from at least one other group by cultural forms, practices, or way of life. Members of a group have a specific affinity with one another because of their similar experience or way of life, which prompts them to associate with one another more than with those not identified with the group, or in a different way. Groups are an expression of social relations; a group exists only in relation to at least one other group. Group identification arises, that is, in the encounter and interaction between social collectivities that experience some differences in their way of life and forms of association, even if they also regard themselves as belonging to the same society.
>
> (Young, 1990: 43)

In this passage, Young twice suggests that differences between groups originate from experience. There is no suggestion that these experiences are interpreted or their significance contested. Rather, these references to experience imply an unmediated connection between an event or process and a shared, group identity which springs from it. This suggests that all group members will have the same reaction to the same experience; it does not imply variation within a group. Young's account thus to some extent echoes the purity of insider identity assumed by 'community' itself.

Another difficulty in Young's discussion is her consideration of who defines a group/community. Like many writers on community, her focus is on the definition of the group and its members (Young 1990: 43–8). She argues that groups define themselves, that groups are relational and therefore also define each other, and that individuals are in part also constituted through group membership. But this begs the question of the power relations which structure definitional practices. What if a group, a marginalized group, struggles to rework its designation as other into a different identity for itself: how is it legible to the powerful? Lauren Berlant (1994) worries, for example, that the legibility of a marginalized group can lead all too easily to its assimilation into the dominant culture (see also Hoffie, 1991). What Berlant (1994: 155) calls 'the taxonomizing chaos generated by identity politics' may parallel the professionalizing discourse of the US state apparatus so that there is a need to consider 'the problems of translation that arise in the movements of expertise and self-representation between sub- and dominant cultures' (p. 127). In the context of cultural managerialism and fierce competition for funding resources, taxonomies of difference may work only to stabilize identities once again. The result would be once more the fantasy of pure identity which the move away from community was meant to avoid. As Phelan (1993) suggests, perhaps it is not the case that to become visible and nameable, to express an identity, is a necessarily radical strategy. To be named, to be discursively recognizable, may itself be a tactic which always already concedes too much. To be named is to make sense, to be made sense of; it is to be positioned in the realm of the legible, the knowable, the translatable. It is to be made vulnerable to knowledge; to be produced through discourse; to be produced.

For Berlant (1994), the subversive move is therefore to refuse to define. For her, a truly radical politics must entail a 'something', a political project 'riskily underdefined', 'a suspension of the will to knowledge' (Berlant, 1994: 145, 152). She is certainly not alone in advocating a politics of which refuses to make everything accessible to discourse. Certain critical theorists of both postcoloniality and sexuality are making similar arguments. Homi Bhabha (1994), for example, suggests that the disruptiveness of postcolonial literature lies in its subversion of the rules of discursive recognition (and, like Berlant, he reads Toni Morrison as exemplary here); and Joan Copjec (1994) has insisted that the sexual should be theorized as disrupting the discourses of power/knowledge rather than as being constituted through them. These are theorists advocating a politics of surplus, excess and lack as most likely to produce a radical politics by destabilizing power/knowledge, and they do so by insisting that there is something excessive to the defining rules of discourse.

The notion of community has also been subject to this critical desire for discursive excess, most persistently by Jean-Luc Nancy. Drawing on various post-structuralisms in order to problematize the notions of subjectivity which undergird hegemonic visions of 'community', he argues that it is crucial to distinguish between two kinds of community. The first of these he terms the myth of community, which, he argues, is produced by 'the will to realize an essence' (Nancy, 1991a: xi). The essence which produces the myth of community is the essence of the human subject as 'the absolutely detached for-itself, taken as origin and certainty' (Nancy, 1991a: 3): a subject who speaks 'full, original speech, at times revealing, at times founding, the intimate being of community' (Nancy, 1991a: 48). This argument parallels those of other critics of community. Like them, Nancy makes clear the political failure of this myth of community, whether deployed by Nazism to authenticate both Volk and Reich or by Soviet Communism to legitimate labour as the essence of the human. Myth, for Nancy, is nothing less than the will to power. The myth of the community is the myth of rational, transparent, transcendent subjects denying difference from themselves.

Nancy argues that challenging that will to power requires a different understanding of community. He calls this other community 'inoperative community'. The central distinction between the myth of community and inoperative community is that the latter is understood in the context of a surplus to discourse. Nancy (1990a) argues that while the calcifying definitions which are the productivity of discourse cannot be abandoned, nor should they be taken as the totality of communication. Discourse has an other, a supplement, which subverts its definitional power. In trying to think the inoperative community, a thinking necessarily at the limits of thought, he comments:

> Perhaps, in truth, there is nothing *to say*. Perhaps we should not seek a word or a concept for it, but rather recognize in the thought of

> community a theoretical excess (or more precisely, an excess in relation to the theoretical) that would oblige us to adopt another *praxis* of discourse and community.
>
> (Nancy, 1991a: 25–6)

Inoperative community, he argues, cannot be thought of in terms such as signification, object, representation, production, commentary, explication, interpretation. Rather, inoperative community is communication *per se*: it is performance, 'dialogicity itself' (Nancy, 1990a: 227), something which occurs only as it is said or done. In his argument, communication does not occur between pre-existing subjects. Instead, communication is both the origin of human beings and our limit: each becomes only in communication with an other. In the moment of communication, therefore, everything that is transcendent disappears: 'infinitely announced, the other puts an end, unceasingly, to the identification and to the assumption of the absolute, perfect understanding' (Nancy, 1990a: 246). Thus communication is both discursive and extra-discursive, for the moment of communication always contains a move beyond discourse and the positions it invites us all to occupy. According to Nancy, it allows us to 'escape the relationships of society ("mother" and "son", "author" and "reader", "public figure" and "private figure", "producer" and "consumer"), but [we] are in community, and are unworked' (Nancy, 1991a: 41). Nancy describes these unworked positions beyond discourse as 'singular'. They are in excess to the discourses of the subject, and also to the myth of community. Instead, they perform inoperative community. Inoperative community constitutes a resistance 'to all the forms and all the violences of subjectivity' (Nancy, 1991a: 35), as well as to all the forms and violences of 'community'. For Nancy, then, a radical understanding of community must be an inoperative one.

Nancy also hints at the spatiality through which inoperative community is constituted. Inoperative community is not a territory but an areality, he writes. It is the space between singular beings in a moment of enunciation, for communication 'consists in the appearance of the *between* as such' (Nancy, 1991a: 29).

> There where there was nothing (and not even a 'there' – as in the 'there is no there there' of Gertrude Stein), something, some *one* comes ('one' because it 'comes', not because of its substantial unity: the she, he, or it that comes can be one and unique in its coming but multiple and repeated in itself). Presence *takes place*, that is to say it *comes into* presence. It is that which comes indefinitely to itself, never stops coming, arriving: the 'subject' that is never the subject of itself.
>
> (Nancy, 1991b: 7)

This is a space not only multiple, composite, heterogeneous, indeterminate and plural, as some geographers have imagined the spatiality of intersecting communities or groups of identity. It is a space the dimensions of which cannot completely be described, defined, discoursed. This other space constantly unworks the certainties of representable spaces and the certainties of identity given form by them. Inoperative community, then, in its doubleness of discourse and surplus, articulates an inoperative space at once representable and obscure.

The next two sections interpret the interviews I undertook with workers in community arts projects in the context of Nancy's arguments about both resistance and space. I hope to demonstrate how, in this particular context, discourses of community – indeed, discourse itself – are both deployed and unworked. I want to suggest that an inoperative community shadows these workers' understandings of both community identity and of the work they do with these 'communities', and that this structures the imagined geographies through which they map their work in the city. It is also articulated through the language they (do not) use to speak about their work, and this chapter will end by briefly considering some of the methodological implications of the non/discourse of these workers.

INOPERATIVE COMMUNITY I: LANGUAGE, VOICE AND TRANSLATION

The founding premiss of community arts projects is that access to the arts is not equally distributed across society, and that this is not only a consequence of the uneven spatial distribution of arts facilities, of the cost of tickets, and of building design, among other things; it is also caused by a number of powerful assumptions about who can really appreciate and practise the arts. As one worker in a project for people infected or affected by HIV/AIDS in Edinburgh put it:

> y'know, all that thing about people thinking that y'know – they have a certain place in the society and that goes y'know, that, that, their relationship with the arts and culture in that society, they don't think it's for them.

Instead, community arts workers 'recognize the right for anyone to be involved in a cultural activity', as a worker at a project for people with disabilities said. This right can be exercised in a number of different ways. The projects with which this chapter is concerned, however, all aim to facilitate the arts skills of people excluded from the arts so that they can produce their own works, often collectively. As a tape-slide worker said of her early involvement in community arts in Craigmillar in Edinburgh, 'there was this really strong sense that like releasing the creative power in everybody was like a really important thing to do'. It was important because, for people marginalized not only from the arts but from society as a whole, arts work can 'encourage people themselves to take control and to do stuff and to develop skills', in the words of a Muirhouse worker. This process is described (and problematized, as this section will suggest) as one of 'empowerment'.

Nearly all of the community arts workers I spoke with are employed in projects funded by the Urban Programme, administered by the Scottish

Office. This programme provides funding for projects operating in what the Scottish Office defines as areas of multiple social deprivation, and which produce cost-effective and tangible results in their area. These arts workers are therefore working in localities defined as deprived, and these are mostly interwar and postwar council housing estates on the outskirts of Edinburgh: Craigmillar, Muirhouse, Pilton and Wester Hailes. In Edinburgh, the dominant culture's accounts of these areas in Edinburgh oscillate between pathologizing them as criminal places of drug abuse, violence and AIDS, and sanctifying them as places where, despite all the odds, the human spirit of creativity can yet flourish. Thus while newspaper reports vilify them, the Scottish Arts Council praises their thriving arts practices and 'refute[s] the idea that an individual's ability or need to produce creative work are limited by where he or she lives' (Scottish Arts Council, 1995: 2). Both these accounts construct the margins of Edinburgh as the other to its centre, its difference either erased entirely, since the same creative humanity is seen as existing there exactly as elsewhere in the city, or rendered as the absolute opposite of all that civilized Edinburgh holds dear. Dominant discourses of the city map the city's edges ambivalently, but never allow the possibility that there may be the location of any kind of alterity. Edinburgh is constructed through the myth of community, its powerful centre taken as origin and certainty and its margins merely their reflection.

The community arts workers I spoke with were clear that power in Edinburgh works in part through the ability to name, to define and to describe with authority. Several were very critical of the press, for example, especially those in Craigmillar, Muirhouse and Wester Hailes, for the way it gave these places 'a bad name'. Others however commented more generally on the oppressive consequences of attempting to define particular groups of people, either by place or by identity. The worker with people with disabilities remarked on the problems associated with defining a territorial 'community' on the grounds of its 'deprivation':

> I think there's huge problems in terms of defining geographical communities, because it's, it's, it's, in a way it contributes to the image of that particular area. So if you say well we reserve our resources for the people of Craigmillar, and you build this in Craigmillar, all your way of doing is to participate in – the fact that people from Craigmillar live in Craigmillar, they should stay in Craigmillar and they shouldn't come from anywhere else.

A parallel argument was made by the HIV/AIDS project worker in relation to definitions of identity. She recalled what she thought was a particularly important debate at a community arts conference she had attended:

> A bit like y'know people were worried that artists were coming in and almost using communities y'know – using the problems that they had, focusing on the problem, and that, y'know that was something that we did discuss a wee bit, like do you go into communities where people are

> defined by their problem, y'know, cos you work on an issue that is mental health or HIV and AIDS, and what's that all about.

Community arts workers, then, are suspicious of the consequences of hegemonic definitions. For them, those who can define are those who are in power, and definition then becomes part of the way power itself works. Trying to challenge these definitions is understood as difficult, moreover, precisely because it entails the problem of translation mentioned by Berlant (1994). At a meeting between people involved in community arts projects and officials from city and regional councils held in November 1995, for example, community arts workers and participants insisted that two languages were being spoken and only one being heard: those of the officials. They demanded instead 'translators' as intermediaries so that both sides could hear each other.

There is no way simply to avoid the definitions of discourse (Fuss, 1989). This is not least in this context because those who are seen as dominating the discursive arena are also those who control the funds which allow community arts workers to resource their projects. Applications for funding must be submitted in a language familiar to, and acceptable to, those in power, to have any hope of succeeding; it must be translated appropriately. For some workers, this meant paying attention to current policy enthusiasms: what one arts worker described as the 'flavour of the month'. Other workers aimed to appeal to those in power by demonstrating their success in their terms. The Muirhouse worker commented on his attempts to secure future funding for the Muirhouse arts centre:

> We're selling it on the – the way you have to sell things these days is the numbers, numbers on seats, the percentages and all that stuff.

For all community arts workers, this situation demands that they are fluent in the left-liberal discourses of community development and empowerment. They refer to 'that, buzzword empowerment, y'know', and deploy 'application-speak': 'I suppose, to use application-speak, it's like sort of to portray Muirhouse as a kinda active community.'

But their fluency is not uncritical. 'Buzzword', 'application-speak', 'jargon', 'spiel', 'catchphrases': all these terms were used to distance the speaker from the language they were using. Sometimes this distancing went further. A Wester Hailes worker commented:

> the idea of community development, where you encourage folk, the trendy word's empowerment y'know, through, through their participation in the decision making process, blah blah blah, they become empowered as individuals.

Blah blah blah. The predictability and certainty of this description of the process of empowerment are undermined here, by a worker all too aware of the difficulties and threats to that process. Most often, 'application-speak' is criticized as bearing no relation to the practice of community arts workers.

The HIV/AIDS project worker separated 'terminology' from 'practice' when she commented that 'I don't have a y'know, have a huge hang up about, I think the work, the type of work you're doing, y'know, your practice is important rather than, rather than, and of course terminology is important', for example. This distinction between a project's work and the language which describes it is also apparent in some of the workers' manipulation of the latter. A Pilton arts worker said about a trip to Spain her project got funding for:

> a, a, well we called it a cultural exchange, it is a cultural exchange [. . .] but it was y'know a cultural visit to Barcelona to go and do some research on mosaic, and in a way it was bit of a, kind of a bit of a junket that one, but it was very good.

As the worker with people with disabilities remarked, 'we can all play with words'. The need to use the correct words to get funding is acknowledged, then, but so too is a distinction between this language and the practices of community arts.

The consequence of this distinction is that application-speak is both used and refused, as in this comment by a worker at a community print resource in Pilton:

> I mean I see the resource area as a kinda – I hate these sort of bud, buzz words, but like enabling people, using equipment, it's about demystifying the technicalities of trying, basically [??] photography or screen printing or desk top publishing.

The language of funding bodies is used for its radical possibility – empowerment, after all, is a worthy goal, and so is enabling, and demystifying – and some terminology is needed to speak at all. But the vocabulary of that language is also qualified, parodied, critiqued and refused, because the powerful are using it to non-radical ends. The discourse of community arts workers then is marked by a kind of doubledness. The words are there but the meaning is elsewhere. What Berlant (1994) calls the problem of translation between sub- and dominant cultures is being negotiated here, not by translators this time, but by the deployment of a discourse which has a surplus threaded through it. The words of the powerful, in all their calcifying definitional predictability, are paralleled by a dissenting usage which allows for something different besides the dominant meaning.

If these workers critique the language of the powerful, however, they also recognize that their own language is often different from that of the people they work with. The places they work belong not to them but to their residents: 'it's their life, it's their community, it's their culture'. Their preference therefore seems to be to explore their projects' practices, and to withhold describing those practices as representations of local identities. As well as the surplus in their language then, there are also silences articulated by a certain refusal to translate. Two silences at least seem to hover around the sound of their words.

The first of these surrounds the making of things. A product of a particular project is presented in and of itself as an achievement, regardless of what it 'means'. Its importance is not in its meaning but in its facticity; enormous importance is given to the sheer existence of the object itself. Community arts workers repeatedly stress the process of making an object as in and of itself a radical moment of empowerment and of community. Another video access worker said:

> You can get a big kick out of just watching a film that you've made. You can get such a kick out of it, because you've made something. There is a sense of achievement and that's probably another reason why people do it.

The Pilton arts worker also said that one of the most important consequences of participating in an arts project was the fact that participants got 'to create something that they're proud of'. The workers celebrate this making: the Pilton video worker described one project as:

> a success in a lot of ways in terms of y'know the fact that we actually managed to produce a feature-length drama. Er, and that, for example the *Scottish Screen Digest* has just been published and it's got y'know 'Feature Film Production in Scotland since 1990', and there it is, 'The Priest and the Pirate' y'know and a little star saying 'video feature'. So it's, y'know in a sense it's quite nice that it's up there with y'know the the sort of mainstream films.

In none of these discussions did community arts workers ever attempt to interpret the object produced. Their meanings were never decoded for me. The products of their projects remained veiled in this way, even when I was asking about specific performances, videos, photography projects or tape-slide shows. The workers refused to explain the significance of these products for me in any terms other than the processes of their making, or their impact on audiences. The products were never represented, never described as artifacts awaiting interpretation for community arts workers. Instead, they were always placed in the context of performances, both in their making and in their audiencing. These products were understood as moments of 'communication', not as representation, and could not therefore be described in the context of an interview.

A similar silence somewhat paradoxically haunts one of the most common images community arts workers use to describe the aims of community arts: enabling project participants to 'find a voice'. This phrase was used by almost every arts worker I spoke with. A project tries 'to get folk a place to have their voices', 'giving people who don't have a voice or, a voice', 'to develop their skills, their confidence, their, their voice'. This voice speaks both as an individual and a community, connected. Another video access worker said:

> it then gives people a voice in order to express themselves, to say things

about what they want to say about their community, which I think's really important.

It is a voice which performs and in that moment of communication it may make 'kind of powerful statements that will kind of speak direct to people'. Yet community arts workers very rarely comment on what that voice might say. An arts worker in an environmental project in Craigmillar, for example, was typical when he said while discussing one project that community arts could make statements, but then proceeded to discuss something else:

> [the project] seems to fit in a lot of good things. Community involvement, community arts is a viable thing to do in terms of confidence and all the social aspects, but it can also make statements, it definitely can ... erm ... I'm not going to take this as a flagship thing but a small success in a way was when we were at Niddrie Marischal.

It seems that the critical point of this voice is speech itself, not its content. The Wester Hailes arts worker explained:

> So whereas y'know, a positive image for some folk might mean everything's rosy, for us it means exhibiting the talents, exhibiting a critical view of the world that folk can develop, and it's showing that the folk have a voice in a way, that's another than the, y'know, kinda tabloid coverage which tends to happen.

Again, the workers here are referring to an aspect of local folks' participation without describing it. For the Wester Hailes worker as for Nancy (1991a: 73), 'it is not a matter of a message: neither a book nor a piece of music nor a people is, as such, the vehicle or mediator of a message. The function of the message [...] does not take place in [inoperative] community'. Instead, participants' voices are evoked as a mode of communication which resists the power-ridden process of description, including those of community arts workers.

In this section, I have argued that community arts workers have a keen sense of the reproduction of power through language. Currently, both people and places are seen as defined and confined by powerful institutions, including the mass media and funding bodies. The critical tactics adopted by these projects in the light of this analysis can be understood in terms of the politics of translation. The existence of other voices is acknowledged and an equitable process of translation demanded. The language of the powerful is deployed in order to win resources to facilitate the development of these other voices, but in a voice which carries an excess to that language. And these other voices are never given content by the workers, never translated, just as their own dissent produces not a new vocabulary but a disavowal of the old one. Instead, against the grain of the definitional terms all discourses must employ, community arts workers rub uninterpreted objects and unrepresented voices. Following Nancy, and in contrast to the arguments of Kelly (1984), this definitional uncertainty may be one of community arts' most subversive tactics.

INOPERATIVE COMMUNITY II: PROCESS, PERFORMANCE AND PLACE

The previous section suggests that the tactics of community arts workers could be understood as an effort to render community inoperative: to striate the dominant culture's discursive myths of community with excessive objects, surpluses to meaning, contentless voices. These tactics are a refusal of the discursive legibility of the areas, practices, products and participants in which these arts projects are located. The transparent space of the myth of community is thus fractured and shadowed, its discursive terrain disrupted. This section considers in more detail the spatiality of this resistance.

Community arts workers understand power in terms of language – who can speak, whose language effects action – and their tactics of resistance are therefore in part, as we have seen, directed at the politics of voice and translation. Their tactics are those of excess and surplus. But their tactics are also about a politics of praxis beyond discourse and its myth of community. It seems to me that what offers some kind of guarantee to community arts workers that there will always be an excess to their work is their articulation of the praxis of their projects. The practices of being involved in community arts projects are presented in such a way that they always imply the possibility of escaping the discursively constituted relationships of society, just as Nancy suggests inoperative community does. 'Process' is a key term in these workers' understanding of what their work involves. 'Process' refers to how participants learn skills and create art when they become involved in a project. The tape-slide worker, who very often uses 'feeling' as a benchmark for deciding whether is something is right or not, said her project was developing 'processes, working with people, to kind of, to build up people's confidence and to kind of feel, let them feel like they were kind of generating kind of images for themselves'. The worker with people with disabilities argued that process was at its heart: 'the main focus is on the common process of being involved'. Different projects prefer different practices of process. But what many shared was a sense in which the point of process is to produce more process, more participation. The tape-slide worker said:

> If you get them to a point [. . .] where they go 'oh so that's what it was all about, oh, right, now we are beginning to understand', then you're doing quite well [laughing]. 'Oh yeah, well so maybe we could do another one.' And then it's just really a start.

Participation entails more of itself. It regenerates itself, and this is its purpose. The process of participation is therefore never quite complete. It is a performance constantly reconstituting itself.

The notion of process as something constantly practised and constantly shifting structures community arts workers' understanding of all aspects of their work. It constitutes their sense of the people involved in community arts projects, for example. For some arts workers in Edinburgh, the process of participation is a process of expressing a pre-existing and knowledgeable self. Here, for example, is a worker in an arts project in Pilton, recounting what she says to her project's participants when she takes them to the National Gallery of Modern Art in the city centre:

> y'know your opinion of something is just as valuable as somebody else's, whether they know a lot more about it than you do or not, because the art's on the wall in the gallery, and you are the person looking at it, and your response to it is just as important as anyone else's, and I see that as a key to getting people – having a bit more sense of self worth, and y'know really start listening to their, their own minds, instead of sort of being all confused, so y'know that's important.

However, many of the workers I spoke with elaborated at length how difficult it could be to get people listening to themselves, and they argued that this difficulty was itself the most profoundly oppressive aspect of the marginalization of communities. The HIV/AIDS project worker paraphrased what she thought many people from marginalized 'communities' thought about themselves:

> 'Why should anyone want to hear about what I've got to say', or even, not even in a wider sense, but y'know, it's a lot about people not really valuing their experience, their life experience, and not thinking they've got anything important to say.

Given this analysis of the absence of self-worth, community arts workers consistently stressed that their projects were hoping to produce self-confidence. An arts worker in Craigmillar argued that 'through art, through getting a chance to do it, people's confidence can increase, and self-esteem can increase obviously', and a video access worker commented that 'it's part of our philosophy of promoting videos as a means of y'know like raising self-esteem and confidence in the community'. Participation in the processes of a community arts project is thus dynamic because the processes are not always seen as enabling the reflection of experience or the expression of an anterior identity: they can also be understood as developing a self. Participants are changed by their involvement in community arts. The Wester Hailes worker talked about people 'who are with us every single day of the year, and moving on, in terms of their own development'. Participants do not remain as they were, and the process of developing self-confidence is thus also a process of developing a different self. One video worker based in Craigmillar, whose interest in video was started through his participation in a community arts video project, described the effects of his participation thus:

> It's the self-esteem, y'know you suddenly realise you're more important than you thought. It takes a while to realise it, it takes, it's not just something that comes out the blue. It's like you build a house, you're building a house for yourself, you're building yourself up, and you know you've done it step by step, and you know you've achieved

> something and you know you've went along this road, building and building and building, and you know all the work you've done, and you know all the skills you've developed and it's really, y'know it's a real thing to give you strength.

His self here is imagined as strengthening and changing through his participation in a community video project: his participation made him how he now is.

The same Craigmillar video worker argued too that the sense of self of his audiences could also be fortified by watching community videos: 'it's a sort of nourishment, it's a food then, they're happy to have it and it keeps them going y'know'. In watching a performance, moreover, the self also changes because its relationship to others alters. The Muirhouse worker described the annual community pantomime his project facilitates:

> It was a real, it was really good. I just about died doing it. [??]. It was worth it. It really was quite special. At the end of it like everybody up on their feet and leaping about, it was just, really getting off on it like, really working and that got people involved in other things and from seeing that [??]. It was really good, y'know. It picks people up, a sort of snowball effect.

Such a performance is understood as good because it energizes and connects all its participants: local performers, arts workers and audience. Indeed, this sense of connection between audience and performance defined community arts itself for one community drama practitioner:

> I mean you hope that at the end of the day what your [??] has got is entertainment value and it's got artistic quality, but I think that the relationship between the audience and or what ever it is, whether it is a book or what have you, is different in quality, so that the audience in some way has to become part of that process, which either is shared by the way that you, y'know present what you're doing, the actual content, that it tells its own story, that you see how this has come to be or what have you – or some sense in which the audience has local ties, or y'know what have you. There is a kinda, there is a definite sort of relationship there.

The worker here leaves the nature of that 'definite sort of relationship' undefined; another moment at which the process of community arts is situated beyond the legibility of discourse. But here it is also possible to suggest that what this excessive shadow on the legible spatiality of the myth of community is: it is a spatiality of action and performance, the constituent elements of which shift even as they also emerge through the process of being performed.

It is from this kind of performed connection that 'community' begins to emerge for community arts workers. Community is represented by them precisely as a number of connections. These can be connections forged by the sociability of community arts, something all the workers were anxious not to underestimate. A video access worker said:

> some people, a lot of people they weren't even interested in video, a lot of them were just interested in talking to somebody.

Connections can also be institutional. The Craigmillar arts centre worker, for example, suggested it was the network of community organizations which made Craigmillar a community:

> I see it [Craigmillar] as a fairly, fairly well defined, I see it is as a fairly well defined community [...] I mean the Festival Society has just gonna, gone through a sort of strategic planning exercise, and they were, their like kinda by-word, or whatever the phrase is, is 'representing Craigmillar', and I think that's, I think we do, and I think most people see us as that.

Other workers suggested that any kind of connection produced a community. A youth video worker commented:

> I think any any mechanism that brings people together, y'know even if it's a local café or – I mean, er, or a community arts project or mothers and toddler groups, or anything that brings people together and erm helps to develop relationships is gonna be a great thing. So – um, I don't, I don't – I mean that's that's the only – I mean if you do, if you do bring people together and break down barriers that exist between them then you can think of events or activities that achieve some kind of local prominence, y'know like galas and these things are really important cos they pull people together.

This pulling together is understood as a process which, like the participation of individuals, gains its own momentum. A video worker in Pilton described one of his projects:

> Now the video was a means of identifying what the problems were, but it was also a means of creating solidarity amongst the group, so, so it had like two, at least two different effects, and the group grew bigger and bigger by involving people in the actual video, so that you had a production which is gonna say certain things about er what the group's needs are but it's also drawing people in in terms of participation. Y'know, so like people want to belong, people want to get their point of view across in the video and then they've got a, like once they become part of that production, then they've got a much greater stake in the campaign and in arguing for what what those people need.

And, again like participating individuals, the dynamism of connection changes the form of what is connected. A Wester Hailes arts worker commented:

> we, we've got a carnival band that has become one of these established groups that we see the community development being encouraged, the

> formation of an activity, you encourage the participants to control of the destiny of that activity and it becomes another permanent group in the community, another resource if you like.

Community then is about connection, but its connections are understood as in process. Connection itself changes what there is to be connected.

The processes of community arts practices are understood as changing their participants, then; they grow and develop and connect. They shift and change and interrelate in a continual flux. Community arts workers understand performances as making links between its producers and its audience. All participate (hence many community arts workers make no distinction between the producers and the consumers of community art [Raven, 1989]). A performance produces a relation, and, as Nancy suggests, this relation is understood as communicating subject positions not produced by discourse. The tape-slide worker, for example, spoke of the impact of a production which her project had facilitated in terms of how it had reached beyond the usual divisions in Craigmillar to speak both to local residents and to the local redevelopment partnership:

> and even, like the Craigmillar show, erm, we screened it to business community people, who're like part of Friends of Craigmillar, and again it just had a really powerful impact on them. And it's, it's partly the fact that, that kinda local people have put it together, and, but it's just so direct, it's just, it just so directly um speaks that you cannae argue with it, y'know?.

This performativity dissolves the binary division between the powerful and the powerless which the discourse of community arts also deploys, to suggest that 'outsiders' may indeed be able to hear what community 'insiders' have to say. As a video worker commented, 'in the end um the most important thing is for people to see what other people are doing and for other people to be able to learn from other people's experiences'. Everyone becomes an other. These performances step outside the positions structured by discourse, to enable the demands made by the marginalized to pass beyond the territories to which they have been consigned. They produce what Nancy describes as 'singular beings':

> singular beings [...] are what they are to the extent that they are articulated upon one another, to the extent that they are spread out and shared along lines of force, of cleavage, of twisting, of chance, whose network makes up their being-in-common. This condition means, moreover, that these singular beings are ends for one another [...] the whole community is not an organic whole.
>
> (Nancy, 1991a: 75–6)

This community of performed connection, then, may be one articulation of the inoperative community theorized by Nancy. Its spatiality is not the 'there' of territorializing, mythical community, but is instead a coming together of excesses to discourse.

Further, this spatiality of performance is also a spatiality which has a different relationship to alterity than does the spatiality of the myth of community. Nancy argues that the latter erases all radical difference from itself in its assumption of absolute, perfect understanding; while inoperative community constructs difference differently, in its unworking of social identities. The myth of community articulates the Western desire to transcend its own specificity, and Nancy argues that this desire for transcendence produces the disavowal of difference on which the myth of community is founded. In contrast, Nancy argues that it is necessary to recognize difference, which involves, fundamentally, recognizing the absolute difference of death itself; drawing on Bataille, he argues that 'death irremediably exceeds the resources of a metaphysics of the subject' (Nancy, 1991a: 14). Inoperative community thus acknowledges death and difference: 'community is revealed in the death of others; hence it is always revealed to others. Community is what takes place always through others and for others' (Nancy, 1991a: 15). And indeed, several of the arts workers I spoke with mentioned death – or madness, 'this woman came and said, "if I don't write, I'm going to go mad, please help me"' – as a reason for the participation in community arts projects. Participation was a means of warding off the threat of complete disappearance, but disappearance then marked the process of participation. This was most explicit in the case of the arts worker with the HIV/AIDS project. One of the participants in a tape-slide project became ill during its development:

> it was quite a difficult period for the group, but it, the work seemed to help that kinda process of him dying. It helped him and it helped the group. Here you are, he left this piece of work, which has been seen by a lot of people, a lot of his family have seen it, and – it's really, it's a significant piece of work.

Other projects also suggested that participation entailed a recognition of some kind of death. The arts centre worker at Craigmillar described the strengthening process in these terms:

> you can see people, who maybe, have been at quite a low ebb, but then – y'know gaining strength, gain strength and, and just strength to be able to live.

And the Craigmillar video worker also remarked:

> a lot of people would be in a lot worse position if it wasn't for community activity, but then community activity comes out o' people's – desire for survival, I mean we group together to survive, and they know from experience the best way of doing things is to work together to survive, so it's just a manif, a manifestation of survival y'know in the best possible way.

This fear of death, or the fear of some other kind of erasure, fuels the desire for participation and for communication. 'A[n inoperative] community is the presentation to its members of their mortal truth' (Nancy, 1991a: 15). Dissolution threatens, but also necessitates, communication.

The spatiality of inoperative community then is a spatiality of both absence – mortality – and presence – performance. It is a spatiality whose complex flux is both constitutive and dissolutive, a paradoxical space in which selves are performed and do not exist. It is a spatiality strung out of links between relational terms whose performance always reconstitutes the performers and whose practice is always marked by its erasure. This is a spatiality which haunts the legibility of discursive space; an inoperative spatiality not entirely describable.

A FINAL FEW WORDS

In this chapter I have argued that the critical politics of community arts projects in Edinburgh may not conform to the model of a marginalized, radicalized and unified 'community' articulating an oppositional worldview in resistance to the powerful. Their critique is more complex, and this is because they connect material oppression to cultural marginalization, and demand that both must be ended. They work towards this end by placing as much emphasis on voice and performance as on what they call 'issue-based' work. The performances they facilitate often do address issues of material deprivation; but in the conversation of these arts workers there is also always a space striating these performances for something beyond the discourses of both the powerful and the community development radicals, including themselves. These products and processes are understood to be as much about performance as substance; as much about communication as discourse; as much about absence as presence. The performances of product and process may, they argue, produce an excess beyond the hegemonic definitions of community and identity. They critique such definitions even as they must use them, and they also refuse the power of discourse when they never describe the products and processes of the projects: in their interview performances with me they did not represent or interpret those performances but were silent around them.

If their tactics rework what might be meant by a politics of resistance, they also restructure the space in which such a politics takes place. This space is not the territorialized uniformity of 'community'; but nor is it entirely the multidimensional matrix of identity politics. Both these geometries remain representable and therefore too open to assimilation by powerful definitions, although both may strategically be deployed to win funding (Rose, 1997). Instead, arts workers' emphasis on performance evokes a space of contingent connections beyond the constraining positions of discourse. It is a mobile network of groups and individuals performing themselves and changing shape as they do so. This dynamic works at many 'levels', a phrase repeated by worker after worker: 'although it seems quite straightforward in itself, there's all these different kinda layers'. The Muirhouse worker described the effects of participation on the participants thus:

> it will work for them on a number of levels. Whether that is about, y'know, encouraging people to go on to do further training, or give them the confidence to go and seek like employment, all sounds very, very vague, but I mean in real terms I mean I think it does work on those levels, like and it does, people get involved socially and actively within something like that and find themselves building up their skills, building up their confidence and that, that translates to other kinda avenues.

Thus the form of community is imagined through a fluid and multi-dimensional space, but one which remains 'vague', radically undermined by uncertainty. That uncertainty is a symptom of the death that inoperative community acknowledges, and which also structures this space. The threat of erasure of self and of community is articulated as death or madness but also, for several arts workers, as an absence of space: a 'vacuum', with and as 'nothing'. This absence, this lack, striates the performance of inoperative community, renders the space of inoperative community partially indescribable for these workers. What words, after all, can represent 'nothing'? What space? Only silence and emptiness – there is no there there.

This non/space of a politics of non/representation has certain methodological implications. It suggests that a radical politics is not always articulated through words, but that silences and absences also constitute critique. It suggests that coding and categorizing should not, but also cannot, be the only methods for invoking this politics. It suggests that the assumption of knowable identity which underlies calls for methodological self-reflection may not be an adequate strategy, on its own, for refusing the authority of the author. But it also suggests that, in the moments of communication during an interview, there are possibilities for escaping the oppositional roles of 'interviewer/interviewee' or 'academic/activist'. There is a possibility of communicating something else beyond the positions power consigns us to. If, as Nancy (1990a: 160) says, 'I am only "I" if I can say "we"', then those positions never quite contain us. We need others, and the practices of community arts workers in Edinburgh suggest that communication with others we understand as different from us can be performed differently from the terrible exclusions of the myth of community.

ACKNOWLEDGEMENTS

The research on which this paper is based was funded by the Economic and Social Research Council, grant number RR000235698. Thanks to all the interviewees and to Sue Lilley for her transcription skills.

10

RESISTING RECONCILIATION

the secret geographies of (post)colonial Australia

Jane M. Jacobs

it is paradoxically in hiding that the secrets of desire come to light.
(Cocks, 1989, *The Oppositional Imagination*)

In her book *Generations of Resistance* (1991), now in its second edition, Lorna Lippmann documents the long history of Aboriginal and Torres Strait Islander demands for justice in the nation of Australia. Her book is part of a large revisionist historiography which has brought to the sphere of public history previously repressed narratives of the making of the Australian nation. These once hidden histories tell of the violence, racism and oppression of colonial settlement. They also explode the myth of Aboriginal acquiescence to colonial authority by detailing the long and diverse history of Aboriginal resistances: the guerrilla warfare, the civil rights marches, the legal challenges, the pastoral worker walkoffs, the mining blockades, the homelands movements and the street protests. This revisionist historiography celebrates those moments when Aborigines broke out of the subordinate positioning conferred upon them by the colonial order and entered onto the historical stage as protagonists. That there is an array of such counter-colonial events to be recovered points to the always partial and frail completeness of colonial domination in nations like Australia.

This revisionist history also has a geography. Indeed, the re-narration of the Australian nation is regularly articulated in and through space. Take as an example the Aboriginal 'Tent Embassy', a small and somewhat dilapidated tin shed painted in the black, yellow and red of the Aboriginal flag, which sits on the groomed lawns of Capital Hill, immediately opposite Old Parliament House. In 1972 Aborigines and Torres Strait Islanders established the Embassy as part of their struggle for land rights. Then it was a confronting appropriation and re-deployment of one of the structures of modern nationhood – an embassy – which stood for one of the fundamental rights – the right of sovereignty – which had been denied to indigenous Australians. Establishing a Tent Embassy was a political strategy which took, to the political heart of the nation, Aboriginal complaints about the historic denial of their pre-existing rights over the land which came to be known as Australia. What was then considered by many to be an audacious act is today accepted as part of the making of modern, (post)colonial Australia. The Tent Embassy is now a registered site on the National Estate of Australia's cultural heritage and tours of the capital city regularly include this site of Aboriginal resistance on their itinerary. In short, modern Australia wants its colonial history to include a story of resistance.

There are many other sites in this emerging national geography of Aboriginal resistance. One recent account of Aboriginal resistance documents in detail the history and geography of *Six Australian Battlefields* (Grasby and Hill, 1988). The book ends by encouraging readers to visit these sites in the spirit of a 'pilgrimage' to a previously unknown past. Precise directions are given by the authors for not only do many of these places 'no longer appear on the map' but the traces of resistance which might inhabit such landscapes are not always self-evident (Grasby and Hill, 1988: 271). For example, the modern pilgrim to the site of the 'battle of Parramatta' – where in 1797 the Eora people, led by Pemulwuy, faced the armed forces of the new government of New South Wales – today finds a geography of colonial triumph:

> The whole area today is a sylvan setting for ... lazy afternoons in the summer sun. The quick-running brown stream is flanked on one side by gentle grassy banks, on the other by thick trees, and the only battle sounds to be heard are the click of bowls from the greens of the Parramatta Masonic Club across the way.
>
> (Grasby and Hill, 1988: 273)

If Australians are to make pilgrimages across the revisionist maps of the nation then they may as well pack a picnic lunch.

My purpose in this chapter is not simply to embellish this revisionist historiography and geography by providing yet more stories of Australian Aboriginal resistance or placing more points on an alternative map of the nation, as important as these oppositional formations are. Rather, I would like to propose that within modern Australia this starkly oppositional framework for thinking about resistance is problematised by a range of initiatives which have attempted to redress the uneven structures of power and privilege established under colonialism. Since the 1970s, Australia has self-consciously engaged in a range of revisionist processes which have attempted to acknowledge Aboriginal rights (in land, as citizens) and to incorporate the Aboriginal experience of colonisation into the public history of the nation. In the last decade of the twentieth century these various initiatives have been harnessed into the more comprehensive national objective of 'reconciliation'. Reconciliation aims to 'heal the wounds' produced in Australia's colonial past (Council for Aboriginal Reconciliation, 1994a, vii).

This chapter, then, concerns itself with the place of resistance within this new regime of healing. It explores this issue firstly through an examination of the formal processes of reconciliation and what reconciliation might mean in nations struggling towards a 'post'colonial future. It then examines one specific, place-based reconciliation project: the construction of an 'Aboriginal' walking trail in the city of Melbourne. This trail is one of a number of similar trails being marked into the contemporary Australian landscape (see Jacobs, 1996). Through these trails the goal of reconciliation is given performative expression. Melbourne's Aboriginal walking trail demonstrates the ways in which these reconciliatory gestures, far from settling the nation down into a state of calm co-habitation, can activate a range of unexpected refusals and resistances. So the concern of this paper then is not with the possibility of reconciliation creating political space in which there may be no need for resistance. Rather my concern is with examining the way in which the performance of reconciliation actually produces diversely positioned effects which unavoidably complicate the straightforward oppositional arrangement within which resistance has been conventionally framed.

SHAME JOB:[1] RECONCILIATION IN MODERN AUSTRALIA

While in the 1960s resistance was one of the resources of political hope, today it seems that reconciliation has taken centre stage. Throughout the world previously intractable political conflicts are being redirected into the process of reconciliation. It may be that this move towards reconciliation is a form of millenarianism with the *fin de siècle* taking its effect as a desire for the resolution of conflict and the hope of a calm beginning for the twenty-first century. This is certainly evident in the timetable for Australian reconciliation. The Council for Aboriginal Reconciliation was established in 1991 and has until the year 2001, also the centenary of Australian Federation, to fulfil its vision (Council for Aboriginal Reconciliation, 1993: 1). Indeed a recent government report argued that 'without meaningful progress in the reconciliation process there will be no truly national celebration' in 2001 (Centenary of Federation Advisory Committee, 1996: 13). Reconciliation is not then a description of a pre-existing state but a political naming which is attempting to mark out a postcolonial rallying point for the nation. In this sense it is a performative formation which requires what Judith Butler (1993: 208) describes as 'phantasmatic investment'.

The officially stated goal of the Council for Aboriginal Reconciliation is: 'A united Australia which respects this land of ours; values the Aboriginal and Torres Strait Islander heritage; and provides justice and equity for all' (Council for Aboriginal Reconciliation, 1993: 1). Despite such emphatic definitions, a number of recent commentators on the reconciliation process in Australia have noted the ambiguity of the term (see Hoorn, 1992 and Nicoll, 1993). On the one hand, it can suggest a reconciliation between parties – a sense of both parties reconciling *with* each other to produce 'a united Australia'. On the other hand, it might also mean the participating parties reconciling themselves *to* each other and *to* the limits of their historical predicaments: for example, Aborigines reconciling themselves to the losses experienced under colonisation; settler Australians accepting some sense of responsibility for the violence of colonisation. Hoorn (1992) suggests that in the Australian context reconciliation is weighted towards the latter sense, and that Aborigines are the ones who are doing the real work of reconciling themselves *to* the implications of the nation's colonial past. It is certainly true that the meaning of the word 'reconciliation' does include the sense of parties submitting or resigning themselves. A politics of reconciliation does then imply some form of surrender.

In its ordinary use, the term reconciliation encapsulates a narrative in which parties which were at one time estranged then become friends (Hardimon, 1994: 85). The Australian project of reconciliation conforms to this narrative and is a most self-conscious effort to install a sense of a unified modern nation. Indeed, reconciliation is one tactic (along with multiculturalism) of a much broader political strategy, instigated by the Labor government of the early 1990s, to create what it dubbed 'One Nation' (Keating, 1992 and see Brennan, 1995). In the first major report of the Council for Aboriginal Reconciliation, entitled *Walking Together: the First Steps*, it noted that the initiative 'has provided a marvellous opportunity to all Australians to be participants in a worthwhile nation-building exercise' (Council for Aboriginal Reconciliation, 1994a: ix). Reconciliation is presented as a strategy for all Australians, be they dispossessed Aborigines or non-Aboriginal settlers, to feel that they belong in the nation. It was, as Hardimon (1994) notes, Hegel who proposed that Versöhnung (reconciliation) played a crucial role in allowing the modern, 'alienated' individual to feel 'at home' in the world. But it was also Hegel who observed that reconciliation did not simply mean replacing enmity with love, estrangement with friendship, or conflict with harmony. This may be the outcome of reconciliation, but the process itself is not an erasure of all the negativities in the social world. Rather, Hegel proposed that the sense of being 'at home' in the world depended on coming to terms with both the positives *and negatives* of that world.

This is most certainly the tone of reconciliation in modern Australia. The emergence of a policy of reconciliation can be traced to what was Australia's first officially sanctioned review of the history of colonisation, the Royal Commission into Aboriginal Deaths in Custody (1991). In addressing the problem of an inordinately high level of deaths among incarcerated indigenous people, the Royal Commission moved well beyond the walls of prisons, police cells and remand centres. It was the entire history of colonisation which was blamed for the current levels of Aboriginal deaths in custody. The final recommendation of the Royal Commission dealt not with the conditions of Aboriginal incarceration, but with the idea of the nation reconciling itself to its colonial history. The Royal Commission advised that:

> all political leaders and parties recognise that reconciliation between the Aboriginal and non-Aboriginal communities in Australia must be

> achieved if community division, discord and injustice to Aboriginal people are to be avoided.
>
> (Australia. Royal Commission into Aboriginal Deaths in Custody, 1991: 65)

The Commission then proceeded to advise that there be bi-partisan public and political support for the process of reconciliation. The passing (with full parliamentary support) of the Aboriginal Reconciliation Act in 1991, and the attendant establishment of the Council for Aboriginal Reconciliation, began the formal political response to the directives of this Royal Commission.

The official policy of reconciliation operates as an overarching container for a variety of processes and events which are attempting to make amends for the past. Perhaps the most historic of these reconciliatory events was the 1992 decision of the High Court of Australia to acknowledge that Aborigines and Torres Strait Islanders did have prior rights over the lands which came to be known as Australia, and that these rights were not necessarily extinguished by the colonial designation of Australia as *terra nullius*. In handing down their 'Mabo decision' Justices Deane and Gaudron remarked that it might make some amends for the 'national legacy of unutterable shame' (*Mabo* v. *Queensland* (No. 2) 1992 175 CLR: 104 per Deane and Gaudron JJ). They continued to note that: 'The nation as a whole must remain diminished unless there is an acknowledgment of, and retreat from, those past injustices' (*Mabo* v. *Queensland* (No. 2) 1992 175 CLR: 109).

A similar message was conveyed to the nation by then Prime Minister Paul Keating in his landmark speech in 1992 to launch the International Year of the World's Indigenous People. He called for all Australians to acknowledge the 'true' history of the nation. In front of a predominantly Aboriginal audience, Keating admitted the 'guilt' of colonial settlers:

> We took the traditional lands and smashed the traditional way of life. We brought the diseases. The alcohol. We committed the murders. We took the children from their mothers. We practised discrimination and exclusion. It was our ignorance and our prejudice. And our failure to imagine these things being done to us.
>
> (Keating, 1993: 5–6)

The political framing of reconciliation focuses specifically on re-arranging the 'truth' of the nation. Race relations in colonial nations like Australia has formerly been structured around a different 'truth' which entailed certain misrecognitions of the nature of Aboriginal culture: as 'primitive', 'uncivilised', 'without property', 'without rights'. Such constructions of Aboriginality were fundamentally important to a nation which sought to legitimate colonial settlement (the dispossession, the stolen children of assimilation and so on). Furthermore, the stability of colonial dominance depended, in part, on certain erasures and repressions of the way it really was; in short, the production of what the journalist John Pilger has called *A Secret Country* (1984). Reconciliation is an official entry into the process of disclosing previously repressed aspects of the nation's history and setting these 'secrets' into a national framework of 'truth'. Reconciliation, then, provides an official infrastructure for excavating things once made 'invisible' (a people, their sovereign rights, the ugly history of colonisation, the sites of resistance) so that they might become a visible part of the sanctioned history and geography of the now reconciled nation. The goal of reconciliation may be some form of harmonious co-habitation, but it will be reached by way of an often painful process of realigning what is known about the nation: by performing a 'new truth' of the nation. Of course, what might come to stand as the new truth of the nation is not in any way given. What can and cannot be said in the emerging truth of the reconciled nation will not of course transcend existing power relations. Rather, this new truth will come into being through the contingency of those relations (Butler, 1993: 207).

Reconciliation is an official strategy of correcting the national sense of self. It attempts to bring the nation into contact with the 'truth' of colonisation – and this includes the attendant emotional 'truths' of guilt, anger, regret and hurt – in order that there might be a certain 'healing'. It is perhaps not surprising that, rather than settling things down, this retrieval of the 'truth' of the nation is setting a whole range of things in motion again. Indeed, for many non-Aboriginal Australians the performance of reconciliation is producing anxieties about their place in the future of the nation and the validity of their current understandings of the national past. The restructuring of national narratives and material rights which has occurred under the banner of reconciliation has resulted in the appearance of a virulent new racism in which it is claimed Aborigines now have too much – too much of the nation's history, too much land, too many special rights and services. Rather than reconciliation restructuring the parameters of national knowing into a new space of calm co-habitation, it is actually producing a most contested politics of knowing and rights. Reconciliation may have as its goal a transcending of a more familiar oppositional politics, but it is at the same time generating new political articulations characterised by a range of significant reversals and inversions.

It is in trying to understand these unpredictable effects which surround the performance of reconciliation that a less commonly used meaning of the term resistance may be usefully deployed. This is the psychoanalytic (specifically Freudian) understanding of resistance as a patient's refusal to move to a point which will enable healing to occur. Resistance, in this context, is a form of defence against the anxiety which might be produced by recognising some repressed 'truth' or confronting the repressed emotional traces of a past trauma. It is also a resistance against that knowledge which might badly shake or force an abandonment of the existing order of things (Cocks, 1989: 69). By and large this psychoanalytic meaning of the term resistance is neglected in studies of a cultural politics of resistance. Resistance is more generally associated with an oppositional politics and is almost exclusively associated with those groups which have been marginalised by dominant structures of power. In the case of Raymond Williams (1977), for example, resistance belonged to 'residual' or 'emergent' formations which were somehow 'outside' – but then came back into – dominant structures of power. The more explicitly psychoanalytic engagements with resistance which have emerged in recent postcolonial writings have worked to unsettle such rigid binary frameworks. In the work of Homi Bhabha (1990a, 1990b), for example, a starkly oppositional politics is opened out into a diverse and more fragmented range of subversive formations which are associated precisely with the way in which the coloniser and the colonised are not discrete categories. For Bhabha it is the unruly circulation of colonial constructs of self and other which work to subvert colonial authority. Bhabha's concept of hybridity, for example, operates as a subversive strategy because it entails the colonised contesting colonial authority through seizing the 'given symbols' of that authority (Bhabha, 1992: 63 and 194). Bhabha's charting of a psychoanalytically grounded postcolonialism depends upon understanding the destabilising anxiety that mimetic and hybrid formations produce for the coloniser (1994). He does not, however, consider that alongside of such open and productively anxious responses there may be others which actively try to foreclose the possibility of destabilisation; that is, another type of 'resistance'.

RECONCILED SPACE

As reconciliation requires a restructuring of the nation's knowledge of itself, it is not surprising that one of the primary responsibilities of the Council for Reconciliation is pedagogical. The Council has a mandate to educate wider Australia about Aboriginal culture (past and present) and to remould the story of Australian 'settlement' into a story of 'colonisation', with all its attendant grimness and with its new heroics of survival. The Minister for Aboriginal and Torres Strait Islander Affairs, who guided reconciliation into public policy, saw it specifically as a way to:

> educate non-Aboriginal Australians about the cultures of Australia's indigenous peoples and the treatment of Aboriginal and Torres Strait Islander people by European settlers and their descendants. This sad history – which includes the dispossession and dispersal of Aboriginal people, confinement in reserves, removal of children from their families and the destruction of much Aboriginal culture – needs to be recognised as a primary cause of the current disadvantaged position of Australia's indigenous peoples. All Australians need to understand this country's past and the place of Aboriginal people in it.
>
> (Tickner, 1991: 21)

The Minister of Aboriginal and Torres Straight Islander Affairs was careful to mention that this re-education of the nation was not intended to produce guilt, but an empathetic infrastructure upon which reconciliation might be built.

From the outset it was intended that the official, national commitment to reconciliation articulate itself 'in smaller, practical, localised terms' through community-based initiatives (Council for Aboriginal Reconciliation,

1994b: 5). Every Australian was called upon to establish their own 'personal agenda for reconciliation' in the hope that this would ultimately restructure the 'national ethos' (Council for Aboriginal Reconciliation, 1994c: 26–27). In December of 1993 a community out-reach organisation called Australians for Reconciliation was established to coordinate and encourage grassroots involvement in reconciliation. Using the project title of 'Working Together', Australians for Reconciliation have encouraged a wide range of community-based initiatives. One significant part of their work has been to encourage local government authorities and community groups to engage in projects in which the goals of reconciliation are given performative expression.

It was in this spirit of a local initiative of reconciliation that, in 1994, the City of Melbourne commissioned one Aboriginal and one non-Aboriginal artist to work together to produce the 'Another View Walking Trail'. The trail is subtitled 'Pathway of the Rainbow Serpent', but it traces neither a traditional mythological pathway nor a pre-contact story. The trail winds through the streets of the central business district of Melbourne, passing some seventeen sites which include newly installed artworks produced by the artists as well as existing monuments originally erected to celebrate the triumphant moments of the colonial settlement of Australia. The trail is marked by small plaques set into the ground and has an accompanying brochure, available from tourist information booths and the local authority, with a text written by Aboriginal historian and writer, Robert Mate Mate.

Like most Australian cities, Melbourne's gridded street plan reflects the colonial desire to replace the 'disorder' of an unknown new land with the ordered spatiality of something familiar (see Carter, 1987: 46 and 204). Historians of early settlement in the region have noted that the impact of colonisation had a 'sudden and dramatic' effect on the Kulin, the Aboriginal people who originally inhabited the area. The area surveyed for the City of Melbourne had been an important meeting ground, but soon became 'virtually barred to them' (Presland, 1994: 104). Like other recent urban place-making projects with an Aboriginal theme, this trail intended to (re)Aboriginalise the city space (see Jacobs, 1996). It did not necessarily uncover pre-existing sites of significance to Aborigines, but charted a new geography of Aboriginality in the city.

The artists' statement made the association between the trail and the project of reconciliation explicit. And just as the national programme for reconciliation stresses a re-narration of the nation so too does this trail serve to bring its viewers into contact with the 'true' history of the nation:

> only when the past is confronted and accepted ... can we move on and build a creative and harmonious future ... This project asserts that Aboriginal and non-Aboriginal Australians have a shared history, and that in order to have a shared future we need to acknowledge success and failure in the past in our contemporary reading of history.
>
> (Roy Thomas and Megan Evans, quoted in City of Melbourne, 1995a: 3)

This is a theme repeated in the conceptual theme of the Rainbow Serpent which, according to the artists, symbolises 'dying the death of not knowing' and a rebirth 'into the sphere of awareness' (City of Melbourne, 1995a: 1). In another City of Melbourne publication, the link between national (re)birth and 'truth' is again stressed: 'A nation comes of age when it is able to face the truth about its past'. The 'truth' which is being told on this walking trail is of the 'bitter past' of Aboriginal and European relations in Australia (City of Melbourne, 1995b: 64). How then is a new 'truth' expressed on this reconciliatory walking trail and how might we think about its relationship to a politics of resistance? Does it productively subvert non-Aboriginal authority or does it simply produce non-Aboriginal resistances to the new nation?

THE PATH TO RECONCILIATION

The new artwork on the trail is varied. Some of it represents scenes of traditional life and stories of traditional importance to the Kulin people, the original inhabitants of the Melbourne area. For example, a painting mounted (somewhat precariously) on the side of a major road bridge depicts the Kulin story of the creation of men and women. Elsewhere on the trail, in a garden named after the matriarch of Imperialism, Queen Victoria, a pavement-set mosaic depicts the Rainbow Serpent coming out of the ground. On one of Melbourne's most fashionable and expensive shopping streets, known somewhat ambitiously as 'Little Paris' by local tourism boosters, a brass depiction of the Seven Sisters Dreaming has been tastefully embedded in the pavement. Each of these artworks is based on a re-interpretation and re-deployment of traditional Aboriginal stories and practices. We can think of this as a form of strategic essentialism, where there is a 'performative invocation' of identity (Butler, 1990b: 325) which takes up some of the (more positive) characteristics by which Aborigines have been constituted as a 'subject' under colonialism (Butler, 1992a: 109–110). In their content these artworks stage a return to tradition, but this return is in pursuit of a modern objective. Tradition is activated as part of a strategy of subverting the sure footing of colonial Melbourne by reclaiming an 'Aboriginal' originary narrative for the space which became the city.

At other sites on the trail newly commissioned artworks stand as counter-points to pre-existing monuments which celebrate colonial conquest. The statue of the maritime explorer Matthew Flinders, who had surveyed the southern coast of Australia in 1802, has him gazing triumphantly out across the prow of his ship-to-shore boat as it is dragged onto the new land by two seamen. Matthew Flinders now also gazes out across a small cross-shaped box embedded in the lawn at the base of the statue. The red-tinted transparent perspex lid of the box is etched with text. It begins: 'IN THE NAME OF' and continues in a repetitive chant 'progressjusticecivilisationEnglandHisMajestyprogressjusticecivilisationEnglandHisMajestyprogress ...'. Inside the cross-shaped box, spent brass cartridges and bleached bones rest on a blood-red lining. The installation is most certainly intended to produce a subversive counter-narrative to the smooth, masterful story of settlement enshrined in the statue of Matthew Flinders. Yet this small installation, what one commentator called a 'mini-monument', is barley visible to the passer-by or even to those who might choose to rest on the quaintly designated 'Ladies only' seats nearby (Stevens, 1996: A15).

Some of the new artwork on the trail is located at what might best be thought of as modern 'sites of significance', places which represent the troubled history of Aboriginal and non-Aboriginal relations. It is far from surprising that a number of these sites are associated with the law: Parliament House, the Old Melbourne Gaol, the Melbourne Remand Centre, the Supreme Court. As the recent investigation into Aboriginal Deaths in Custody demonstrated, it has been the regulatory framework of the law which has produced the sites of greatest conflict between Aboriginal and non-Aboriginal Australians. Indeed, as I will argue later, this tension between the law and Aborigines is a persistent feature of this 'reconciliatory' initiative.

At Parliament House, where the trail begins and ends, a pavement mosaic re-interprets the painting 'Ceremony' (circa late 1890s) by William Barak, an Aboriginal artist from the Woiworung people. It was at this site, in 1936, that the last recorded Aboriginal 'corroboree' was performed. At the Old Melbourne Gaol (no longer a functioning gaol but a heritage-listed tourist destination) the artists have erected four metal poles in the style of traditional Aboriginal burial poles (City of Melbourne, 1995a: 6). It was here on 20 January 1842 that two Tasmanian Aborigines, (Jack) Maulboyhenner and (Robert) Devay, became the first prisoners to be executed by public hanging in the new colony. In the early days of colonisation, these two men had been appointed as go-betweens for the Protector of Aborigines, George Augustus Robinson. They had assisted in the removal and round up of Aborigines in Tasmania and Victoria and even had their portrait painted to mark their status within the emergent colony. But after leaving the employ of the Protector, Maulboyhenner and Devay led a series of Aboriginal attacks on whalers in the Westernport area of Victoria, a retaliation which cost them their lives. The poles erected outside the Old Melbourne Gaol – referred to in the guidebook for the trail as 'the colonial place of suffering' – are screen printed with text and figurative images. The text includes extracts from the sensational reports of the hanging – the shouting and laughing crowd which had gathered to experience what one chronicler of the hanging called a 'pleasant fine morning's fun'. Figurative images of Maulboyhenner and Devay are stencilled onto barely transparent perspex sections on the poles. These are reproductions of the official portraits of Maulboyhenner and Devay, commissioned during the time when they were more comfortably aligned to colonial intentions. Contained within one deliberately distressed perspex case is a noose, the trace of the deaths which the poles symbolise.

There is little doubt that this Walking Trail does articulate 'views' which counteract and subvert the 'known' history and geography of the city of

Melbourne. It takes symbols of dominant culture (such as the statue of Matthew Flinders) and destabilises their authority by presenting oppositional narratives of the events they set in train. It reclaims city space by inserting into the built environment strategic performances of Aboriginal 'tradition'. It memorialises those who had heroically resisted colonial authority. These features of the trail can all be aligned with a politics of resistance, be it the starkly oppositional forms Raymond Williams might have hoped for or the more fragmented subversions that recent postcolonial theory leads towards. Yet the Aboriginal 'resistance' which is expressed through this trail is also problematic. For example, the expressions of an 'Aboriginal view' were not solely the work of Aboriginal-identified artists or of an Aboriginal-led initiative. The trail was the result of the creative collaboration of an Aboriginal artist and an Anglo-Celtic artist under commission from the City of Melbourne. And the trail itself was the product of the self-conscious desire of Melbourne's local authority to give over (some) city space to an 'Another' (read 'Aboriginal other') expression. The 'politics of resistance' which is expressed in this trail is a result of a modern coming-together of non-Aboriginal desires and an Aboriginal politics.

Such a coming-together should not, however, be misrecognised as the smooth space of reconciliation. As I have noted in relation to other such trails, these reconciliatory place-making gestures rarely involve Aborigines gaining any legal, propertied control of urban space. The Aboriginalisation of urban space rarely runs *that* deep in the context of the highly valued and much sought after space of the city. Rather, it occurs within the confined opportunities offered by sympathetic local authorities who have incorporated difference into their visual imaging of urban space. The Another View Walking Trail is one of many community arts strategies which have been commissioned by local authorities in order to create symbolic markers of cultural diversity within the much desired idea of a multicultural/postcolonial Australia, 'One Nation'. The various subversions and re-arrangements of history and space which occur through these sites are articulated within a facilitatory framework which operates at the discretion of local planning authorities and the various funding bodies which support their efforts. If the Another View Walking Trail is designed to 'pave the way for a future which we can all share' (City of Melbourne, 1995b: 67), then it is often not Aborigines who are mapping the course that such pathways take.

RESISTING RECONCILIATION

The question of whether such trails can in fact act as a path to reconciliation, and whose reconciliation might be performed through such trails, are not issues about which one need only speculate. The Another View Trail, for example, generated a controversy which was linked precisely to its primary goal of adjusting the 'truth' of the nation. The idea of a new range of city artworks which were to disclose the 'bitter' history of Aboriginal and non-

Aboriginal relations met with considerable resistance from influential Melburnians. One press report dubbed it the 'guilt trail' (Stevens, 1996: A15).[2] But it was not only the content of the planned artworks which created disquiet. It was precisely the geography of these planned artworks, where they were to be located, which generated such vocal opposition. Influenced by these complaints, the City of Melbourne withdrew five of the artworks planned for the trail on the grounds that they were 'too controversial' and 'extremely confrontational' (City of Melbourne Public Art Officer quoted in McKay, 1996: 1). The City of Melbourne wanted to Aboriginalise urban space in order to perform reconciliation. The generation of offence and non-Aboriginal complaints did not accord with its parameters of what reconciliation meant in the eyes of the local authority. It withdrew those artworks it deemed had gone 'that bit too far' because, in the words of one City spokesperson, it 'would really have worked very much against the overall project' (City of Melbourne Public Art Officer quoted in McKay, 1996: 1).

While the artworks which were successfully incorporated into the trail do give 'Another View' of Melbourne, there is perhaps a more familiar, less reconciliatory, view which is encapsulated in the story of these absent artworks. The most complete and thorough erasure happened in the case of a pavement-set mosaic which was to be located outside the Supreme Court. The planned artwork attacked the very foundational principle of the Australian–British tradition of law – the idea of justice, that Enlightenment ideal which was so much a part of the legitimating drive of colonial expansion into so-called 'uncivilised' lands. The proposed mosaic was to be located at the base of an existing statue of Justice which stands outside of the Courts. It took up the theme of the sword of Justice depicted in the main monument and was to represent instead an Aboriginal head impaled on that sword. Members of Victoria's judiciary, led by Victoria's Chief Justice, Justice John H. Phillips, vocally opposed the mosaic on the grounds that it was an inappropriate addition to the stately and authoritative space of the Law Courts.

It may seem strange that the powerful institutional framework of the law should be so unsettled by the prospect of a relatively small-sized mosaic installation. In many ways this proposed mosaic provided a critique of the relations between Aborigines and the law similar to that expressed in the death pole installation at the Old Melbourne Gaol. But there the death poles addressed past injustices and sat alongside a building which represented the law-in-heritage. The mosaic at the Law Courts did not address an injustice safely contained in a historical event. It spoke to a far more persistent and ubiquitous notion of injustice – one which extends to the present and right into the ordered confines of the Law Courts themselves. Being reminded of the 'injustice' of the law was possibly not what the modern Australian legal fraternity wanted to hear. In the past thirty years the law as it relates to Aboriginal Australians has changed significantly. For example, it was 'the law' (and indeed only the law) which held the power to reverse the fallacy of *terra nullius*, land unoccupied, and recognise Aboriginal prior rights to

Australia. It was 'legal justice' which ushered in the Mabo decision and the Native Title Act.

The Another View mosaic proposed for the Law Courts was unconcerned with this 'new', possibly more just, law. Perhaps if the proposed mosaic had celebrated this reformed law rather than reminding the legal fraternity of the less glorious aspects of its effects on Aborigines, then maybe the judiciary might have been happy to give over a fragment of their public forecourt to 'another' view. Perhaps if the proposed mosaic had shown the rainbow serpent rising up to meet the sword of justice to produce a marriage of old Lore and new Law, then this site might still be part of 'another' view of Melbourne. But the proposed mosaic was far more 'confrontational' than that. The 'truth' it told was not one that the legal fraternity wanted to hear. In the official interpretation which accompanies the trail there is no mention of this proposed location nor of the more recent 'bitter history' of Aboriginal and non-Aboriginal relations which adhere to this space of erasure. The space of Justice has remained pure and its place in the vanguard of the making of the nation has not been sullied by what the law no doubt saw to be reconciliation taking the wrong path.

All of the other artworks to be censored from the final trail had a fatal combination of 'controversial' content and 'inappropriate' geography. Each contained an explicit reference to Aboriginal deaths. But again it was not only content which sealed the fate of these artworks. All of them were to be located near to existing monuments to the colonial period. For example, a brass map of the site of Aboriginal massacres in the state of Victoria was to sit alongside the memorial marking the original grave of John Batman. Batman was not only one of the founding fathers of Melbourne but was the first and only recorded colonist who, on the 6 June 1835, signed a treaty (which in the following year was disallowed) with the Aborigines of the Port Phillip area. This particular memorial had already felt the hand of the revisionist history. The original text on the monument had referred to the land he selected for settlement as 'unoccupied'. In the wake of the Mabo decision an additional plaque was added to the memorial statue acknowledging that prior to colonisation 'the land was inhabited and used by Aborigiana people'. The installation of a map of the early massacres of Aborigines would have further destabilised Batman's place in the current history of the nation as one of the 'good' colonisers.

In some cases the artwork was censored but the proposed site remains marked on the trail. This is the case with the Queen Victoria monument, a typically Gothic structure celebrating the triumph of the British colonisation of Australia. Below the always imposing figure of Queen Victoria are allegorical representations of the 'pure' Enlightenment ideals which motorised British Imperialism: wisdom, progress, history and justice. Deprived of the 'Another View' artwork, the site functions on the trail to bring the visitor into contact with the 'lie' of these imperial principles. The guidebook text offers an Aboriginal re-interpretation. Here 'History' becomes a specifically Eurocentric 'history' which excluded Aboriginal experiences. 'Wisdom' becomes a debilitating ignorance of those Aboriginal knowledges of land which were often fatefully disclaimed by colonists. The 'Progress' which resulted in the destruction of Aboriginal cultures and of the land itself is questioned. And the text points to the contradiction between the colonial idea of 'Justice' and Aboriginal experiences of that 'justice'. This re-narration of this monument self-consciously subverts its authority. But there is little at this site to mark this subversion. A small plaque embedded in the pavement surrounding the Queen Victoria monument registers to the passer-by that this is not simply a monument to the 'pure' ideas behind the making of the colonial nation but also stop number 12 on the Another View Walking Trail. The destabilising irony produced by the Aboriginal appropriation of this monument is knowable only to those visitors who come equipped with the brochure containing the interpretative text written by Aboriginal historian Robert Mate Mate. In short, only those tourists who want to know the new truth of the nation will see this 'other' view.

The censorship of the Another View Walking Trail is a version of a far more widespread resistance to the disclosures and accommodations which are required in the process of reconciliation. For example, the 1996 Federal election produced a form of open anti-Aboriginal racism not seen in contemporary Australian political life since the 1950s. A number of maverick candidates ran campaigns based on the proposition that Aborigines (the most disadvantaged sector of the entire Australian population) 'squandered' government funds and had too many special provisions. This then is a racism based on a perception that Aborigines now have too much, more than the 'average' Australian. This is postcolonial racism.

CONCLUSION

It may be thought that the movement towards reconciliation establishes a state in which there is no room – or need – for resistance. But reconciliation does not do away with resistance at all. Indeed, that there was Aboriginal resistance to colonisation is one of the 'truths' that Australia is being asked to incorporate into its new national knowing. The revisionist histories of resistance with which this chapter opened are a requisite part of the performance of reconciliation. But the recent history of Aboriginal and non-Aboriginal relations which surround the Another View Walking Trail suggest that while reconciliation may not dispense with resistance it does prefer it in certain forms. For example, reconciliation prefers resistance to be something which happened *then* but is remembered *now* – as in the case of the Old Melbourne Gaol burial poles which memorialised the long-ago hanging of two Aboriginal heroes of resistance. Reconciled (non-Aboriginal) Australia wants the grand moments of colonial triumph to be chastened by historically contained memories of Aboriginal opposition. As was demonstrated through the opposition to and censoring of the Another View trail, there is far less comfort with reconciliatory performances which unsettle the sacred infra-structure of colonial history or remind modern Australia that it is still far from postcolonial. It is no coincidence that the sites which were censored from the trail were not those which depicted a traditionalised version of Aboriginality. The 'truth' which non-Aboriginal Australians are most comfortable with is a romanticised and primitivised Aboriginality, one which by-passes the very history with which reconciliation is supposed to bring them into contact.

There is little doubt that the idea of resistance, as it was once theorised, has undergone significant transformation. For many on the left there remains a nostalgia for the hope when resistance would cohere and become a tectonic revolutionary moment. But, for most, this possibility is understood as only a hope, a modernist fantasy of the other kind. Tectonic forms of resistance still occur: colonised groups still organise into anti-colonial national struggles, workers still do (occasionally) unite rather than bargain. But it is just as likely that 'resistance' will be articulated through a range of more fragmented formations in which stark oppositions give way to more complex subversions. Resistance may still be articulated as being grounded in incommensurate difference, as is the case when tradition is taken up into contemporary political struggles. So, for example, the Another View Walking Trail mobilised traditionalised constructs of Aboriginality – the corroboree, the rainbow serpent, the seven sisters dreaming. But these traditional articulations were produced under modern conditions; that specific co-incidence between the desire of the local authority to Aboriginalise city space and Aboriginal political objectives. Of course resistance may be based, as Bhabha notes, around the unsettling anxieties produced by a seizure of the signs of colonial authority – the appropriation of the sword of justice, the re-mapping of Victoria as a bloody site of massacres, the re-telling of the stories of colonial 'settlement'. Such was the anxiety produced by at least some of the proposed artworks which tackled these themes that they generated another, but nonetheless significant, form of resistance on the part of non-Aboriginal Melburnians. Indeed, it was not simply that this alternative truth was to be articulated but it was very much about 'where' this truth was to be told. The policing of the disclosure of 'truth' which occurred in the Another View Walking Trail was a spatial issue. That is, there was a regulation of the geography of the 'other' view which ensured that at least some of the grand sites of colonial authority remained pure. This suggests that reconciliation may not stop certain uncomfortable 'truths' being told but that these truths cannot come too close. Reconciliation may be recovering 'new truths' but non-Aboriginal resistance to these truths may prevent them from entering into a productively promiscuous relationship with the existing order of things.

What some feminist circles have described as 'backlash' politics now so fully inhabits all dimensions of the contemporary political environment it is essential that the contours of this other form of 'resistance' be charted. This is not to abandon the project of documenting and supporting counter-hegemonic formations. Nor is it to suggest that the postcolonial project of shaking up a binary-based oppositional politics is irrelevant. But it is to suggest that the contemporary moment requires us to think about the consequences of a politics in which some of the most powerful sections of society are claiming they they are the ones who are besieged, they are the ones whose rights are being denied. The psychoanalytic idea of resistance – that mechanism which keeps things hidden – might help in understanding the regimes of desire which underpin this virulent backlash politics and the service it performs for the persistence of hegemony.

ACKNOWLEDGEMENTS

I would like to thank the following colleagues for their timely inputs to this paper: Steve Pile, Fiona Nicoll and Katherine Gibson.

NOTES

1 It is worth noting that in Aboriginal English the term 'shame job' is commonly used as a humorous and ironic confirmation that one has transgressed expected (and often mainstream) norms of behaviour.

2 The issue of whether non-Aboriginal Australians – and especially those who are descendants of the early colonisers – should feel or even accept 'guilt' for the losses Aborigines experienced under colonisation has produced considerable debate, including specific intervention from the right-wing think tank, the Institute of Public Affairs (see Brunton, 1993).

11

IDENTITY, AUTHENTICITY AND MEMORY IN PLACE-TIME

Michael Dear

> Gradually, it has become clear to me what every great philosophy has so far been: namely, the personal confession of its author and a kind of involuntary and unconscious memoir.
>
> (Friedrich Nietzsche, *Beyond Good and Evil*)[1]

> You cannot write anything about yourself that is more truthful than you yourself are. That is the difference between writing about yourself and writing about external objects. You write about yourself from your own height. You don't stand on stilts or on a ladder but on your bare feet.
>
> (Ludwig Wittgenstein, *Culture and Value*)[2]

When I was a very young child, one of the high points of my day came when Chris Kinsey finished delivering bread to various homes in the mining village of Treorchy, in Wales. He used to carry the pungent slabs in a voluminous basket attached to the front of a bicycle, and when the basket was empty, he would grab me and dump me inside it. Then he would ride off at breakneck speed through the narrow streets back to the bakery, making my eyes water. I preferred those white-knuckle excursions to what was another routine event in my daily childhood round, when my father would pull me aside to talk politics. His sermons used to take place at the oddest hours, because he usually worked shifts and his 'free time' was (to me) painfully unpredictable. His perennial favorite was the tale of how Winston Churchill sent in the troops to break up the miners' strike during the depression. Everybody in the Rhondda hated Churchill, even after the Second World War, and my father was determined that I should understand my political birthright. It mattered not a jot that his history was a trifle shaky. I also passed a lot of time with a cousin, Marian, who was quite a bit older than me. She spent most of her short life in a psychiatric hospital. Relatives euphemistically whispered that Marian was 'in the sanatorium,' or even more cryptically, 'in Bridgend' (after the hospital's location). But I knew she was mad. It never bothered me; she was simply the most entertaining relative one could ever wish for.

When I was an undergraduate, my moral tutor (yes, they were called that!) invited me into his office during my freshman year specifically to say: 'You are exactly the kind of student we don't want at this university.' He never bothered to explain why. It was true that I was a teenager (thus automatically worthy of intergenerational scorn), but I suspect his censure had more to do with matters of class and nationality. I had been born into a family of Welsh coal miners and was destined to follow the tradition; he was a middle-class Englishman in an English university.

As a graduate student in England during the late 1960s, I spent a lot of time marching around London's Grosvenor Square calling out the names of soldiers killed in Vietnam into a megaphone directed toward the Embassy of the United States of America. Later, completing my doctorate in Philadelphia, I discovered that my working-class and anti-war credentials (though belonging to different places and times) brought an unexpected legitimacy to many of my arguments in progressive circles. These were the same credentials that almost got me expelled as an undergraduate.

Many decades have gone by. Since my childhood in Wales, I have lived for long periods in Australia, Canada, England, and the United States of America. Throughout this time, my scholarly work has resolutely remained focused on the mentally disabled and the homeless, always seen through left-tinted lenses. I cannot shake off the commitments from a particular time and place; nor do I want to. They are who I am.

The postmodern emphasis on difference has placed an intense spotlight on the subject, body, and personal identity. As authors struggle to engage the political in themselves and their works,[3] the questions of identity and authenticity have arisen in two related but nevertheless distinct ways. In the first place, *disguise* is a consequence of what happens once a work is received by others. It is not uncommon for multiple layers of disguise to become the 'person' or the 'thought' irrespective of the intentions of the author or the text (Derrida, 1984). Secondly, *disguise* is an integral component in the work of many artists and scholars, whether it is deliberate or unconscious. Many intellectuals (and others) have chosen to exploit the ambiguities inherent in disguise by deliberately occupying the margins of discourse.[4]

The significance of personal positioning cannot be doubted, for, as Olsson has observed: 'who questions his body, thereby questions his culture.' Adorno also placed the question of human subjectivity at the center of his critical inquiry into authenticity and domination.[5] The problems and pitfalls inherent in a postmodern politics of identity have been acutely pinpointed by Jacques Derrida. His *Otobiographies* (1984) invites us to consider what is at stake when an author signs a work, because it is impossible to control its fate by any power of the author once it is released upon the world.[6] In the case of Nietzsche's writings, Derrida is less interested in the fact that Nietzsche's works were appropriated by proto-Nazi ideologues than with exactly why his texts lent themselves to such interpretations in the first place.[7] He concludes that there are many competing versions of Nietzsche, none with an absolute claim to the 'truth', but all made possible by something in the structure of his writings (Derrida, 1984).

In this chapter, using a series of examples from the Nazi era, I explore the relationships among place, time, identity, authenticity, and memory. The two forms of disguise I have just mentioned form the framework for this inquiry. First, using the examples of Martin Heidegger and Paul de Man, I examine the ways in which people are inevitably the products of particular places and times, but also how interpretive contexts (or 'cultures of criticism') overwhelm and reconstitute the ideas and the person, no matter how specific or forthcoming the texts are. Next, I explore the way in which individuals deliberately manipulate their identities by interrogating the lives and writings of Albert Speer and Philip Johnson. Finally, I show how the commemoration of memory, at later times and in different places, distorts the identity and authenticity of individuals, peoples, and ideas. In this way, it becomes possible for fictions to supersede truths. Such cultural distortions of place and time have special relevance for the practice of politics in a postmodern era, which has placed new emphasis on the subject, difference, and polyvocality in discourse.

I would like this essay to be viewed as an analysis of the political culture of changing place-time, especially as it relates to postmodernity. My inquiry is situated at the intersection between what would conventionally be described as 'political geography' and 'cultural geography.' But this essay is part of an ongoing re-writing of 'political geography' in the light of the agenda established by contributors to Keith and Pile's 1993 volume, *Place and the Politics of Identity*. And in its emphasis on culture, this inquiry shows how the production of knowledge is never innocent of the place-time circumstances in which it is created and (later) received. As such, it is about the relationship between altered human subjectivities and cultural transformations, appealing to an agenda already prefigured in the collection of essays edited by Pile and Thrift (1995b).[8]

AUTHENTICITY AND THE CULTURE OF CRITICISM

> If philosophy is simply understood as a search for truth and politics as the pursuit of power, the two appear to have very little in common. In reality, however, both are concerned with the production, use, and control of truth, with generating, channeling, and manipulating streams of power – though in admittedly very different ways – and from this comes their closeness and their conflict. Philosophy and politics are, in fact, inextricably tied together, but their relationship is also precarious and unstable.
>
> (Hans Sluga, 1993: vii)

Martin Heidegger remains one of the most influential philosophers of the twentieth century. He was also a Nazi. Paul de Man was one of the founders of deconstructionism. He also wrote anti-semitic newspaper articles during the Second World War. Using these two examples, I now examine in more detail the relationship between individuals, their work, and the time-space specificities of the contemporaneous cultures of criticism.

The case of Martin Heidegger catapulted to a new public prominence in

1987 following the appearance of *Heidegger et le Nazisme* by Victor Farias (1989). Elements of the debate have been concisely and meticulously reconstructed by Thomas Sheehan (1988), who described Heidegger as 'a provincial, ultraconservative German nationalist,' who was from 1932 a Nazi sympathizer. Three months after Hitler came to power, Heidegger became rector of Freiburg University. He also joined the Nazi party. Heidegger's record, as rector and teacher, is deeply distasteful. Soon after his 1933 appointment at Freiburg, he wrote that his goal was to bring about 'the fundamental change of scientific education in *accordance with the strengths and the doctrines of the National Socialist State*.'[9] He introduced the Nazi cleansing laws into Freiburg, so that Jews, Marxists, and non-Aryan students were denied financial aid. He secretly denounced colleagues and students. He declined to supervise Jewish doctoral students, and betrayed his mentor, Edmund Husserl.[10] Hugo Ott's (1993) political biography portrays a man who was disloyal, self-obsessed, mendacious, and spiteful. In 1945 an internal de-Nazification committee at Freiburg charged Heidegger with the following: 'having an important position in the Nazi regime; engaging in Nazi propaganda; and inciting students against allegedly "reactionary" professors' (Sheehan, 1988: 47). After a protracted debate, during which his health suffered badly, Heidegger was in 1949 declared a Nazi fellow traveler and was prohibited from teaching. By 1951, however, the political climate had altered sufficiently to allow emeritus status to be granted, thus enabling him to teach once again at the university. Heidegger's postwar defence of his political actions leaves little doubt that he remained convinced of their rectitude.[11]

The case of Paul de Man is more ambiguous. Between 1940 and 1942 the young de Man wrote 169 articles for *Le Soir*, a Belgian newspaper that had been taken over by collaborationists following the 1940 Nazi occupation. During the same period he wrote 10 other articles for a Flemish journal, *Het Vlaasmsche Land*, which was also under German control. These pieces were mainly book reviews, concert notes, and general literary and cultural criticism. In 1942, de Man resigned as critic of *Le Soir* for reasons that remain obscure. When Belgium was liberated from the Nazis in 1944, de Man appeared before a military tribunal but no collaborationist charges were brought against him.[12] The attack on de Man principally hinges on a single article, 'Jews in Contemporary Literature,' published as part of a special section on anti-Semitism in *Le Soir* of March 4, 1941. In it, de Man argues that European literature would not be weakened if Jews were to be placed in a separate colony. The crucial passage in this article reads:

> En plus, on voit donc qu'une solution du problème juif viserait à la création d'une colonie juive isolée de l'Europe, n'entraieraît pas, pour la vie littéraire de l'Occident, de conséquences déplorable.[13]

Since so much hangs on these words, I shall provide only the most literal of translations:

> In addition, one sees therefore that a solution of the Jewish problem that aims for the creation of a Jewish colony isolated from Europe, would not entail, for the literary life of the West, deplorable consequences.

With its isolationist sentiments, it is easy to see why this sentence is the principal target of de Man's detractors. However, it is also likely that this is the single instance in de Man's complete œuvre where such sentiments are so plainly stated.[14]

How do we read, see, and listen through the horrific images associated with Nazism? Should we now discard Heidegger and de Man, just as we routinely discount the rhetoric of our political enemies? To begin to answer these questions, let me first identify some important distinctions between the time- and place-specificities of the two cases.

Heidegger's involvement with Nazism is beyond doubt (although, as Sluga points out, many philosophers on both sides were all too willing to align themselves with their respective national interests – see Sluga, 1993). Wolin concludes that philosophers will never again be able to read Heidegger without taking account of his odious politics (1993a: 273). Sheehan draws a similar conclusion: 'Heidegger's engagement with Nazism was a public enactment of some of his deepest and most questionable philosophical convictions' (1988: 38). Although there is little to be gained from revisiting Heidegger's political past in order to pass judgement yet again, Sheehan argues that his philosophy must now be reappraised:

> One would do well to read nothing of Heidegger's any more without raising some political questions. ... [One] must re-read the works for what might still be of value, and what not. To do that, one must read his works ... with strict attention to the political movement with which Heidegger himself chose to link his ideas. To do less than that is, I believe, finally not to understand him at all.
>
> (1988: 47)

Some of the works, such as the commentaries on Plato and Aristotle, may be unaffected by Heidegger's Nazism. Other positions will require a complete review. Sheehan identifies the problem areas with a forensic precision: 'Above all, I believe we can ill afford to swallow ... his grandiose and finally dangerous narrative about the "history of Being" with its privileged epochs and peoples, its somber insistence on the fecklessness of rational thought, its apocalyptic dirge about the present age, its conclusions that "only a god can save us"' (1988: 47). The point, according to Sheehan, is not to stop reading Heidegger but to begin demythologizing him.

Paul de Man's lesser offense has unleashed a violent attack on (and subsequent defense of) the man, his work, and his politics. One of his principal detractors, Stanley Corngold, is absolute in his denunciation: 'I believe that de Man's critical work adheres to and reproduces, in literary masquerade, his experiences as a collaborator.'[15] In other words, de Man's role in the invention of deconstruction is nothing more than an elaborate subterfuge to whitewash a duplicitous past. Other critics have ransacked de Man's private life for evidence. For instance, Georges Goriely asserted that de Man 'was "completely, almost pathologically dishonest," a crook who bankrupted his family. "Swindling, forging, lying were, at least at the time, second nature to him."'[16] A more temperate and ultimately more persuasive critique was offered by Jacques Derrida, an Algerian Jew and friend of de Man. Derrida confesses that when he was shown de Man's wartime journalism: 'My feelings were first of all that of a wound, a stupor, and a sadness that I want neither to dissimulate nor exhibit' (1988: 600). He then proceeds with an analysis of the texts, making a strong argument for the undecidability of de Man's position. Even when de Man discussed the solution of a Jewish literary colony isolated from the rest of Europe, Derrida asserts that this 'could not be associated with what we now know to have been the project of the "final solution"' (1988: 632). He also points out that de Man had publicly explained his behavior in a 1955 letter, written shortly after his arrival in the USA; that fame came late in life to de Man; and that further announcements about a distant past would have been 'a pretentious, ridiculous and infinitely complicated gesture' (Derrida, 1988: 638). Derrida insists that we view the totality of de Man's work before condemning him on the basis of his wartime journalism.

Cultures of criticism can be transitory and volatile, but they are always highly time- and place-specific (one obvious example is *fin-de-siècle* Vienna).[17] The divergent receptions that have greeted the revelations about Heidegger and de Man provide insight into contemporary influences on our collective critical psyche. These influences include:

1 the timing of the revelations (Heidegger's Nazi affiliation was well-known for many years; de Man's anti-Semitism is newly revealed);
2 the magnitude of the crime (Heidegger's unrepentant Nazism is extensively documented and quite unequivocal; de Man's offense may be reducible to a single long-lost essay);
3 the nature of the offense (Nazism is a twentieth-century phenomenon and almost universally reviled; anti-Semitism has its roots in previous millennia and still has many adherents);
4 the perpetrator's stature (Heidegger is too important and central a figure to dismiss; de Man is a somewhat lesser figure); and
5 the critic's agenda (what is to be gained from this attack? why is it being launched now?).

The fluidity of the cultural climate following the Second World War explains the rapid fluctuation in Heidegger's fortune, and the proliferation of disguises surrounding his works. In 1949, he was punished for his Nazism; yet by 1951 the climate had changed sufficiently to allow for a limited rehabilitation. Since then (and especially since his death in 1976) there has been an almost continuous reappraisal of his contribution, albeit in an atmosphere of heightened skepticism and critical intentionality. Paul de Man's journalism was written in a war-torn Europe in which German hegemony seemed inevitable (see Kaplan, 1993). Some have blamed this context for his youthful anti-Semitic lapse. According to Miller, de Man 'stupidly wrote the deplorable essay in order to please his employers and keep his job' (1988: 685). Following his arrival in the United States, de Man's academic reputation developed slowly, to be secured only by the rise of deconstructionism during the past decade. When his wartime journalism received widespread publicity in 1988, a chaotic flood of condemnation was released. The intent of many critics was to establish de Man as a duplicitous person who, by extension, had devised a duplicitous theory. *In extremis*, critics attacked de Man's deconstructionism as an elaborate invention to blur the meaning of his political writings (although less taxing subterfuges are surely available to an intelligent man intent on concealing his past).

How do these events reflect on our present culture of criticism? It is clear that different standards are being applied to the legacies of the two men. Heidegger, the unrepentant Nazi and major twentieth-century philosopher, is being handled for the most part with care and respect. De Man, part of our urgent present even though he died only seven years after Heidegger, is being vilified for a lesser offense. Why this double standard? Why this unseemly rush to condemn de Man? I have already mentioned that status, timing, and the nature of the offense influence critical reactions at different times and in different places. But the full answer lies deep in our critical culture. Some have argued that the crescendo of distortions surrounding de Man are not solely or even principally directed against him. According to Miller,

> it is fear of this power in 'deconstruction' and in contemporary critical theory as a whole, in all its diversity, that accounts better than any other explanation for the unreasoning hostility, the abandoning of the canons of journalistic and academic responsibility, in the recent attacks on de Man, on 'deconstruction' and on theory in general.
>
> (1988: 685)

In a similar vein, Derrida angrily criticized those who use the revelations about de Man to condemn deconstructionism, accusing them of using the same 'exterminating gesture' they claim that de Man made during the war (1988: 651).

If these interpretations are correct, there is something poisonous in the present culture of criticism. Postmodernism in general, and deconstruction in particular, have introduced unsettling levels of ambiguity and uncertainty into our discourses. Their advocates have emphasized that we cannot move forward with the same imperious certainty that we previously espoused. Instead, we must live with interpretive uncertainty, with a world we will never fully understand. Critics of de Man, postmodernism and deconstruction appear to find such ambiguity intolerable. Their counterattack has spilled over into the personal life and politics behind the philosophy, discrediting the person in order to bury the thought.

The Nazi regime was evil. Heidegger's Nazism was wrong; so was de Man's (putative) anti-Semitism. But such moral lapses do not nullify the philosophy and criticism of either. As Pierre Bourdieu argued, Heidegger's thought cannot be reduced to an ideological product of his sociopolitical circumstances; but neither can it be treated as having no relation to its historical context (see Zimmerman, 1988: 1116). The same caveat should apply to Paul de Man. *The proper way to read these (and other) texts is neither in isolation from the author's politics nor as reducible to them.* The ramifications of this conclusion are enormous. Hans Sluga closes his meditation on the Heidegger controversy by recommending that we 'rethink the whole question of philosophy's relation to politics' (1993: ix; see also chapter 10). The specific place-time circumstances within which a work is written and subsequently interpreted can no longer be regarded as innocent.

IDENTITY

> I believe that ... a writer is not simply doing his works in his books, but that his major work is, in the end, himself in the process of writing his books. The private life of an individual, his sexual preference, and his work are interrelated, not because his work translates his sexual life, but because the work includes the whole life as well.
>
> (Michel Foucault; quoted in Miller, 1993: 19)

Everyone leads a life replete with contradiction, including the Gucci Marxist and the materialist evangelist. Such contradictions will come as no surprise to anyone with even the slightest experience of human nature. The significant question is what are we to do with our knowledge of difference and inconsistency? We can usually recognize when the limits of our personal tolerance have been reached; but what about the myriad ambiguous cases? How do we strike a balance between admiration and antipathy, between skepticism and persecution? These questions are especially pertinent when individuals deliberately reconstruct their legacies, as in the examples of the disguises of Albert Speer and Philip Johnson.

Albert Speer was an architect who fell in with Hitler's inner circle during the 1930s.[18] Rapidly establishing himself as one of Hitler's favorites, Speer garnered increasingly prominent commissions including the plan for rebuilding Berlin as the world capital at the end of the Second World War. After the death of the German armaments minister in 1942, Hitler appointed Speer to that position, giving him authority over an industrial army of 14 million workers, many of them slave laborers and concentration camp inmates. The war crimes tribunal at Nuremberg found Speer guilty of exploiting slave labor, but innocent of complicity in genocide; he was sentenced to twenty years in Spandau prison. Speer himself denied all knowledge of the particularities of the Nazi genocide, claiming that pressure of work and Hitler's personal charisma prevented him from inquiring after potentially discomfiting details.

In the conclusion of his study on *The Last Days of Hitler*, Hugh Trevor-Roper identified Speer as 'the real criminal of Nazi Germany' – not because he was a murderer, anti-Semite, or fanatic, but because he was a sophisticated, intelligent man who personified the educated Germans' disdain for politics.[19] This same aura, however, carried Speer through Nuremberg with his life intact; Bradley F. Smith noted that his Western judges looked with sympathy on such a 'clean-cut and apparently repentant professional man with strong anti-Soviet tendencies' (Smith, 1981: 248; quoted in Craig, 1995: 9). For much of the remainder of his life, Speer devoted himself to self-examination through friendships with *inter alia* a Protestant chaplain, a Catholic monk, and a Jewish rabbi.

In a penetrating account of Speer's life, Gitta Sereny (1995) focuses not on what Speer did but on what he knew and why he acted the way he did. She accepts Speer's own account of his early infatuation with Hitler and Nazism. Referring to his first lunch with Hitler, Speer says to Sereny:

> Can you conceive of what I felt? . . . Here I was, twenty-eight years old, totally insignificant in my own eyes, sitting next to him at lunch . . . as virtually his sole conversation partner. I was dizzy with excitement.
>
> (1995: 103)

Speer was not without selfish professional and political ambitions, however self-effacing he appears in this portrayal. Reflecting on his life after release from Spandau, he frankly confesses:

> During the twenty years I spent in Spandau prison I often asked myself what I would have done if I had recognized Hitler's real face and the true nature of the regime he had established. The answer was banal and dispiriting: My position as Hitler's architect had soon become indispensable to me. Not yet thirty, I saw before me the most exciting prospects an architect can dream of.
>
> (Speer, 1970: 32)

Sereny also stresses the power of the personal bond between Hitler and Speer, which kept him in thrall long after he recognized Hitler's criminality.

The almost palpable ambiguities in Speer's life are anticipated in the epigraph to Sereny's book, where she quotes W. A. Visser't Hooft (leader of the World Council of Churches, 1948–66):

> People could find no place in their consciousness for such . . . unimaginable horror . . . they did not have the imagination, together with the courage, to face it. It is possible to live in a twilight between knowing and not knowing.
>
> (Sereny, 1995: vii)

Historian Claudia Koonz suggests that truth for Speer was an 'elaborate intellectual game,' and that Sereny's dissection reveals a 'slick opportunist who brilliantly served Hitler, outwitted the justices at Nuremberg and found power in postwar West Germany' (Koonz, 1995: 12). These are Koonz's words. Sereny herself concedes that Speer lied on occasion but that 'truth' is evasive, each turn of the kaleidoscope producing different variations on the pattern of truth.

In 1977, after decades of dissembling and subterfuge, Speer ultimately conceded:

> to this day I still consider my main guilt to be my tacit acceptance (*Billigung*) of the persecution and murder of millions of Jews.
>
> (1995: 707)[20]

Speer died from a stroke on December 14, 1981.

The 1994 publication of Franz Schulze's biography of architect Philip Johnson reveals a great deal about the human penchant for disguise. Schulze himself underplays Johnson's 1930 flirtations with Nazism and his attempts to form a right-wing political party (lacking a definite political program, 'their only firm decision was that members of their party would wear grey shirts').[21] Paul Goldberger is more critical of what he calls the architect's 'ghastly political escapades' which reveal a man of stunning intelligence but equally stunning amorality, 'not evil but lacking any clear center' (1994: 14). Johnson himself determinedly separated aesthetics from morality. When he attended a 1933 Nazi rally he responded less to the political content and more to Hitler's charisma and 'all those boys in black leather' (Ballantyne, 1995).

Hitler's armies entered Poland on September 1, 1939. Soon after, the German Propaganda Ministry invited Johnson to follow the Wermacht to the front. Later he reported:

> The German green uniforms made the place look gay and happy. There were not many Jews to be seen. We saw Warsaw burn and Modlin being bombed. It was a stirring spectacle.
>
> (in Schulze, 1994: 139)

About this same time, five articles by Johnson appeared in the magazine *Social Justice*. According to Schulze:

> The first, published July 24, attacked Britain and 'aliens' for turning France into 'an English colony': 'Lack of leadership and direction in the [French] state has let one group get control who always gain power in a nation's time of weakness – the Jews.'
>
> (in Schulze, 1994: 138)

Schulze tries to account for the lack of critical attentiveness to the young Johnson's dubious politics. He notes Johnson's unassailable position in the architectural firmament; that (unlike Heidegger and de Man) he is still very much alive; and that his later activities in causes identified with Judaism may be interpreted as effort toward atonement (see Schulze, 1994). Johnson also enlisted the help of powerful defenders. Abby Aldrich Rockefeller, one of the New York Museum of Modern Art's founders, is reported to have defended Johnson thus: 'every young man should be allowed to make one large mistake' (in Schulze, 1994: 143). And biographer Schulze himself somewhat disingenuously concluded:

> In any case, to the extent that his actions can be made out, they were decidedly unheroic, meriting little more substantial attention than they have gained.
>
> (in Schulze, 1994: 139)

Apart from his politics (disreputable), his status as an architect (derivative, opportunistic), and his role as iconoclast (unmatched in twentieth-century architecture), reviewers of Schulze's biography have dwelled much on Johnson's sexuality. Critic Allan Temko argues that Johnson's 'unabashed homosexuality' complicates everything we know about him.[22] He was deeply excited by the rallies that Speer staged for Hitler, and Temko describes Schulze's 'major discovery' that Johnson felt a 'sexual thrill' watching the destruction of Polish villages (Temko, 1994: 1 passim). Martin Filler has written critically of the amount of prurient attention devoted to Johnson's sexuality which (he claims) has distracted attention from the virulence of his political activities, writings, and correspondence. He expresses disbelief in the face of Johnson's ability to avoid accountability, even though his past has caught up with him (Filler, 1994: 46 *passim*). Johnson captured some facets of his kaleidoscopic personality in a 1953 Smith College talk:

> I studied philosophy as an undergraduate, instead of architecture. Perhaps that is why I have none now. I do not believe there is a consistent rationale or reason why one does things . . . I am too far gone in my relativistic approach to the world really to care very much about labels. I have no faith whatever in anything.
>
> (in Schulze, 1994: 271)

Schulze politely sanitizes Johnson's credo thus:

> Whatever the irreducible core of Philip's personality, it lay beneath multiple layers of motivations manifest in an almost unnatural facility at the intermingling of activities and interests, not all of them discernibly consonant with one another.
>
> (in Schulze, 1994: 105)

As he grew older, Johnson's studied ambivalence allowed him to champion successive emergent architectural fashions, including postmodernism and deconstructivism. He attracted increasing controversy, not least because he recognized it as one way of attracting attention (Schulze, 1994: 273). His fundamental amorality is perhaps best revealed after his 1980 commission (with John Burgee) to build the Crystal Cathedral in Garden Grove, Los Angeles County. The Cathedral is the center of widely televised evangelical Christian ceremonies. In 1990, invited to share Robert Schuller's pulpit at the dedication of the church's new bell tower, Johnson was asked:

> 'Tell us, Philip, what went through your mind when you designed this

> beautiful building?' ... Philip responded by drawing close to the preacher and speaking in a subdued, almost inaudible voice. 'I thought I knew history. I thought I knew what the Gothic spires of old stood for. The Romantic period and the thirteenth-century periods were the highest periods, in spiritual Christianity. I thought I knew how to combine these things to create a great tower. I was wrong. I could not have done this – I have to say it humbly, and I don't ever feel humbly, but I do this morning – I got help, my friends. I think you all [voice breaks] know where that help came from.' Schuller beamed in transcendent acknowledgement. 'Philip Johnson we love you!' he cried.
>
> Later, when asked by an interviewer who knew him well enough, 'Philip, how *could* you?', Philip briefly buried his head in his hands in mock shame, then grinned and replied, 'Wasn't that *awful*!'
>
> (in Schulze, 1994: 341–2)

Andrew Ballantyne concludes that Johnson was prepared to go further than most to obtain a commission; that he was more cynically and effectively manipulative than other architects. He quotes Johnson: 'I wanted to do the job. I got very religious.' According to Ballantyne, it was as simple as that (1995: 7).

MEMORY

> I used to think I would be rewarded for good behavior. Therefore if I wasn't understood, I must not be understandable; if I wasn't successful, I must try harder; if something was wrong, it was my fault. More and more now I see that context is all. When someone judges me, anyone or anything, I ask: *Compared to what*?
>
> (Gloria Steinem, 1994: 282–3, emphasis in original)

Memories are recollected for a myriad reasons: we recall only good times we spent with lost loved ones, turning our thoughts away from painful images of their death or departure; or we reconstruct meaning in our lives by constantly reordering past experiences into a narrative of coherence; and we seek to atone for past wrongs, struggling for redemption and reconciliation. The problems of remembrance and forgetting – of disguise – are rarely more acute than in the case of institutionalized *commemoration*.

In 1995, a dispute emerged over the portrayal of Jewish victims of the Holocaust, at Yad Vashem, the Jerusalem-based museum and memorial (Haverman, 1995: A-6). Alongside stark photographs of carnage and mayhem is one image of unidentified Jewish women about to die; the women are stripped of their clothes and have arms folded across their breasts presumably to hide part of their nakedness and/or to keep warm. A group of Orthodox Jews in Jerusalem has objected to the display of dead or near-dead people without their clothing, on the basis that the victims are being degraded yet again by their nakedness. Defenders of the photographs have objected to the sanitization of Nazi cruelty that would be implied by the removal of the images. If their demands are not met, the Orthodox complainants have threatened to establish their own museum to memorialize the victims 'properly and objectively,' in the words of one leader, Rabbi Moshe Zeev Feldman. Efraim Zuroff, Israeli director of the Simon Wiesenthal Center responded: 'You would think that on this one thing there would be a willingness to stand silently together. . . . Do you think the Nazis distinguished between Jews when they killed them?' (Haverman, 1995: A-6). More than half a century later, in the new state of Israel, Jews are in conflict over whose version of the Holocaust should be preserved.

Distaste and resentment also simmer under the surface of negotiations to commemorate the Holocaust in Berlin (Kramer, 1995). In 1987, talk-show host Lea Rosh announced that she would ensure that Berlin would have a memorial to the six million Jews who died. She was granted permission to erect the memorial on a prominent site by Chancellor Helmut Kohl. As Jane Kramer underscored, Rosh is not a Jew (Kramer, 1995: 50). She changed her name from 'Edith,' and then took up the cause of commemoration. Berlin historian Reinhard Rürup called the proposed site 'inappropriate' as a memorial to a single victimized group, the Jews, when in fact many groups suffered under the Nazis, including gypsies, homosexuals, and mentally retarded people (Kramer, 1995: 51). Rosh responded that she would see to the erection of memorials for other victims, once the Jewish memorial was completed. This proposal did not appeal to Rürup. He predicted that when other persecuted groups stepped forward to press their cases, all memory would be lost to the 'banality of conflicting claims' (Kramer, 1995: 52). The question of the Berlin memorial has since stalled. Some Jewish critics have entered the debate, objecting to any Holocaust memorial in the place-time of contemporary Germany, on the grounds that it would imply a German mediation of Jewish suffering and identity.[23]

The question of memory extends even to national identity. The complicity of ordinary Germans in Nazi atrocities is a highly contested topic. Most commentators have denied the existence of a collective complicity with the Holocaust. In contrast, Daniel Goldhagen dismisses the ideas that ordinary Germans disapproved of the Holocaust and its methods, or that they participated in genocide only under duress. Instead, Goldhagen (1996) insists, the vast majority of Germans shared Hitler's anti-Semitism and willingly collaborated in its brutal implementation.

Hitler, it is generally conceded, never disguised his intention of ridding Germany of the Jews. Yet his murderous vendetta had to await the beginning of the Russian campaign when circumstances and geography (specifically the annexation of Poland) contrived to facilitate his program (Craig, 1996: 6). Central to this effort was 'the camp,' a generic term that included the concentration camps, extermination camps, detention facilities, work camps, transit camps, and ghettos – together described by Goldhagen as Nazi Germany's largest institutional creation. He shows that the killings, usually assumed to take place out of sight in gas ovens run by the SS, actually involved ordinary German people often in full view of other ordinary German people. Goldhagen focuses on three specific instruments of death: police battalions, work camps, and death marches. The police battalions (*Ordnungspolizei*) were recruited haphazardly, minimally trained, and then sent out to round up Jews for transport to camps, or to shoot them. Members of such units were not compelled to take part in these activities. In work camps, Jews were forced into unproductive tasks with inadequate food and rest, and constantly abused by brutal guards until they died. (At Majdanek work camp, the mortality rates were not significantly lower than at extermination camps such as Auschwitz.) Forced marches involved food deprivation, beatings, and shootings. They continued even during the last phases of the war, demonstrating (for Goldhagen) the intensity of Germany's commitment to genocide.[24]

The history of the extermination camps betrays an equally broadly based enthusiasm and commitment to their grisly goal.[25] Initial gassings of prisoners were affected at the front with crudely re-rigged vehicles; later, stationary gas vans were delivered custom-built to occupied territories. And finally, gas chambers capable of mass extermination were built, and refined into evermore capable killing factories (see Kogon, Langbein and Rucherl, 1993). One ambitious young physician, Dr Freidrich Mennecke, wrote enthusiastically to his wife about the latest technology at Dachau: 'There are only two thousand [prisoners], who will be quickly done, as they can be examined only in assembly-line fashion' (in Kogon, Langbein and Rucherl, 1993: 41).

THE POSTMODERN IMPERATIVE

> The arrival at a theory of identity is also an arrival at a certain theory of space, apparatus, body and structure.
>
> (Catherine Ingraham)

Where do identity and authenticity lie in the mutable folds of place-time? If they exist they are nowhere at rest. They float like fragments of a continent on a molten sea – always in motion, sometimes disappearing under one another like tectonic plates, only to return to the surface in a reconstituted form. The geological metaphor is apt. It reminds us that very little is new: we are merely sifting through the metamorphosed foundations of previous eras.

It is an elemental folly to erect artificial barriers between knowledge and the personal politics of place-time. Indeed, Norris emphasizes that writers who include autobiographical details in their 'scientific' writings may assist in overcoming this false opposition and 'inaugurate a reading attentive to the various points of exchange, of inter-textual crossing and confusion, between life and work' (Norris, 1986: 213). Person, text, and politics cannot be separated; a *contextual* reading of Heidegger (or anyone else for that matter) is unavoidable. Unfortunately, we seem neither well-equipped nor even well-disposed to undertake the subtle, informed readings that contextual analysis now requires. But as a beginning, I believe that there is an imperative favoring the polyvocality of postmodernity. Static place-time norms are the consequence of domination and subordination,[26] those invisible glues that a critical social science seeks to dissolve.[27] Without communication across ideologies, places, and times, all is silence.

ACKNOWLEDGEMENTS

I am very grateful to Michael Brown, Dennis Crow, Michael Keith, Steve Pile, and Jennifer Wolch for thoughtful comments on an earlier version of this essay.

NOTES

1 Friedrich Nietzsche, *Beyond Good and Evil*, section 6; in *Basic Writings of Nietzsche*, ed. Walter Kaufmann, New York, The Modern Library, page 203.

2 Ludwig Wittgenstein, *Culture and Value*, translated by Peter Winch, Chicago, University of Chicago Press, 1980: 33e.

3 See, for example, Nick Blomley's appealing essay, 'Activism and the academy' (1994: 383–5).

4 There is an important and ambivalent tension between those who choose to occupy the margins and those who feel excluded from the mainstream. On marginalism, see the collection of essays by Russell Ferguson, Martha Gever, Trinh T. Minh-ha, and Cornel West (eds), *Out There: Marginalization and Contemporary Culture*, Cambridge, MIT Press, 1990. The essay by Gayarti Chakravorty Spivak, 'Explanation and culture: marginalia' (pages 377–93), is especially relevant to my text.

5 Theodor W. Adorno, *The Jargon of Authenticity*, Evanston, Northwestern University Press, 1973. Trent Schroyer has written a perceptive foreword to this translation by Knut Tarnowski and Frederic Will.

6 Derrida, 1984. Also see Christopher Norris, *Derrida*, London, Fontana, 1986, for a critical summary of Derrida's argument (especially chapter 8).

7 Friedrich Wilhelm Nietzsche was born in 1844. His father died when he was five, and the young Nietzsche was forced into the company of grandmother, mother, sister and aunts, a situation he apparently did not much care for. He turned to philosophy after studying theology and classical philology, but was never content at the University of Basel, where he had taken up a professorship. He left the university after a few years and devoted himself to a life of writing, isolation and sexual asceticism. Friedrich's sister Elizabeth, a dominant personality, traveled in 1886 with her husband, Bernhard Förster, to Paraguay in order to establish an Aryan colony called Nueva Germania. Upon her return to Germany, the dreadful Elizabeth took control of her brother's work and reputation: it was she who largely rewrote Nietzsche's posthumous masterwork, *The Will to Power*, investing it with her own proto-Nazi views; it was she who invented and promulgated the cult of his philosophy, inserting it into the new order emerging out of war-ravaged Europe. She died just before the outbreak of the Second World War and was given a full Nazi funeral, which was attended by Adolph Hitler. No other woman, with the possible exception of Cosima Wagner, was so celebrated in the cultural world of pre-war Nazism. And no-one did more to secure (her vision of) her brother's reputation.

8 See especially chapters 1, 2, and 18.

9 Quoted in Sheehan, 1988: 39; the emphasis was Heidegger's own.

10 As well as in the Farias volume cited earlier, the basic case against Heidegger has been assembled by Hugo Ott, *Martin Heidegger: a Political Life*, translated by Alan Blunden; New York, HarperCollins, 1993. Many of the original texts by Heidegger in support of Nazism are reprinted in Wolin, 1993b.

11 See Heidegger's final interview, published posthumously in *Der Speigel* in May 31, 1976; translated as 'Only a God can save us' by Maria Alter and John Caputo, in *Philosophy Today*, 20(4/4): 267–85. The translation is reprinted in Wolin, 1993b.

12 The basic facts in the case of Paul de Man are contained in David Lehman, *Signs of the Times: Deconstruction and the Fall of Paul de Man*, New York, Poseidon Press, 1991 (although the reader is warned that this account is basically hostile toward de Man and deconstructionism).

13 Paul de Man, *Wartime Journalism 1940–1942*, edited by Werner Hamacher, Neil Hertz, and Tom Keenan, Lincoln, University of Nebraska Press, 1989. De Man's words were published seventeen months before the first deportations of Belgian Jews to the death camps; see J. Hillis Miller, 'NB', *Times Literary Supplement* June 17–23, 1988: 685. While it is unclear whether de Man intended this opinion to imply a literal or a figurative ghettoization of the Jews, it is hardly possible to read his statement as an invitation to genocide.

14 Derrida draws this conclusion after reviewing de Man's writings; see Jacques Derrida, 'Like the sound of the sea deep within a shell: Paul de Man's war', *Critical Inquiry*, 14, 1988: 590–652. He concludes: 'in the sum of the total articles from that period that I have been able to read, I have found no remarks analogous or identical to this one' (page 631).

15 Stanley Corngold, Letter to the Editor, *Times Literary Supplement*, August 26–September 1, 1988: 931.

16 Quoted in James Atlas, 'The case of Paul de Man', *The New York Times Magazine*, August 28, 1988: 37.

17 See, for instance, Carl E. Schorske, *Fin-de-Siècle Vienna: Politics and Culture*, New York, Knopf, 1980; and Allan Janik and Stephen Toulmin, *Wittgenstein's Vienna*, New York, Touchstone, 1973.

18 Speer's life is recounted in a remarkable autobiographical account, *Inside the Third Reich: Memoirs*, New York: MacMillan, 1970.

19 Hugh R. Trevor-Roper, *The Last Days of Hitler*, New York: Collier, 1962. This is how Trevor-Roper accuses Speer:

> Nevertheless, in a political sense, Speer is the real criminal of Nazi Germany; for he, more than any other, represented that fatal philosophy which has made havoc of Germany and nearly shipwrecked the world. For ten years he sat at the very centre of political power; his keen intelligence diagnosed the nature and observed the mutations of Nazi government and policy; he saw and despised the personalities around him; he heard their outrageous orders and understood their fantastic ambitions; but he did nothing. Supposing politics to be irrelevant, he turned aside, and built roads and bridges and factories, while the logical consequences of government by madmen emerged (page 302).

20 Professor Richard J. Evans concludes that 'Albert Speer lived his lie to the last.' See Richard J. Evans, 'The deceptions of Albert Speer', *The Times Literary Supplement*, September 29, 1995: 6.

21 See Andrew Ballantyne (1995: 7).

22 Homosexuality was also a factor in the intellectual agenda, career, and manner of Michel Foucault's dying. Critic James Miller sees Foucault as an essentially private moralist who begins and ends his career by attempting to orient himself in relation to society and his own desires. His early life was dominated by coming to terms with his homosexuality, but the mature Foucault focused on social boundaries and their transgression. Powered by the Nietzschean injunction to become what one is, Foucault sought potentially transformative 'limit-experiences', and under the influence of Georges Bataille, the limits of erotic transgression, especially sado-masochistic practices. According to Miller, Foucault's conscious efforts at scripting the self allow us to directly observe the relationship between the man and his work: '*all* of Foucault's books, from the first to the last, comprise a kind of involuntary memoir, an implicit confession.' Indeed, Foucault himself invited such investigations. Not surprisingly, the somewhat prurient and sensational overtones in Miller's study have been denounced by Foucault's supporters as homophobic gossip. Others, less sympathetic to the man and his work, have used the revelations to mount what has been called a campaign of 'intellectual liquidation' – claiming, in essence, that because Foucault's acclaim was founded on a lie, his work could now legitimately be discredited (see Miller, 1993). Aspects of this reaction are discussed by Alan Ryan (1993: 14). The phrase 'intellectual liquidation' is used by Colin Gordon (1993: 27).

23 In Atlanta, Georgia, there is an analogous dispute between the family of Dr Martin Luther King Jr and the US National Park Service over how to commemorate the civil rights' leader's memory (see Smothers, 1955: 1–7).

24 Clive James has written that Goldhagen overstates his case; James emphasizes the coercive powers of the Nazi state, Hitler's obsession with Jews, and the multiple acts of local resistance by Germans as evidence of the absence of a monolithic anti-Semitism in Germany during the Second World War. See Clive James, 1996: 44–50.
25 See Feig, 1981; Fleming, 1984.
26 See Miller, 1993: 35.
27 Olsson, 1991: 54.

12

LOCAL CULTURES AND URBAN PROTESTS

Shlomo Hasson

The rise of urban protest movements in disadvantaged neighborhoods and immigrant housing estates has attracted the attention of geographers, sociologists and political scientists. These movements, as several researchers suggest, have been closely associated with the creation of new social spaces in which ethnic groups and underprivileged social classes challenge the dominant political, economic and cultural system, assert their identity and seek to advance their civil and social rights (Keith and Pile, 1993b; Hasson and Ley, 1994). This chapter seeks to explore the process whereby the movement's actors form their identity and shape the nature of protest. It is argued that the actors' schemes of interpretation play a crucial role in affecting the nature of urban protest, and that everyday life experience gained in specific places shapes the actors' schemes of interpretation. Based on the study of eight urban social movements that emerged in Jerusalem's immigrant housing estates during the 1970s and 1980s, I will show that in spite of the similar socio-economic conditions shared by the participants in the movements, their protest tended to diverge in different directions. To account for this variation I will examine the schemes of interpretation developed by members of these organizations and the specific local cultures underpinning these interpretations.

URBAN SOCIAL MOVEMENTS: THEORETICAL DISCUSSION

An urban social movement is a concerted attempt of urban groups to further or secure a common interest either in the ecological, cultural or political sphere through collective action outside the sphere of established institutions. In accounting for urban protest reference has been often made by scholars adhered to the collective behavior approach to structural processes and problems, such as large-scale immigration, or modernization processes, or social mobility.[1] According to these theoreticians, the structural processes may lead to the destabilization of social frameworks and of the rules and norms associated with them, thus causing social unrest and protest.

The Marxists, and especially the structural-Marxists, tended to see the social movements as the result of structural forces, but these forces were interpreted in the Marxist codex as a structural contradiction between the forces of production and the relations of production. The structural-Marxist approach contended that beside the major conflict between the forces and relations of production in the work place, there emerged a secondary arena of social exploitation within the built environment itself. The capitalists' motive of profit accumulation, which results in low level of physical development and poor housing conditions and social services, contradicts the citizens' will for a better and safer environment. The unavoidable result has been the rise of a series of urban movements that conceive the environment not as an asset to be traded in the market, but rather as a place wherein bio-physical and cultural reproduction is to be secured (Castells, 1977).

Despite the differences in theory and ideological–political outlook, Castells' earlier structural explanation was very similar to those presented by the theory of collective behavior developed in the 1960s and 1970s. In both cases there were deep structures that, according to the writers, transcended the human agent and affected the movement's action; hence the primary importance accorded to the analysis of these structures. The search for deep structures continues in Castells's book *The City and the Grassroots*, although this time in a much more subtle form (Castells, 1983). In this book, Castells seeks to theorize the rise of urban social movements by relating them to deeper societal structures: the capitalist mode of production (form of surplus appropriation in capitalist societies), the industrial mode of development (which concerns the increasing productivity of labor and capital through technological developments in the spheres of production, management, and control), the information mode of development (which concerns the increasing technological knowledge and its concentration in a small technocratic elite), and state power (which implies control over the means of violence). Each of these constitutive structures, Castells remarks, produces new dominant social oppressors: capitalists, managers, technocrats, and bureaucrats who are to be challenged and confronted by the urban social movements (Castells, 1983: 319–321).

The Marxist account for the rise of urban social movements has been sharply criticized by Weberian theorists. Relying on the works of neo-Weberians such as Rex and Moore (Rex and Moore, 1967; Rex, 1968), Pickvance (1976, 1977) and Saunders (1980) pointed to internal divisions within the working class that prevented an overall class mobilization, and offered a narrower basis for social organization. These lines of division were defined not in accordance with the social position in the production process, but according to the position in the goods market. As the neo-Weberian school claims, the existing distinctions within the working class as far as ownership levels of goods (housing or other durable goods) are concerned, and differences in life opportunities and social interests stemming from these distinctions, are the concrete causes that determine the development of urban social movements. In other words, the urban social movements do not express a struggle between social classes in the Marxist sense, but rather between different sectors of consumers that have different life-opportunities, and accordingly also different socio-political interests.

It has often been noted, however, that structural conditions and even specific experiences, do not necessarily breed urban resistance. They merely denote the fact that the probability of political action is higher in one place or at a certain time than in another. The more recent cultural–political literature, especially those studies associated with the resource mobilization approach, suggests that to realize the potentialities embedded in the societal structures, the social actors have to assume an active role in interpreting the context of which they form a part. Values, norms and beliefs thus come to play a significant role in informing and shaping the nature of urban resistance. Hall (1982) showed how social movements are engaged in the 'politics of signification'. Snow and Benford (1992) conceptualized this signifying work with the verb 'framing'. As they clarify the concept frame 'refers to interpretative schemata that simplifies and condenses the "world out there" by selectively punctuating and encoding objects, situations, events, experiences, and sequences of actions within one's present or past environment' (Snow and Benford, 1992: 137).

The interpretative scheme, or frame in Snow and Benford's (1992) terminology, plays a fourfold role. First, it fosters a sense of identity by underscoring the seriousness of injustice inflicted upon a certain group. Second, it functions as an explanatory mechanism by identifying the processes, actors and events responsible for the group's conditions. Third, it sets ideals concerning future goals and aspired outcomes. Finally, it suggests a line of action to be taken in order to get from the present to the future. In so doing the interpretative scheme articulates discrete events, places, actors and experiences into a unified and meaningful whole.

But what are exactly these interpretations, and how are they related to urban social protest? Peter Marris makes a valuable theoretical contribution at this point by clarifying the interrelationships between meaning and action. According to Marris 'the choice of language affects the choice of action' in as much as metaphors may shape the strategies of social action (Marris, 1980: 6). Marris's observation is important in that it relates orientations and strategies of action not only to material conditions and social relationships, but also to the metaphors through which people represent these conditions and relationships to themselves. Specifically, Marris draws a distinction between two types of metaphors: 'the structural' and 'the reproduction of relationships' metaphors, and analyzes how they affect social action. The structural metaphor, as Marris contends, 'was introduced by the Left, and insisted upon, because it represented the interconnections between neighborhood, city, nation and the international economy as an indivisible set of relationships'. Consequently, it is asserted that 'neighborhood problems could only be solved by a national or even international strategy for controlling the process of capital accumulation' (Marris, 1980: 229). By sharp contrast, the reproduction of social relationships metaphor assumes a pragmatic position. Whereas the structural metaphor takes social conformity to the rules of the material context for granted, the reproduction of relationships assumes that social relations and social action are affected by the specific interpretations attached to the material context, and by the choices made by human agents. 'However constrained these choices are', Marris claims, 'we rarely perceive them as the inevitable outcome of our situation. On the contrary, we characteristically feel the need to justify them, to assert their moral and rational superiority ... These rationalizations are too idiosyncratic, subtle and various to be treated merely as the internalization of a dominant ideology' (Marris, 1980: 236–237).

It is not hard to see at this point how the metaphors shape the nature of social action. The structural metaphors, Marris claims, 'by their very nature, make radical changes appear impossible ... If instead we think in terms of the reproduction of relationships rather than their structure, this image of powerlessness becomes transformed' (Marris, 1980: 245). This line of argument has been further developed in the 1990s by Homi Bhabha (1990c) and Trinh (1992) who focus on counter-narratives of self-identification that challenge and resist hegemonic myths and symbols. In so doing, these counter-narratives recreate the nature of place, turning the periphery into a 'site of hybridity', an 'in-between space' where the meanings of cultural and political authority are negotiated and imbued with new meanings.

The theoretical question which still requires an answer is how the interpretative schemes (and by implication the counter-narratives) come into being. The treatment of this question by the interpretative approach has been far from satisfactory. Geertz (1973) and Clifford (1986), for instance, provide penetrative insight into the modes of interpretation of different cultural groups, but stop short of clarifying the specific contingencies within which these interpretations were formed and developed. It is precisely at this point that this study seeks to go beyond interpretation and expose those place specific ingredients that inform interpretation and subsequently affect social action.

URBAN SOCIAL MOVEMENTS AND THE SOCIETAL STRUCTURES IN ISRAEL

The 1959 riots in Wadi Salib, an Arab neighborhood in Haifa populated by Sephardic Jews (Jews who were born in Muslim countries and their Israeli-born descendants) after the War of Independence in 1948, marked the first outbreak of social discontent in Israel. The Ethnic Coalition of North African Immigrants, an organization created during the outbreak of protest, cited ethnic discrimination against North African immigrants as the main reason for the riots (Eisenstadt, 1967: 253). At the beginning of the 1970s the Israeli Black Panthers appeared on the public scene. This movement was also born in a former Arab neighborhood, this time Musrara in Jerusalem, populated after 1948 by new immigrants, mostly from Morocco. In a series of violent demonstrations the movement demanded to eliminate all discriminatory practices against Sephardic Jews and to supply better housing, employment, and education for the Sephardic community (Cohen, 1972: 93–94). In 1973 the nucleus of the Ohalim movement – Ohel Yosef – consolidated in the immigrant housing project of Gonen (Katamon) Tet in Jerusalem.[2] The members of Ohel Yosef at first opposed the activities of the Black Panthers and looked for what they defined as positive activity to improve the life of their community by initiating self-help services from below. In the years 1977–1981 Ohel Yosef members and supporters worked to establish Ohel councils in other Jerusalem neighborhoods: Shmuel Hanavi, Nahla'ot, Baka, and on Stern Street in Kiryat Hayovel. In 1979 the governing councils of the different Ohel organizations merged into a city-wide movement, the Ohalim movement, aimed at advancing the interests of disadvantaged neighborhoods and ethnic groups all over the country. The city-wide movement was responsible for activity on the supra-neighborhood level, while activity within each neighborhood remained in the hands of the individual Ohel councils. The 1980s saw the appearance of several other protest organizations in Jerusalem: the Dai (Enough) movement in Musrara, which made headlines by squatting in a neighborhood basement and establishing a 'settlement' there; Shahak (Improvement of Community Life, henceforth ICL) in Ir Ganim, which protested local housing conditions; Tsalash (Young for the Neighborhood, henceforth YFN), also in Ir Ganim; and the Katamon Higher Committee, which represented residents in Katamon Gimel, Daled, Hey, Rashbag and in the Pat housing project (Hasson, 1993).

Structural conditions

A basic characteristic of all the protest movements in Jerusalem is their common location in the generational, social, and geographic context. The generation is that of people who were born in Israel or who came to the country at a young age, and who at the time of their protest activity were in their twenties or early thirties. Members of the group have often described themselves as 'the second generation' of immigrants. Their location in the social context is that of being members of families who came to Israel from Muslim countries and who, upon arriving in Israel, constituted a well-defined social group at the lower levels of the class structure. Subsequently, the Israeli class structure was characterized by a deep *ethno-class division* between the older Ashkenazic (European-born Jews) group and the Sephardic one. This division, primarily felt during the 1950s and 1960s, and to a lesser extent in the following two decades, was characterized by socio-economic inequality among Sephardic and Ashkenazic Jews and by unequal access to political power. From a geographic perspective, the ethno-class division was translated through state housing policy into spatial closure of the Sephardic newcomers. Spatial closure denotes the geographic segregation of the lower strata of society in separate housing projects through a deliberate housing policy. Such a segregated pattern implied substandard housing conditions, a low level of social services, limited informal social contacts with the absorbing society, and the prominence of social inequality. These factors bred social resentment and facilitated territorial mobilization.

This polarized structure is clearly reflected in the social geography of Jerusalem. At the periphery of the city, adjoining the pre-1967 border areas lies a belt of housing projects and formerly settled Arab neighborhood inhabited by new immigrants. Part of the belt stretches west to east along the city's southern rim; it includes Ir Ganim, the Katamons, and southern Baka. The belt then turns to the northeast and includes Shmuel Hanavi and the formerly Arab neighborhood of Musrara. The proportion of immigrants in these areas who were born in Muslim countries was between 70 and 90 per cent in 1961 and 1972 respectively, while the figure for the older neighborhoods was about 10 per cent (Hasson, 1977: 70).

Every location in the various contexts had a great influence on the sociology of knowledge of the members of the second generation. The generational context determines, according to Karl Mannheim (1965: 291), the range of experiences, and prepares the members of the young generation for a defined way of thinking and historical activity. The Marxist argument emphasizes the influence of the class context on social consciousness and on class interests, while Mannheim's theorization (1936) in this field elaborates on the emergence of a counter-utopia that challenges the ruling ideology. As for the geographic context, Giddens argues that segregation of residential areas has a far-reaching impact in shaping the system of beliefs, attitudes, and ideas. In other words, the geographic closure affects social consciousness, life-style, and values (Giddens, 1980).

Shared experiences

Even though the distinction between the three contextual locations is important for analytic purposes, in reality all three collaborated to produce a common range of experiences, perceptions and social responses. One of the experiences common to all participants in the urban movements was what sometimes has been referred to as physical and social disadvantage, or ethnic discrimination and social deprivation. This experience was the first and primary one upon which other experiences, such as socio-economic mobility, were later founded. As a first childhood experience, it served as a central point of reference for the protest activities. There was a clear attempt, in every protest activity, to counter this first experience, and a somewhat vague attempt to produce an alternative social model. Over and over again members of the movements pointed to their marginal position within Israeli society. Leaflets of the Black Panthers asserted that Israel is 'a state of Black and White people'. Referring to the group early experiences, the leader of the Ohalim movement said: 'All we knew at that time was that we were marginal people that could be arrested and rearrested'.

It is not, however, only the experience that matters, but also, as Mannheim (1965) notes, the temporal order of the experiences. For the second generation, the first experience was a kind of trauma experienced during childhood – at the beginning there was physical and social deprivation – and all the rest came afterwards. The Black Panthers described the institutions of young delinquents where some of them matured and gained their first experiences. Others described in detail the history of broken families, unemployed parents and the dependency on social welfare. 'My father', claimed the leader of the Ohalim, 'was a violent person who terrorized the neighborhood', and reports of local community workers approve this description. Members of the YFN reported on the poor level of schooling in the local, Hassidic school of Habad. In this sense members of the second generation differed from their parents, the immigrants of the first generation, for whom similar experiences came much later in life. Hence the unrelenting effort to negate the first experience, by protesting against the state identified with it, and the attempt to form a new socio-cultural situation and personal identity.

The negating tendencies were expressed on different levels of meaning and activity and reflected the second generation's different needs and conceptions. On the concrete level, the negation took the form of activity aimed at criticizing and modifying the existing material conditions – housing, employment, services – and replacing them with others. It was this concrete negation of physical–material disadvantage that united different population groups around the leadership of the second generation, and that created the local base of support, activists, local resources, and local legitimacy for the activity of the movements. For instance, the demonstrations held by the Panthers in 1971 attracted from several hundred to seven thousand people. The squat campaign at the Pat housing project in 1976, organized by Ohel Yosef, meant to protest against bad housing conditions in Gonen Tet, involved some one hundred young couples, almost none of whom were members of the movement. The involvement of the young couples in such activities reflected their protest against substandard physical conditions, a protest to which the movement gave socio-political expression in the form of squatting.

Beyond the concrete negation of material disadvantage, there was another form of negation, shared by all the movements, that expressed itself at the symbolic level. At its center was a socio-cultural message, according to which there emerged in the immigrant neighborhoods a new generation, a new entity different from the previous generation and from Israeli society's stereotyped image of them. This critique was leveled against what members of the movements perceived as something taken for granted by mainstream Israeli society – the existence of two cultures or societies: a higher culture outside the immigrant neighborhoods and a lower culture within them. In other words, the experience of physical and social disadvantage over time became associated with an experience of social stigmatization. It was precisely this stigmatized identity that the members of the second generation tried to reject and replace, by producing new social symbols or by reviving old ethnic traditions. Characteristic of this tendency is the claim made by one of the Ohalim leaders that 'They never believed that power would emerge from this ghetto.'

THE DIVERGENCE OF PROTEST

The movements' location in the social, generational and geographic contexts created an a priori tendency for urban resistance. This resistance aimed at social liberation in the two senses posited by Berlin (1969): negative – that is, negation of the current economic, political, and cultural conditions – and positive – the creation of alternative models in which residents become involved in shaping their social and physical conditions. What seems to be extremely intriguing about these organizations is the diversified patterns of protest they display despite their common location in the social, generational, and geographic matrix.

Three major types of urban social movements emerged in Jerusalem during the 1970s and 1980s: urban social movements, neighborhood protest movements, and coproductive organizations.

Urban social movements

The urban social movements of the Black Panthers and the Ohalim sought to introduce a radical change in the existing reality of everyday life and the policy that stood at its base. They demanded the government to provide affordable housing solutions to young couples in poor neighborhoods, to provide employment and to reduce the ethnic–social gap between Ashkenazic and Sephardic Jews. They also demanded recognition of the cultural richness of the Sephardic community, and participation in decision-making processes. The social, cultural and political messages were aimed not only at the neighborhoods in which the movements arose, but also at other areas with similar socio-economic conditions. To attain their goals and exert pressure on the state, the movement's participants resorted to violent demonstrations and squatters' campaigns.

The call for social, political and cultural reform of urban–social policy did not imply a rejection of the basic system of social values underlying the existing policy. It was argued that the basic values of democratic and welfare system on which the State of Israel was founded had been repudiated and distorted while the state moved from the normative–declarative to the political–concrete levels. Committed to the central value system, the urban social movements failed to develop any alternative ideology or social myth that would lead people to abandon their previous commitments and attitudes and put them on track of a radical change in the structure of society.

Neighborhood protest

The Neighborhood Protest movements of Ohel Yosef, Dai and ICL (Improvement of Community Life) functioned in narrow residential areas. Their goals were improvement of the housing conditions of local residents, raising the level of local services, and strengthening of local political representation. These movements addressed neighborhood residents and dealt with local issues. The Ohel Yosef council, for instance, was engaged in developing local services in Katamon Tet, in the development of local creative talents, in the education of youth, and in the improvement of local housing conditions. The Dai movement was interested in preventing older Sephardic families of Musrara from leaving the neighborhood, which members of the movement thought was the result of a deliberate gentrification policy undertaken by the state through the Project Renewal Program. Local patriotism was a significant characteristic of the Dai movement, revealed in expressions like: 'The residents of the neighborhood will not allow the neighborhood to be destroyed under the guise of Project Renewal.'

Members of the neighborhood protest movements developed local services from below and occasionally confronted the bureaucracy and the politicians. Although engaged in protest and confrontation, members of these movements did not seek to produce a radical change in the social, political and cultural system. Unlike the urban social movements, urban resistance in this case was a tactical one, aiming to achieve public resources.

Coproductive movements

The coproductive movements of Shmuel Hanavi, the YFN (Youth for the Neighborhood) and the Katamon Higher Committee dealt with concrete local problems in a pragmatic manner. They organized neighborhood clean-up campaigns, initiated cultural activities, organized learning centers for children and youth, and participated in improving the local housing conditions and physical infrastructure.

These movements based their activity on coproduction with the state. Coproduction was manifested in developing working relations with politicians and with senior administrators, and in sending movement's representatives to joint committees with the local and central government. Confrontational strategy was rare, being used only in exceptional cases. In pursuing the coproduction strategy, movement's members relied on personal connections with senior administrators and politicians. The leader of the YFN movement explicitly condemned the confrontational strategy assumed by the urban social movements of the Black Panthers and the Ohalim, and declared: 'We naturally took a solid, drab and quiet approach.'

Summary

The urban protest of the second generation was expressed in three distinct types of urban movements: urban social movements, neighborhood protest movements and coproductive movements. The movements differed in their engagement of territory and place. The urban social movements sought to transcend the neighborhood, and to produce some universal changes in the political, economic and cultural systems. The neighborhood protest movements challenged the system, but functioned only on the neighborhood level, especially in the area of the supply of collective means of consumption. The coproductive movements, although born out of protest, wished to bring about limited change in the environment while acting in concert with the state. Following Castells's early (Marxist) writings, the first type of movement may be called genuine urban social movement; the second type, a reform movement; and the third type, a control movement (Castells, 1977).

THE INTERPRETATIVE SCHEMES

The general question arising at this point is: why do protest movements diverge? This question becomes particularly perplexing in the Israeli case, since all the movements, discussed so far, developed within the same social, generational, and geographic context. I would like to suggest that in order to answer this question, one must look at the fundamentally different interpretations given by members of the second generation and the social organizers accompanying them to the social dilemma in which they find themselves.

Relating to the socio-geographic conditions, members of the second generation took two significantly different interpretative schemes. One located these conditions within a deep, invisible context – that is, gave them a structural interpretation that went beyond the empirical facts – while the second relied on an empirical interpretation that was confined to tangible and concrete objects within the neighborhood.

Structural interpretative scheme

The Black Panthers, the Ohalim movement, Ohel Yosef, and the Dai and ICL movements had a tendency to place their local problems in a structural context out of which the meaning of the problems was derived. For example, a concrete local problem, such as overcrowded housing, was associated with a deeper socio-political context, presented as an outcome of basic social conflicts (the ethnic gap or class conflict), and thus turned into a universal problem. The Black Panthers, for instance, argued:

> You're screwed not because you were born, God forbid, screwed, but because they screw you. Let's assume that you are an Iraqi, Yemenite, or Moroccan-born manual laborer and father of a large family. It's easy to guess, more or less, your past. When you came to the Land of Israel – they threw you in a transit camp. You received exploitation wages, and more serious – the fruits of your labor were eaten by them – the work managers, the factory owners, the bosses. To this day they pride themselves in building the country, paving the roads.... Today they hold senior positions in the country you built. And you – the real worker, the real builder – were screwed in the end, because you didn't come to Israel from Moscow or Leningrad. So why should you get a decent apartment?
>
> (Placard, August 1971)

The Black Panthers and the Ohalim movement grounded their local problems in the context of ethnic and class conflict and tied the two together inseparably. In both movements there was a tendency to emphasize ethnic discrimination as a factor that determined class position, and the class conflict had an ethnic tinge. The interpretations that relate to the deep structure point to its weaknesses and call for a fundamental change in the system of social, cultural, and political relations. Advancing such changes would obviously lead to conflict and confrontation with the state.

Neighborhood protest movements, such as Ohel Yosef, Dai, and ICL, also tended to develop universal interpretations linked to deep structure. At a demonstration against settlements in the West Bank the leader of ICL argued:

> The struggle for peace is primarily the struggle of the exploited groups in the country.... The ICL movement is a movement of disadvantaged Sephardim and Ashkenazim who see the connection between foreign and defense policy and the economic and social situation in the disadvantaged neighborhoods.
>
> (Speech by the chairperson of the ICL movement at the disadvantaged neighborhoods demonstration in the West Bank, January 5, 1983)

The local problems in Katamon Tet, Ir Ganim, and Musrara were thus linked by the leaders with deeper societal problems, such as class conflict and the allocation of resources for settlements in the occupied territories.

Empirical interpretative scheme

Coproductive movements, such as the Katamon Higher Committee and the YFN movements, made no attempt to explain the problems they faced in terms of prejudice or ethnic discrimination, or in terms of class conflict. They made no attempt to ideologize or theorize about local problems by looking to deeper levels. Their interpretation had an empirical character and was limited to visible phenomena.

The empirical interpretations, which focus on surface phenomena, identify problems on the visible physical and geographic level. Solving these problems can be accomplished peacefully through a coproductive action taken within the geographical area in which they were discovered. Hence the basic difference in the essence of the social action undertaken by the various movements.

LOCAL CULTURES AND THE CONSTRUCTION OF THE INTERPRETATIVE SCHEMES

The question arising at this point is how the two distinct interpretative schemes were constructed. The answer to this question is complex and requires a simultaneous and non-reductive examination of a number of factors: the character of the local groups involved, the specific biography of groups and individuals around whom the movements were formed; the cultural interaction between the groups and their environment, whether the neighborhood environment or the wider socio-political environment outside the neighborhood and the learning processes within the organizations.

The character of the local groups involved

The interpretation of the socio-geographic conditions encountered by the movements' participants was influenced by the socio-cultural characteristics of individuals and social groups involved in the organizations. Among the urban social movements of the Black Panthers and the Ohalim the presence of delinquent youth and unemployed was most conspicuous. The group that founded the Black Panthers had originally been a street gang, while the members of the dominant group in the Ohalim movement had in the past belonged to a criminal network. Members of the groups had been unemployed for a long period, and had difficulty adjusting to work and accepting authority.

A completely different social background characterized the groups that founded the coproductive organizations of Ohel Shmuel, YFN and the Katamon Higher Committee. Ohel Shmuel was established in 1978 by a group made up mostly of high school students, some of whom studied in day schools and some in night programs. They brought in a group of adults, from the older generation, and along with them identified local problems and suggested possible solutions. Most members of YFN had high school education and were professionally established. The leaders of the Katamon Higher Committee were all gainfully employed, some of them in senior positions, such as a district sanitation inspector and an assistant television producer.

Members of the neighborhood protest organizations of Ohel Yosef and Dai represented two radically different social extremes. Some of the leaders of Ohel Yosef, for instance, had high school education while others had been members of a criminal gang. Dai was made up in part of remnants of the Black Panthers, while other members, such as the movement's leader, a student at the Bezalel art academy, were socially established and had university degrees. Most members of the ICL leadership came from the relatively well-off groups in the housing project and none of them suffered any basic disadvantage, be it housing, income, or employment. The leader of the group was a laboratory technician at the university, while others were teachers or service workers.

The social background, then, provides a partial explanation of the interpretations and forms of protest that were adopted. This is especially true of the two extremes, the urban social movements on the one hand and the coproductive organizations on the other. The deep interpretations were associated with the underclass groups, while the empirical interpretations and the instrumental protest originated in the relatively upper social groups, apparently reflecting these groups' different attitudes towards the center, i.e.

the institutional system, the prevailing rules, norms, and procedures of society. The exceptions were the neighborhood protest organizations, where even the upper social groups adopted a critical position towards the centre.

Personal biographies

The interpretations and forms of protest adopted by the organizations were partially rooted in the personal biographies of individuals and groups within the organizations. The personal profiles of the leaders of the three organizations identified may illustrate this point.

The leader of the Ohalim Movement (an urban social movement) was born in 1955 and was one of 11 children. His father served in the border police and later became permanently unemployed. His mother worked as a cleaning lady and seamstress, but her salary was not enough to feed the family, which needed welfare support. Home life was tense and violent, and the children united around the mother. According to the local community workers, the instability of the family had an enormous effect on the development of the children, especially of the eldest and future movement leader, who was an active participant in home affairs (Interviews with local community workers). In third grade he began displaying signs of aggression and was transferred to a special class, and later to a special school, which he in practice did not attend. At the age of 11 he was first put on trial and was sent to a special educational institution, where he studied for three years and finished eighth grade. By the age of 14 he had several entries in his police record and was put under the supervision of the juvenile rehabilitation service. After a year in the army he was discharged because of adjustment difficulties. On Bar-Yochai Street in the Katamon Tet neighborhood a group of young men of similar background gathered around him and he became the leader of a local gang. In his words: 'I was the leader of a crime gang. I was a thief and terrorized the neighborhood.' In 1975, a local community worker succeeded in persuading him to join Ohel Yosef, which had up until then been run by a mainstream-oriented group. This was the turning point in both his life and that of Ohel Yosef, since until 1975 he and his gang had not interfered with the work of the local organization. His involvement gave the organization a charismatic leader with a predilection for activity outside the established system.

Coinciding with his activity in the local organization, which reached its climax in 1979 and 1980, when he became the leader of the Ohalim Movement, he continued his criminal career, although less actively; as a result he was tried and imprisoned. In other words, in these years he was simultaneously involved in anti-establishment social activity and criminal activity, the two seeming to have joint roots in his early history. Community workers and psychologists who worked with him suggested that his family background, the problematic home environment, and his father's violence had pushed him towards criminal activity and protest against the establishment, since he saw the latter as responsible for the social problems he had witnessed. Against this background, the community workers and psychologists suggested, he developed a hostile interpretation of the role of the state, an interpretation that was fostered later on by outside agents. Nevertheless, as several community workers emphasized, his activity in the movement led to socialization and to his moving closer to the central value system.

The biography of the leader of YFN, a coproductive organization which is at the opposite extreme from the Ohalim Movement, is entirely different. The leader was born in 1958 in Casablanca, and in 1962 his family moved to Israel and was sent to live in the asbestos transit houses in Ir Ganim. He studied for eight years in the local Hassidic school, completed his studies at the prestigious Himmelfarb High School, and served three years in the army's tank corps. Upon completing his military service he entered the Hebrew University and completed his masters' degree in Jewish philosophy. During his university studies he began taking part in public affairs – he was a member of the student council for three years, ran the university's training program for public activists, and served as a consul-intermediary between the university and the neighborhood. His public activity at the university, his education, and his personal criticism of the local boss system in Ir Ganim motivated him in June 1981 to found YFN. His socio-political criticism focused on the patronage system that characterized the rule of the local boss, and the political–economic support of this system granted by the parties. This criticism lacked, however, the hostility and resentment that characterized the leader of the Ohalim movement. It would seem that his exposure to the social system outside his neighborhood aided him in learning the dominant rules and procedures, contributed to the development of a favorable view of the existing system, and consequently fostered a tendency to function within the existing socio-political framework. This may explain his call for 'solid, drab, and quiet activity' and cooperation with the state.

It appears that the biographies of these two leaders and the interaction that developed with the center served as an important factor in shaping their personal interpretation. The leader of the Ohalim Movement, for instance, demonstrates an interaction that rejects the existing system and its rules, while the leader of the YFN movement demonstrates an interaction that accepts this system. The internalization of the prevailing social rules led the latter to a positive interpretation of the system, and in return the YFN was accorded legitimacy and resources by the state. In contrast, the rejection of the rules by the leader of the Ohalim Movement, a rejection rooted, among other things, in his personal biography, led him to a hostile interpretation of the system and made it difficult for him to receive legitimacy and to mobilize state resources.

The biography of the leader of Dai, a neighborhood protest movement, and his interaction with the center has elements of both previous profiles. The Dai leader was a resident of Musrara and had experienced upward social mobility. During the period of his activity in the movement he was a student at the Bezalel art academy, where he had received a scholarship. This trajectory is somewhat similar to that of the YFN leader. But in the case of Dai's leader it did not lead him to adopt the existing rules. On the contrary, it sharpened his criticism, his indignation, and his resentment of the existing system. Like the leader of the Ohalim movement, his resentment of the system had personal roots. When he returned to Musrara after a long absence he tried to buy an apartment and was not able to do so, because, he claimed, the Project Renewal Program did not let him. It was only after a long struggle that, he claimed, 'they threw up their hands and gave in'. This incident exacerbated his sense that Project Renewal was not designed for the benefit of local residents but for people with connections and resources who could pay the prices being charged. His anti-establishment activity thus had a motif of personal revenge against the bureaucracy he had to endure. But the personal motif in the shaping of his world view sprang from a wider, hostile view of the state, a system he identified with what he termed 'the Ashkenazim' (European-born Jews). The Project Renewal Program, identified by the Dai leader with 'its Ashkenazic director,' served, in his opinion, as a front for this system, and so, in his words, 'it has to be thrown out of here'.

It appears, then, that the difference in interpretation and forms of protest are associated with social experience, personal biography, and nearness to the center. The closer the group is to the center in its conceptions, values and norms the greater the tendency for an empirical interpretation and activity within the established system.

The differences in interpretation reflect, then, the fundamental differences between the groups in the extent to which they accept the central values and their recognition of the center's authority. In Israel there is a strong tendency for consensus, which is associated with the ideals of building the nation and gathering the exiles, and with the external threat. The socially unattached and delinquent groups at the root of the urban social movements, like the Black Panthers and the Ohalim movement, challenge the consensus and existing order and sometimes unconsciously advance an alternative order. Such a challenge may serve under certain conditions as a social magnet for another group – made up of radical intellectuals – which has its own alternative utopia.

Yet, on the other hand, this challenge may distance from the urban social movements those traditional elements within the neighborhood and outside it that would prefer a pragmatic orientation and activity within the established system. In other words, the interpretation and patterns of activity that grow out of social experience and personal biography may trigger a negative or positive response in accordance with the kind of cultural interaction that develops between the organization and its environment.

Cultural interaction with the environment

In general it is possible to distinguish between two types of environment with which the organizations had social interaction, and by which they were influenced in terms of the interpretation and forms of protest they developed. One environment was the internal one, composed of neighborhood elements, which held traditional values, and the other environment was the outside one composed of community workers, street gang counselors, non-professional intellectuals, and politicians who held radical values.

The neighborhood environment, which represented to a large extent a tradition that consolidated partly before the immigration to Israel and partly while coping with conditions in Israel, encouraged the creation of a pragmatic orientation. The culture of this environment may be likened to a kind of heritage from the past that was adopted unconsciously and uncritically. It was a kind of ideological hegemony, to use Gramsci's (1971) terms, that shaped the interpretation of, and attitude towards, society. This cultural–political tradition was characterized by the following: ethnic closure, a traditional–religious spirit, a tendency to support the right wing on the national level, a sense of community, adjustment, conformism, dependence on a local figure who represented the community to the authorities, and avoidance of conflict. The nature of this tradition was opposed to radical protest, which assumed a radical–leftist orientation, and tended to support what it saw as pragmatic activity of an apolitical nature, directed at improving the neighborhood. It was strengthened indirectly, and to some extent unconsciously, by community workers who took an apolitical stance and who aspired to identify concrete local problems and encourage the residents to find solutions to them. On the basis of these traditional values and the contribution of apolitical community workers, a set of expectations emerged that saw the neighborhood organization as an instrument for dealing pragmatically with local problems. In other words, the internal tradition in conjunction with professional–pragmatic and apolitical ideology created a system of expectations and demands, and this system served as a filter through which reality and possibilities for action were reflected. An example of this is a letter sent to the leader of YFN by a local resident, containing between the lines a pragmatic–local conception as regards the role of the organization:

> I write to you with a great request as chairperson of the neighborhood council that you help me find the most efficient possible solution for a loan to complete construction. I turn to you with this request since I am a resident of the neighborhood and have a right to assistance from the head of the neighborhood council.
>
> (Letter of a resident to the chairperson of YFN, December 12, 1984)

The YFN organization strove, in its initial stages, to become an organization that would, according to its founders, serve as a model to be imitated by the entire society. The local traditions and expectations were different and demanded activity within the neighborhood in concrete consumption spheres – housing, transportation, lighting, and services. The response to the environment required giving a different interpretation to the problems, to developing a local orientation and a pragmatic approach in coordination and cooperation with the state.

The traditional cultural framework that was based on patron–client relations, on a sense of community, and on conformism, was broken to a certain extent in the case of the underclass groups due to social associations established with liberal–radical groups who came from the outside. The articulation of the challenge to accepted social norms, presented by the underclass groups, with the political–moral criticism carried by liberal–radical agents, who came from outside the neighborhood, was common to the Black Panthers, the Ohalim Movement, Ohel Yosef, Dai, and ICL, and it had considerable influence on the interpretation and orientation of these organizations. Working among the Black Panthers were people from Matzpen (a Trotskyist faction), Oded (a left-wing organization of Sephardic intellectuals), high school students, leftist university students, community workers with progressive orientation, and politicians from the periphery of the establishment. In the Ohalim Movement and Ohel Yosef there was a similar coalition; the activists here were in particular South and North American students with a Marxist approach, and radical community workers.

It appears that in the pre-organizational state, people in the neighborhoods tended to follow the routine pattern they had become accustomed to over time. This phenomenon found expression in ideological utterances such as the ones made by two leaders of the Ohalim movement:

> We were told many times to assume responsibility and to run the neighborhood on our own. But we would say that it was beyond our capabilities and should be seen by the municipality ... The people in the gang were just like me, engaged in crime in poverty. We didn't know about social services and didn't care about the law. We were marginal people who could be arrested and re-arrested.

These utterances clearly indicate routine patterns of behavior and action guided by deeply entrenched beliefs, norms, and values, which help to sustain dependency on state agencies. These expressions are ideological in the sense that they express an internalization of existing power relations. Put another way, social marginality and compliance with existing relations had become part of a set of values, norms, and behaviors that were taken for granted.

The deconstruction of these routinized conventions and social behaviors and the construction of new identity were accomplished by progressive community workers and outside professionals working in conjuction with a newly formed local leadership. These agents developed new sets of images and symbols that challenged old conventions and reoriented ideological configurations and social action. As one of the community workers in Katamon Tet conceded: 'I always told them that, if you want revenge, you can be far more effective through social protest.'

These agents had a considerable influence on the interpretation adopted, since they tried to explain to the members of the movements that the problems characteristic of disadvantaged neighborhoods were part of a larger set of deep contradictions and conflicts that affected broad sections of the public. Leftist elements from Siah (the Israeli New Left), for instance, spurred the Black Panthers into raising the class problem. A placard written by Siah activists, or with their help, before the demonstration of May 18, 1971, stated:

> We call on you, residents of the poor neighborhoods, factory workers, and all the exploited people of the country to join a consistent, uncompromising struggle.

The contact between ICL and Peace Now and with Mapam led to the creation of a political–universal message that confronted the government's settlement policy in the occupied territories. Consequently, members of the movement took part in Peace Now demonstrations and went on to establish a 'tent settlement' as an act of protest against the government's settlement policy in the West Bank and the Gaza Strip. The intellectuals and organizers arriving from outside thus gave a broader interpretation to local problems, passed on messages, directed activity, and in this way raised local consciousness. An examination of the placards, poems, texts of plays, and slogans adopted by the organizations with a deep interpretation shows that their protest was a kind of 'political art,' that is, an articulation of communications–artistic sophistication and political activity, brought in different ways from outside to the local elites.

It would be overly simplistic to label the interaction between the cultural elite from outside and the members of the movements as manipulation, or to see in this interaction an expression of emotionalism or sentimentalism on the part of intellectuals from outside, even though these elements did exist. Apparently, this interaction marked an attempt to deliver to the social movements in disadvantaged neighborhoods messages that originated in high culture, while adapting and translating them into the language of popular culture prevailing in the local environment.

The mediation between the two cultures was carried out through a medium that, with LaCapra (1983: 52–54), may be called the carnival. The carnival, in the framework of which come the demonstration, the protest-squat, the provocative play, and the furious placard, served as a means of attracting attention, of overdramatization, of criticism and mockery of the social system, and sometimes of the participants themselves. The activity turned into a sort of show which displayed the complex messages associated with resource allocation, social and class divisions, and sense of discrimination. The classic expressions of the carnival were the simplistic placard with a humorous slant, the dramatic, show-like activity, and poetry. As one of the community workers who participated in the organization of Ohel Yosef's squatting campaign at the Pat housing project explained, the central idea behind the placards was the use of concise slogans so that the message would penetrate and be accepted. For example, 'Families had children in order to get housing points, but then they changed the point system.' The system of points for housing (by which young couples accumulated points towards a mortgage based on income, number of children, housing density, and other factors) became an object of criticism and was widely and sharply attacked by community workers, one of whom wrote the following poem:

Poor neighborhood resident remember!
In the housing company the point system operates!
For a sick heart or leg problems
They give you a pair of points
And a man in a sleeping bag
Gets another point.
And for each and every soul
And for cracks in the walls
And for most diseases
You might even get double.
The point system operates
The miseries grow;
Overcrowding buries us,
And the neighborhood asks
To hell with it! How must you look
To get two thousand points?!

(Ross, undated: 49)

This protest poem was very popular among the residents of the disadvantaged neighborhoods, and in the first (and last) issue of the newspaper put out by the Ohalim movement, it appeared at the top of the page as a sort of anthem.

The cultural elite that came from outside the neighborhood played, then, an important role in shaping the protest, developing criticism, creating ethnic–cultural myths, and in establishing links with other groups. It reinterpreted the past, singled out objects for criticism, created symbols of group identification, and indicated directions of activity. The question that arises, then, is to what extent were the movements with structural interpretation autonomous in creating their messages and activities, and to what extent were they directed from backstage, sometimes without being aware of it, by the outside environment?

The interaction with the external environment (outside professionals and facilitators) and the adoption of radical interpretations often created a deep rift between the organizations with structural interpretations and their neighborhoods. The radical tradition associated with the structural interpretation, and the orientation and types of activity that derive from this tradition contradicted to a great extent the traditional conceptions of the local community. This created a conflict in which, generally, the neighborhood's traditional view won. As a result, the movements that adopted structural interpretations such as the Black Panthers and the Ohalim movement, did not put down roots in the neighborhood and were conceived, by those who adhered to the local traditional conception, as a stain on the neighborhood. In contrast, there was greater correspondence between the local tradition and the movements that adopted the empirical interpretation and acted within the system. As a result, they won broader support in the neighborhood. This phenomenon was evident in the 1982 elections to the local council in Ir Ganim. The YFN, which had a strong link with the local culture (religious observance, a right-wing leaning, apoliticization on the neighborhood level, avoidance of conflict, an instrumental orientation, and respect for the authority of the local rabbis) won 14 of the 15 seats on the council, while ICL, which had close contact with the external–radical environment, remained foreign to the neighborhood and won not a single seat. In other words, the interpretations and symbols shaped by ICL were not accepted by the residents, while those of YFN won much favor.

Learning from the experience of previous movements

The various protest organizations consciously related to patterns of activity developed by previous organizations, derived lessons, and consequently created new interpretations and patterns of activity. This learning process was carried out through a critical examination of previous experiences, which involved a partial adoption of already existing knowledge and the development of new layers based on the organization's specific experience. In light of their examination of past experiences, they adopted conceptions, expressions, and approaches in a selective way, made selective use of previous symbols and categories, and rejected other symbols and approaches.

The founders of Ohel Yosef in 1973 related in a critical manner to the militant and political activity of the Black Panthers, which in their opinion did not contribute to improving conditions in Musrara. Relying on this critical evaluation, the Ohel Yosef activists defined their goal as diametrically opposite to that of the Black Panthers – constructive activity within the neighborhood in order to develop the local community. As the then leader of Ohel Yosef put it: 'The movement was in some sense a reaction to the Black Panthers. The active part did not favor violent activity . . . we wanted positive activities.' Later, in 1975, Ohel Yosef was radicalized by the entrance of a group of criminals. This radicalization culminated in the establishment of the Ohalim movement in 1979 and was criticized by the leader of YFN, a movement that appeared in 1982: 'I'm very critical of the Ohalim movement … they're the most radical on the map', and in a letter to the mayor he said:

> This should make it clear that our way was not that of the demonstrations and violence so characteristic, unfortunately, of our city. The damage they cause is almost irreparable, at least as far as self-image is concerned. We naturally acted in a solid, drab, and quiet manner.
>
> (Letter by the head of YFN to the mayor, March 1984)

The social critique of previous organizations does not mean a total rejection of the patterns they developed. The chairperson of YFN did not hesitate to adopt symbols and approaches from the past:

> Unfortunately, our members gradually began realizing that in the atmosphere that has been created in Jerusalem there is no practical possibility of serving as a conventional council. The only way to get things moving is still the way we have not chosen, the way that is so opposed to our principles and to our central idea – the way of violence and demonstrations and turning tables over etc. – the way with which our predecessors succeeded.
>
> (Letter from the chairperson of YFN to the mayor, March 1984)

There seems here to be a clear reference to precedents characterizing the militant movements that turned to activity outside the system. The militant approach adopted by these organizations serves here as a source of power, as a political resource that can be used to put pressure on the state.

CONCLUSIONS

This chapter described and analyzed the rise and divergence of urban social movements. It has been shown that although the movements' actors shared similar social, generational, and geographic conditions, social protest diverged and was expressed in different forms of organizations: urban social movements, which sought to advance a comprehensive change in society through militant actions; neighborhood protest movements, which aspired to produce fundamental changes in the neighborhoods while colliding with the state; and coproductive organizations, which wished to bring about limited changes in the neighborhood while cooperating with the state.

The divergence of urban protest did not stem, in my view, from the societal context. It was rooted in the different ways in which different human agents located in varying places framed their experiences: that is, identified the sources of injustice, set their goals and suggested lines of action. In other words, to account for the differentiation of protest, local interpretative schemes are to be studied in depth.

Relating Marris's insights, as illustrated in his study of Neighborhood Development in London, to my observations in Jerusalem, I suggested two major forms of interpretation: structural and empirical. The structural interpretation, prominent among the urban social movements of the Black Panthers and the Ohalim movement, located the local problems (housing deficiencies, lack of social services, dilapidated physical infrastructure, social detachment, and juvenile delinquency) within a deeper context of socio-ethnic inequality. Hence, the conception that any serious response to these problems presupposed a radical change in the relations of power within the city and the state. The empirical interpretation, on the other hand, which characterized the coproductive organizations of YFN, the Katamon Higher Committee, and Ohel Shmuel, viewed the same problems as concrete issues that should be pragmatically confronted through negotiations and exchange with the state. The meaning associated with the social conditions thus affected the position towards the state, and shaped to a large extent the strategies of action.

Some interesting questions seem to arise at this point. To begin with, what were the social mechanisms that informed these radically different interpretations? Why was a certain interpretation adopted and preferred over another one. One cannot conclude the study of protest (and by extension of any social action) by turning solely to interpretation. The interpretations themselves have to be subjected to a further inquiry that examines those contingent, place-specific factors that informed them.

It has been shown that schemes of interpretation are constructed through everyday life experience, for instance by being brought up in a broken or stable family; through group membership, for instance by being a member of a criminal network or of a student group at the Hassidic school; through interactions with different environments, whether local–traditional or outside–radical ones; and through learning from the experience of other organizations, whether critically or approvingly. There is nothing determinate in these contingencies, and they should be treated as informing specific experiences at the local, everyday level.

It may be argued then the different forms through which urban protest expresses itself cannot be directly inferred from the macro-societal conditions (the ethno-class, generational, and spatial contexts), but rather are mediated through the actors' experiences and interpretations. In other words, the context in which the protest grew up is not conceived in terms of creating, demanding, and shaping, but is treated as filled with potentialities that might be realized in different ways, by different agents who live in different places. In encountering the socio-historical context, the social actors do not face it directly. They confront it through specific experiences in specific places, through local knowledge accumulated in the course of daily life experience and through disparate forms of interactions with different environments. Values and norms gained through these specific processes coalesce to shape specific schemes of interpretation. And it is through these schemes of interpretation that the broader context becomes meaningful, and diversified types of social protest are formed.

NOTES

1 See Smelser, 1963; Gurr, 1970; Turner and Killian, 1972; McCarthy and Zald, 1977.

2 The Hebrew word *ohel* (plural: *ohalim*) literally means 'tent', but in this particular case indicates feelings of temporary residence.

13

SPATIAL POLITICS/SOCIAL MOVEMENTS

questions of (b)orders and resistance in global times

David Slater

Across a broad array of analytical domains signs of the spatial surface with increasing regularity. Notions of territorialization and deterritorialization, of flows and fixities, of time–space compression, of inside and outside, of the transcendence and re-inscription of borders, of overlapping territories and terrains of resistance inhabit zones of knowledge that also include growing reference to the global. Terms such as globalization, the global–local nexus, critical globalism and the 'global condition' all reflect a widening sensibility of the need to go beyond national boundaries.[1] At the same moment, there is a strong sense in which the universal and the particular are not only juxtaposed but interwoven and implicated in each other. In an era often characterized by the markers of fragmentation, heterogeneity, plurality, hybridization and difference, the project of Western liberal democracy has acquired an accentuated universality. In the specific context of North–South relations, neo-liberalism with its particular prioritization of market forces, possessive individualism, property and the private has been deployed in ways which have transgressed the national sovereignties of peripheral states and set in train a series of severely detrimental social effects.[2] As Derrida (1994: 85) writes, at a time when an ideal of liberal democracy has been triumphantly proclaimed as the ideal of human history, it must be strongly asserted that 'never have violence, inequality, exclusion, famine, and thus economic oppression affected as many human beings in the history of the earth and humanity'.

Derrida also suggests, in his analysis of the spirits of Marx, that communism could be distinguished from other labour movements by its international character and by the fact that no other organized political movement had ever presented itself as geopolitical, as inaugurating a space that is now reaching its limits. In this global space that Derrida refers to, which is now, post-1989, no longer home to an international communist movement, we have different forms of mobilization, of resistance and movement; movements which have been contextualized in terms of 'nomads of the present' (Melucci, 1989), capturing an important sense of fluidity, and territorial flexibility. Resistances and oppositions have been increasingly seen as independent from any encapsulating universalist discourse. Archipelagos of resistance or reverse discourses that have the potential to be connected across space, but which are also distinct, specific and embedded in local and regional contexts, have emerged in many different societies, encouraging in some cases the use of the term 'new social movements'.[3]

Leaving aside for a moment the question of how social movements are interpreted, in what forms their occurrence and continuing presence are treated theoretically, it is abundantly clear that their existence has been connected to a range of significant themes from development to democracy, from citizenship to culture and from environment to emancipation. These sites of knowledge have their own intersections, but it is a reflection of the expansion of interest in movements that their investigation and critical analysis connect to an ever-growing gamut of themes and issues. As the centrality of class has waned, the revival of interest in civil society, and state–society relations in general has been accompanied by a search for alternative forms of 'doing politics', against a backdrop of normalizing projects of global order and power.

MOVEMENTS AND THE RE-MAPPING OF THE POLITICAL

One potential area of enquiry that has sometimes remained underexposed concerns the imbrication of geopolitics and social movements. Thus, in some discussions of the interconnections between movements and democracy, or development, or cultural change the territoriality or more generally the spatiality of movements, power and politics has been marginalized.[4] It is certainly the case that the way we can think about politics and the political has been connected to movements and to resistances, and it is also evident, as mentioned above, that across a broad spectrum of research and enquiry, there has been an expansion of interest in the spatial. However, I would argue that in the literature on social movements, the difference that the geopolitical can make has not always been taken as a significant issue. I want to argue here that the geopolitical has a double meaning, and can be examined in relation to the territoriality of politics within national boundaries, as well as to the transnational flows and penetrations of different kinds of power. In the context of social movements, struggles for a decentralization of political power within a given national territory, and for a radical re-structuring of the territorial power of the state can be identified as exemplifying the more inner-oriented form of the geopolitical. This is not meant to suggest that there are no links to the transnational or global context of political contestation; rather the primary focus is on the 'inside', whereas there are other instances, other kinds of movements which, although also having an 'inside' are also deeply involved in a transnational arena, as can be seen in the example of some environmental movements. Falk (1993: 39) has referred to this phenomenon as 'globalization from below'. In a parallel way, it is possible to argue that in the analysis of democracy and processes of democratization there is also an inside – the territorialization of democracy within a given nation-state, and an outside – the struggle for a democratization of institutions that operate at the global level, but which have multiple effects within the territorial polities of the countries of the South. Clearly in the cases of social movements and democracy, the inside and outside of the geopolitical are not to be realistically seen as separate, but as overlapping and intertwined in a complex of relations.

Connolly (1991a) has maintained that it would seem that democratic political theorists and international relations theorists have little to communicate to each other about. On the one hand intrastate political theory tends to concentrate on 'internal' issues such as rights, justice, community, obligation, identity and legitimacy, whereas on the other inter-state theory examines 'external' questions of security, alliances, violence, war and subversion. For Connolly, this particular discursive division of labour allows the effects of changes in the contemporary era to fall through the gaps of democratic theory. What is then needed is a realization that democratic politics must extend into global issues since increasingly the most fundamental issues of life are not resolvable within the confines of the territorial state.[5] More generally, it is contended that one of the key requirements of the current period is to supplement and challenge structures of territorial democracy with a politics of nonterritorial democratization of global issues. Here, one can think of a number of movements that re-inscribe the meanings and practices of democratization.

Feminist movements which continue to struggle for the liberation of women from all types of oppression have constructed new forms of organization and solidarity that connect transnationally. In Latin America, for example, the biannual feminist *encuentros* held since 1981 have consistently expanded taking on a variety of themes from autonomy to issues of community leadership (Jaquette 1994). Along another pathway, as Wapner (1995) indicates, transnational environmental groups extend their activities beyond the territorial confines of a given country. In India, for instance, tens of thousands of peasants, landless labourers and indigenous people have demonstrated against a series of dams in the Narmada Valley, and the movement has spread with the formation by local and transnational groups of an activist network that operates both inside India and abroad. In these two examples, gender relations and issues of the environment have become focal points in the re-thinking of the spatialities of democratization. For Connolly these kinds of movements would exemplify the point that nonterritorial democratization can ventilate global issues through the creative interventions of nonstate actors. In their turn, these interventions could potentially reinvigorate the internal democracy of territorial states. However, nonterritorial democratization lacks a secure territorial base and, it can be argued, is less anchored in the terms of its accountability. Democratization beyond a territorial anchorage tends also, as Held (1995) has indicated, to be in the process of emerging as a series of demands and claims that are still relatively nascent in their formation, although this would certainly be less true of the women's movement. In contradistinction, the territorial state, in global times, tends to rest on an increasingly fragile and precarious ground, with pressures from below often opening up fissures in its territorial control, whilst the globalization of financial, economic and cultural power increasingly impinges on the nation-state from above.[6]

It is particularly important to underline the continuing ways in which much social and political theory, either in the context of interpreting social movements, or democracy, or the state, has tended to evade the difficult question of the inside and the outside and their dialectic. In a related intervention, Connolly (1991b) draws our attention to the way in which a certain rather pervasive undercurrent of political theory constructs a boundary between an inside (self) and an outside (other). Behind the boundary we have our own world of community, membership, internal understandings, our morality, distributive mechanisms, democratic accountability, obligations and allegiances. On the other side, outside our own constructed world, there would be alternative worlds of strangers, danger, external principles and uncertain moralities. There would not be very much that connects these other worlds to us politically, morally or temporally. The threat of anarchy and alterity outside would intensify the inclination to regard shared understandings and common principles of membership as adequate norms of political judgement within the state, as well as fortifying the view of the territorial state as the highest unit of political loyalty, identification and democratic participation. In this kind of context, there would be little room for the possibilities of what Campbell (1996) has referred to as 'radical interdependence' across borders.

Whilst Connolly's critique of the split between an inside and an outside, a split which is constitutive of much political theory, is highly relevant for a whole series of issues, and not least for those interpretative modes that classify and contextualize social movements within a self-containing inside, equally it is worthwhile noting that the theory Connolly is evaluating is Western-based theory, and that the forms in which political theory has evolved in the societies of Europe and North America cannot be realistically divorced from the colonial and imperial imaginations and projects that are rooted in those societies. Thus, although the heterogeneities and complex differences within the categories of West and non-West or North and South have to be constantly kept in mind, geopolitical memory, the recalling and representation of those crucial divides in the nature, scope and magnitude of power relations needs to be taken as a central part of our own contemporary project of critical analysis.[7]

If modern politics is a spatial politics, with its crucial condition of possibility being the distinction between an inside and an outside, between the citizens, communities and movements within and the enemies, others and absences without, it is also the case that the characteristic universalisms of so much Western social and political theory have expressed a continuing amnesia towards the geopolitical penetrations, fragmentations and power relations within which modern accounts of universality have been articulated. Furthermore, when modern politics has been viewed as world politics, many accounts have tended to assume that 'world politics' refers not only to some determining structure, but more relevantly to processes that occur in realms somewhere 'beyond' society. In contrast, social movements have frequently been seen as phenomena that occur within society, existing in juxtaposition to those key political structures that give them their essential meaning, namely, states and the states sytem. Moreover, within this inner realm of society, social movements have tended to be interpreted as part of civil society, which in turn has been distinguished as separate from the political affairs of the modern state. In this sense, the ways in which social movements have been analysed has tended to reproduce an approach to politics that confines it to a pre-given realm that is implicitly constructed as non-problematic.

Social movements may well be linked to the political domain through their impact on state policies or the priorities of political parties, but any connection to global politics would be characteristically made through the mediation of the inner political system. This particular perspective tends to treat politics as a domain or separate level from the economic and social, and equally draws demarcating lines between an inner national political realm and an outer domain of world politics. In an earlier series of comments on this theme, I suggested that often a binary division has been drawn between the realm of the political, bounded within the state, and political parties, and the space of the social, framed around the family, the school, religion, the individual, movements. Alternatively, I argued that the political dimension could be endowed with a certain duality, whereby it could be seen as inscribed within the different spheres of the social whole and also as constitutive of the terrain on which the fabric and fate of the social whole is decided (Slater, 1994). Hence, what is and is not political at any moment changes with the emergence of new questions posed by new modes of subjectivity, for instance, 'the personal is political', and different kinds of social relations. This is not meant to imply that the political eliminates the social conditions from which its meaning can emerge; gender, religious belief, the environment, nationality, regionalism and so on, may become political at certain moments, but they are not only political.

A crucial feature of the political relates to the questioning of the socially given, of what appears to be socially natural and uncontested. When 'the given' is not accepted as such but referred back to the 'initial act' that led to its installation, its potential instability is revealed and reactivated (Laclau, 1990: 212). That instability is inseparably bound up with the pluralization of the origin, with the disruption of the implicit notion of a singularity of foundational meaning. The desedimentation of the social entails laying bare its political content, and since the social is expressed through a plurality of forms, the desedimentation of the socially given, in its plurality, reveals the potentially protean nature of the political. In this context, it has been remarked that contemporary social movements have challenged and/or redrawn the frontiers of the political. This can mean, for example, that movements can subvert the traditionally given of the political system – state power, political parties, formal institutions – by contesting the legitimacy and the apparently normal and natural functioning of their effects within society. But, also, the role of some social movements has been to reveal the concealed meanings of the political encased in the social. Social struggles can be seen as 'wars of interpretation' within which the orientation and significance of their demands and revendications are constructed through their practice. It is within a related approach to social movements that Walker (1994: 674–5), in his imaginative intervention, argues that perhaps the most interesting element of social movements concerns the ways in which they may contribute to the 'reconfiguration of the political under contemporary conditions'.

QUESTIONS OF THEORETICAL RENEWAL

Perhaps one of the most fascinating and complex of questions surrounding any treatment of the conjunction of social movements and the political concerns the issue of interpretation itself. For Walker (1995: 311) much analysis of the novelty of social movements is characterized by a quite crucial limitation, which is that 'the horizon of enquiry is already given by historically specific understandings of what it means to speak of community, a class, an interest, an identity, or a movement of action'. As a consequence one tends to encounter questions such as: do social movements constitute a break from or a continuation of class politics; do they reflect a re-invigoration of civil society or an abandonment of the state; do they constitute mobilizations which are free from previous forms of populism; or do they capture a continuation of liberal pluralism? In these kinds of positionings social and political phenomena are not infrequently insinuated with implicitly pre-given and consensual meanings. Walker goes on to assert that for all its sophistication the literature on social movements still seems bound by framings of political possibility that preclude putting into doubt established conceptions of political community or identity. Thus the possibility of the newness of social movements is circumscribed by a 'specific rendition of what it means to be political, and of where the political is to be found' (Walker, 1995: 312).

Looking at some of the recent discussions of social movements[8] one can find examples of the basic thrust of Walker's critical contention, and above all one is struck by a common assumption that politics and political strategy, as well as class and materiality are somehow already pre-given in their significance and location in the broader analytical arena. It is also as if certain categories exist on a pre-discursive terrain, whereby conceptual markers such as class and materiality are constructed as the radically unconstructed. Butler (1990a), in her influential analysis of feminism and the subversion of identity, argued that in the cultural interpretation of sex and gender, the production of sex as the prediscursive had to be understood in terms of the power relations that produce the effect of a prediscursive sex, thereby concealing the very operation of discursive production. Similarly, with certain kinds of arguments about social movements and political change, one still encounters interpretations that implicitly grant a pre-discursive, independent meaning to categories of class, materiality and agency, thus drawing a screen over the way in which those concepts have been discursively constructed. As Butler (1992b: 13) notes, in a subsequent article, agency can be viewed as belonging to a mode of thinking about persons as instrumental actors who confront an external political field, and 'if we agree that politics and power exist already at the level at which the subject and its agency are articulated and made possible, then agency can be *presumed* only at the cost of refusing to inquire into its construction'.

One of the primary questions involved in these kinds of arguments concerns the definition of politics and the political, and their relation to our theoretical understanding of social movements. In many contributions to the debate on social movements, no distinction is made between politics and the political, and as noted above, it is quite often the case that politics is referred to in a way that already presumes a meaning that is consensual and foundational. However, I would not suggest that there is an already fully formed theoretical framework that we can grasp and immediately apply. Our conceptual and thinking spaces are striated by a series of destabilizations and uncertainties that make any such quick alternative framing quite inappropriate. There is a sense in which many of the concepts employed to explain social and political phenomena appear increasingly precarious and partial. In times of re-thinking, re-visioning, re-imagining, the notion of taking 'soundings', or the emphasis on fluidity if not vertigo within the fields of analytical enquiry reflect the presence of a shifting, mobile terrain. Previously staked out domains of knowledge, lined by the contours of linked categories and constructs, are being increasingly destabilized and disrupted by ideas coming out of the border zones that traverse and transgress these older and erstwhile assured domains. One such idea concerns the way we may re-think politics and the political.

Mouffe (1995: 262–3), for instance, in a viewpoint that connects to Lefort's (1988) earlier work, writes that for her 'the political' relates to the antagonistic dimension that is inherent in all human society – an antagonism that can take many different forms and can be located in diverse social relations. In contrast, 'politics' can be taken to refer to the ensemble of practices, discourses and institutions which seek to establish a certain order and to organize social life in conditions which are always potentially subject to conflict precisely because they are affected by the dimension of 'the political'. In this light, politics can be seen as the attempted pacification of the political, or the installation and embodiment of order and sedimented practices in a given society. Depoliticization is the most established task of politics, and also as writers such as Honig (1993) and Rancière (1995) have suggested, it can be argued that key branches of political philosophy and theory have displaced the political as a means to realize the closest approximation of political good in the midst of the disorder of empirical politics. Rancière's suggestive notion of the pacification of the political can be exemplified in a number of additional ways that are specifically relevant to our own discussion.

First, in the context of many societies of the South, wherein social conflicts, material polarization, violence and a growing disillusionment with formal institutions manifest themselves in ways which appear to be steadily more problematic, new policies of 'good governance' and the attempt to introduce Western-style arrangements for democratization can be taken as one form of the external ordering and containment of the political. At the same moment, within some societies of Africa, for example, new networks of solidarity are being established and new mentalities are taking shape, so that, as Monga (1995: 360) indicates, in discussing democratization in Francophone Africa, 'we are now witnessing a complete transformation of the conditions in which politics emerge'. What we have in these instances are simultaneous but deeply contrasting attempts to re-align the relations between politics and the political.

Second, in a sharply contrasting example, the Cuban experience of post-revolutionary order has been characterized by the continuing attempts of a one-party state to neutralize points of potential antagonism by representing government as the synthesis of society. The imperative of order and security has been translated into policies of integration and assimilation whereby difference has been equated with destabilization. In the contrasting case of Nicaragua, an initial attempt to integrate the ethnic minorities of the Atlantic coast into the Sandinista project was subsequently radically altered and replaced by a conversation in which the government came to recognize the rights of difference and autonomy. In the general history of Marxist projects in the Third World, the drive to centralize has been predominantly rooted in an ideological suspicion of the local and the regional, and the pacification of the political has been an immanently territorial project.

Third, in a number of Andean countries, Bolivia, Colombia and Peru being clear examples, in the last fifteen to twenty years there have been a series of mobilizations, protests and movements emerging at the local and regional levels which have challenged the existing territoriality of the state.[9] In particular, new associations have been made between democratization and decentralization and in the struggle against centralism new forms of spatial subjectivity and identity have emerged. These new forms which contest the given territoriality of the political system can be viewed as reflections of the political expressed spatially. In response, central state administrations have introduced a variety of reforms which have sought to contain and incorporate these local and regional resistances. By seeking to re-align and re-structure the territorial power of the central state, a variety of governments in these countries have sought to contain and pacify the geopolitical within their already constituted boundaries.

Overall, in these examples and more generally, the most salient point of my argument is to stress the interactive nature of politics and the political;

to posit their distinction but also their essential inter-connectedness. Hence, the reference to the political does not entail a marginalization of the formal sphere of politics; rather it calls for a distinction between two registers that implicate and involve each other. Politics has its own public space; it is the field of exchanges between political parties, of parliamentary and governmental affairs, of elections and representation and in general of the type of activity, practices and procedures that take place in the institutional arena of the political system. The political, however, as Arditi (1994: 21) has proposed, can be more effectively regarded as a type of relationship that can develop in any area of the social, irrespective of whether or not it remains within the institutional enclosure of 'politics'. The political then is a living movement, a kind of 'magma of conflicting wills', or antagonisms; it is mobile and ubiquitous, going beyond but also subverting the institutional settings and moorings of politics.

In an important sense the idea of the imbrication of politics and the political reflects the continuing debate about the relations between the state and civil society. Let us for a moment refer to a recent passage from Laclau. In an interview concerning the paradoxes of contemporary politics, it is emphasized that the contemporary situation can be characterized by the blurring of the division between the state and civil society. For Laclau (1994: 45), currents circulate between the spheres of state and civil society, 'making illusory the idea of a confrontation or even a delimitation between the two as fully fledged autonomous entities'. As an example Laclau refers to the radicalization of the democratic process, arguing that it would be unacceptable to go along with the view that equates the radicalization of democracy with the deepening of the demarcation of civil society from the state, since in many instances the advance of democracy requires progressive legislation that goes against deep-seated interests anchored in civil society itself. On the other hand, it would also be inapposite to accept the idea that the public sphere is the 'locus of an absolute and omnipotent popular will'; instead, 'democratic politics requires many and complex strategic moves which cut across the two spheres and dissolve the clear-cut differentiation between the two' (Laclau, 1994b: 46).

Laclau's attempt to underscore the importance of an interweaving of moves that merge and dissolve the civil society–state distinction is particularly relevant in the analysis of social movements since not infrequently civil society has been essentialized in a positive frame, as the terrain of the good and the enlightened. The emphasis on imbrication is also pertinent to our discussion of politics and the political since their interaction transgresses the state–civil society divide. The merging and interweaving noted here can be illustrated in the context of the relation between sedimentation and re-activation. Politics, for example, can be thought of as the institutionalization of an order that is designed to overcome or at the least to confine the threatening conflicts of the political – a case of sedimentation. But 'order', or 'governance' is always a series of regulative and sedimented procedures, practices, codes and categories that can never be absolutely fulfilled. This is the case since the political – the possibilities of subversion, questioning, opposition, refusal and resistance can never be fully overcome; the interruption of de-sedimentation, or interventions which constitute a re-activation of the instability that 'order' sought to pacify reflect the inseparability of politics and the political. In this conceptual context, the political is always that irremovable inner periphery at the heart of politics.[10]

TRACING THE GEOPOLITICAL

One question that can be immediately posed at this juncture concerns the potential relevance of the spatial for any demarcation of politics and the political. What difference would the prefix geo- make to our above argument? Inscribing a spatial dimension into the above notion of politics could lead us into a discussion of the internal territoriality of constructed institutional orders, through, for example, a consideration of the local/ regional constitution of national political systems. In addition, a critical examination of the relations between nation-states, located within a posited world system of such units, which have been traditionally regarded as the building blocks of geopolitics, could form a related pathway of enquiry. Also, we might want to go beyond these 'containers', and think geopolitics in terms of global processes that transgress the boundaries of state-units (Held, 1995; Taylor, 1995).

Furthermore, we can go on to denote two connected expressions of the geopolitical, which relate to Arditi's metaphor of the 'magma'; to those underlying, unstable, fluid 'substances' that may break the ordered surface and provoke re-orderings, re-structurings or, in certain moments, transformative ruptures, as created by past revolutionary insurgencies. The first expression of the geopolitical can be defined within the ostensibly inner bounded realm of the territorial state. Here, as briefly alluded to above, there have been examples in a range of peripheral societies of certain kinds of movements that have challenged and continue to challenge established territorial orderings of the state. In some instances, such movements, as has been the case in the Atlantic Coast region of Nicaragua, have been intimately rooted in ethnic identities, whereas in the examples of Bolivia, Colombia and Peru, mentioned earlier, whilst indigenous communities have been differentially involved in the struggles against a centralized state, the local and regional movements of these societies have been unpredictably heterogeneous and have embraced a highly diverse range of demands. Nor is this tentative distinction meant to imply that the 'indigenous' or the 'communal' are somehow uniform. Often, as Agrawal (1995) reminds us, the idea of the 'indigenous' is deployed in a way that masks an important heterogeneity that lies within.

In those specific cases where a concerted challenge has been made to the centralized nature of state power, as has occurred in Bolivia, an eventual legislative response, as reflected in the 1995 Law on Administrative Decentralization, does not bring to a close territorial protest and contestation; in contrast, the continuing interwoven nature of geopolitics and the geopolitical is taken into a new phase. Nor should we assume that the challenges, in the Bolivian case, to the centralization of power are of a singular orientation, since the various departmental *Comité's Cívicos* have articulated their demands in ways which have not always achieved inter-departmental unity.[11] Similarly, at the local level, the Popular Participation Law of 1994, which ostensibly decentralizes power and resources to new rural municipalities, has evoked a variety of responses from grassroots indigenous organizations (Albó, 1996).

In a different societal context, regional movements in 1970s Peru placed on to the agenda the call for a new level of territorial power – the establishment of regional governments in a unitary state. Their impact, and challenge to the given spatio-political order created the conditions for a protracted national debate and eventual legislation to install a new instance of government. However, the influence of these various movements, often embodied in the form of 'regional fronts', was slowly undermined by growing violence, social dislocation and acute instability. Specifically, *Sendero Luminoso*, with its introduction of 'armed stoppages' at local and departmental levels, made it increasingly difficult for regional movements to organize peaceful and effective protests in an atmosphere characterized by violent confrontations between the Shining Path guerrillas and the Armed Forces.[12]

In a particular moment, a guerrilla organization and a series of loosely grouped regional fronts represented two very different expressions of a geopolitical challenge to the existing institutional order. The former challenge came to overshadow all other forms of political contestation, and with a change in government in 1990, followed by Fujimori's 'auto-golpe', the intensity of the threat to the political order was used to justify a re-centralization of power, and a sharp reduction in the importance of newly established regional governments. In the Peruvian case, in sharp contrast to Bolivia, the emergence of a deep association of violence and terror with a movement that initiated its actions in a quintessentially peripheral region of the Andes has greatly facilitated the renewal of a centralist project. Moreover, any counter project aimed at deepening and broadening democratic structures, of 'territorializing democracy' may well be haunted by the sign of 'terror' hidden in the naming of 'territory'.[13]

A second instance of the geopolitical can be thought through in relation to the original constitution of national sovereignties. Significantly, the geopolitical in this context can be used to destabilize some of the meanings previously attached to the political since in many of these conceptualizations the analysis of the relation between politics and the political is seen as being worked out within the confines of an implicitly Western territorial state. Here, there is an assumption of pre-given territorial integrity and impermeability. But in the situation of peripheral polities, the historical realities of external power and its effects within those systems are much more difficult to ignore. What this contrast points to is the lack of equality in the full recognition of the territorial integrity of nation-states. Predominantly, those underlying, mobile, unstable, disrupting currents that can fundamentally shift the terrain of politics are located within the implicitly bounded space of one nation-state which is invariably Western in its origin. Missing is the possibility that externally based forces could also constitute the magma of the political. Such an absence reflects a governing supposition, rooted in modern political theory, that the context is formed by full territorial sovereignty; 'quasi-sovereignty', in contrast, would be applicable to non-Western states (Jackson, 1993).

For the societies of Latin America, Africa and Asia the principles governing the constitution of their mode of political being were deeply moulded by external penetration. Colonialism, for example, represented the imposition and installation of principles of the political that violated the bond between national sovereignty and the constitution of societal being. The framing of time, and the ordering of space, followed an externally imposed logic that did not cease to have effects in the post-colonial period. The struggles to recover an autochthonous narrative of time and an indigenous ensemble of meanings for the territory of the nation have formed an essential part of post-Independence politics.[14]

In what were referred to as 'wars of national liberation' the struggle to breathe new life into the time–space nexus of independence lay at the core of the anti-imperialist movement. At the same time, however, it has to be stressed that the struggles against imperialism in peripheral societies have always assumed a variety of forms, as witnessed within Latin America, where the meanings given to cultural imperialism by *Sendero Luminoso* in Peru contrast markedly to the Sandinista discourse of the 1980s or to today's Zapatistas in Chiapas. Furthermore, of course, contemporary struggles to redefine the geopolitical take place in an era marked, as was noted above, by the hegemony of neo-liberal ideas.

Examined broadly, neo-liberal discourse not only enframes development in its notions of structural adjustment and good governance, but reaches out and gives contemporary meaning to projects for democracy. This attempt to construct a global agenda has been specifically criticized by Parekh (1993: 168) who observes that to insist on the universality of liberal democracy is to impose on other countries systems of government unrelated to their skills and talents, reducing them to 'mimics, unable and unwilling to be true either to their tradition or . . . imported alien norms'; and he germanely adds that the 'cultural havoc caused by colonialism should alert us to the dangers of an over-zealous imposition of liberal democracy'.[15]

INSIDE/OUTSIDE AND ZONES OF RESISTANCE

In the above outline of an argument on the geopolitical, themes of power, inside/outside and movements intersect in a way that can be further developed, and the uprising in Chiapas can be taken as one particularly illustrative example of these kinds of intersections. In this context, it is possible to identify the interweaving of 'levels of analysis', so that the global, the regional and the local can be interpreted as deeply imbricated, with the notion of the 'borderization' of the world underlining the fragility of settled spatial orders.

The armed uprising of between 3000 and 4000 Indians in Chiapas on 1 January 1994, and the seizure of seven towns was timed to coincide with the entry into effect of the North American Free Trade Agreement (NAFTA). One of the first communiqués of the Ejército Zapatista de Liberación Nacional (EZLN) stated that NAFTA 'is a death certificate for the Indian peoples of Mexico, who are dispensable for the government of Carlos Salinas de Gortari' (quoted in N. Harvey 1995: 39). The validity of this vision was subsequently captured in a leaked Chase Manhattan Bank memorandum for early 1995, which argued that the Mexican government 'will need to eliminate the Zapatistas to demonstrate their effective control of the national territory and of security policy'.[16]

Clearly, we have here a pivotal example of the importance of connecting inside with outside, of seeing the global and the local and regional as intimately intertwined rather than as dividing markers of separate and unconnected worlds. In an earlier analysis, the EZLN subcomandante Marcos described Chiapas in a regional/national/global context, outlining an approach that rekindled many previous arguments of the *dependentista* perspective. In a language that evoked Galeano's classic text, Marcos wrote that:

> Chiapas is bled through thousands of veins: through oil ducts and gas ducts, over electric wires, by railroad cars, through bank accounts, by trucks and vans, by ships and planes, over clandestine paths, third-rate roads, and mountain passes . . . oil, electric energy, cattle, money, coffee, bananas, honey, corn, cocoa, tobacco, sugar, soy, melons, sorghum, mamey, mangos, tamarind, avocados and Chiapan blood flows out through 1,001 fangs sunk into the neck of southeastern Mexico . . . billions of tons of natural resources go through Mexican ports, railway stations, airports and road systems to various destinations: the United States, Canada, Holland, Germany, Italy, Japan – but all with the same destiny: to feed the empire.[17]

The presence of a geopolitical imagination that fuses a variety of spatial arenas – the global, the national, the regional and the local – is again strongly evident in an interview with the Zapatista subcomandante, published in August 1995, from which three crucial observations can be highlighted. First, it is argued that current processes of globalization have the potential to break national states, and to accentuate internal regional differentiations, as reflected in the divergence between the northern, central and south-eastern zones of Mexico. Second, with reference to questions of war, it is commented that political confrontation and the battle for ideas has acquired more significance than direct military power, echoing the Gramscian contrast between a war of position and a war of manoeuvre. And third, pivotal importance is given to the role of the means of communication; if a movement or resistance can be made to appear dead or moribund, irrespective of what happens to be the reality on the ground, this constitutes a greater threat than superior military strength.[18] It is in this situation that the use of e-mail has assumed its alternative potential.

I have quoted at some length from one of the key Zapatista leaders in order to underscore the way the thinking within a resistance movement can reflect the interlocking nature of issues that resonate transnationally. Equally, as researchers, such as Dietz (1995), N. Harvey (1995) and Zermeño (1995), have reminded us the Zapatista rebellion is anchored in a long regional history of social struggles and oppositions, which provide it with a deep political sustenance. Furthermore, its leadership expresses a respect for difference and plurality that displays a sharp contrast to previous revolutionary movements, and its recent sixteen popular demands concerning land, housing, work, food, health, education, culture, information, independence, democracy, freedom, justice, peace, security, the fight against corruption and protection of the environment, have been articulated through an emphasis on dialogue and the recent organization of a '*consulta nacional*'.[19]

The wave of massive support that initially greeted the Zapatista insurgency had both an urban and rural component. Moreover, as Dietz (1995: 46) points out, the new alliances of indigenous communities in Michoacán, Guerrero, Oaxaca, Veracruz and Morelos convened regional assemblies in which the recourse to armed struggle was viewed with 'understanding', whilst it was added that the worsened situation of their own regions hardly differed from that which generated the insurrection in Chiapas.[20] The struggles for territorial autonomy, embodied in organizations such as the Independent Front of Indian Peoples (FIPI) and the Indigenous National Convention (CNI), contain both an ethnic and regional dimension that connect with the Chiapas rebellion and reinforce the overall and growing significance of the cultural within the geopolitical.

In this example of the Zapatista resistance and its challenge to the existing institutional order, it is evident that our conceptualization of the geopolitical assumes two linked meanings if we remain within a more internal realm. First, the Chiapas uprising and its condensation of deeply rooted social opposition can be seen as representing a radical questioning of the territorial functioning of the contemporary Mexican state. Its list of demands, and its prioritization of a radical democracy and a just society have been articulated in a context of territory and power, and have established bonds between the regional, the national and the global. Second, it is a movement that through its naming, re-connects to one of the founding moments of the Mexican Revolution. In a continuing act of radical remembrance it subversively re-frames the themes of land, justice and democracy. Through a process of re-activation of contested meanings it presents itself as a moment of resistance which is both cultural and geopolitical. At the same time, the effects of globalization, and through NAFTA the bringing into question of national sovereignty, provide an example of an externally generated geopolitical that crucially impinges on the internal, so that the timing of the uprising and the trajectory of Zapatista discourse cannot be understood outside the interwoven webs of inside and outside. This point can be further elaborated in relation to issues of democracy, justice and the impact of neo-liberalism.

At the beginning of 1996, the EZLN formed the Zapatista Front of National Liberation (FZLN), 'a civil and non-violent organization, independent and democratic, Mexican and national, which struggles for democracy, liberty and justice in Mexico'. In the 'Fourth Declaration of the Lacandon Jungle', the Zapatistas called for a nation of many worlds, and affirmed that democracy will come when the culture of the nation is refashioned from the perspective of indigenous peoples. At the same time, not only have the Zapatistas made common cause with many sectors of the Mexican population in their opposition to neo-liberalism, but more notably they have also extended their strategy to the international arena, calling for an 'Intercontinental Forum Against Neoliberalism'. Meetings are being planned in Berlin, Tokyo, an African city, Sydney and Mexico City. A significant feature of their agenda is to organize a broad-based internationalist culture to counter the culture of neo-liberalism. It includes: 'all individuals, groups, collectives, movements, social, citizen and political organizations, neighborhood associations, cooperatives, all leftist groups, non governmental organizations, groups in solidarity with the struggles of the peoples of the world, bands, tribes, intellectuals, musicians, workers, artists, teachers, peasants, cultural groups, youth movements, alternative media, ecologists, squatters, lesbians, homosexuals, feminists, pacifists' (quoted in Yúdice, 1996: 20). Hence, the Zapatista struggle for democracy, justice and national sovereignty is intimately linked with their opposition to neo-liberalism and to NAFTA. They have made the connections among the global, national, regional and local and do not restrict their geopolitical vision to any one level of analysis or action.

In a comparative sense, if we limit ourselves to the idea of an internal territorial domain, the inner ambits of the territorial state, for example, I want to suggest that the questioning, disrupting, destabilizing effects of the movements I have briefly referred to above can be thought of in terms of three modalities of the geopolitical. First, as strongly evidenced in the cases of Bolivia and Peru, there are regional movements that challenge the existing territorial power of the central state and call for a spatial extension and deepening of the democratic process. The fluidity and heterogeneity of these resistances, and in some instances their elusive transience, have led some investigators to belittle their effects in the 'real world of politics'. Nevertheless, they continue to move and put into question many of our established modes of analysis.[21]

Second, as described in the Mexican case, an armed uprising, a guerrilla movement that is regionally rooted but unconfined to its region, can constitute another internal modality of the geopolitical which crosses the border between the inside and the outside having effects in a connected series of spheres. This is only one example and as suggested earlier on, other guerrilla rebellions, such as the Shining Path movement in Peru, which was also regionally based but not limited to that original region of Ayacucho, constructed a very different set of meanings and practices to confront centralized state power. Moreover, the 1980s also witnessed a series of guerrilla movements in Central America, where the territorial power of internal states was effectively fractured.

Third, the growth in indigenous demands for territorial autonomy in a number of Latin American societies combines an interrogation of existing spatial ordering with a profound questioning of the founding of the state itself. Autonomy was the major demand of Mexico's indigenous peoples when they called for the creation of a National Plural Indigenous Assembly for Autonomy, claiming that for centuries the Mexican government has been trying to integrate them into a homogeneous nation that has never existed. While the call for autonomy is not new to Latin America's indigenous peoples, who have always demanded the right to self-government, today it is a highly charged issue because it is viewed by central governments as a call for secession and the break up of the territorial state. The right to autonomy is contingent on indigenous peoples having the right to their traditional land, which has been a key stumbling block for centuries. Land, in the indigenous cosmovision, is the source and mother of life, and it is strongly argued that a guarantee to territory, with environmental conservation, is crucial for the economic and cultural continuity of indigenous peoples (Collinson, 1996).

These three kinds of challenge to the territorial politics of the state are not to be seen as always separate, since the actual pathways of struggle have sometimes overlapped. They all represent the potential to undermine and weaken the solidity of contemporary political systems, and equally they have sometimes intersected with other social movements concerned with, for example, environmental and human rights issues. The intertwining of inside and outside has obviously varied between these movements, as also has the degree of connection between different kinds of struggle within the same society – archipelagos of resistance where linkages may be tenuous, intermittent or broken would be a more accurate depiction than the notion of territorial coalitions.

In times of acute political turbulence and precarious affinities, the placing of social movements has become increasingly problematic. Calderón (1995: 122) writes of the heterogeneous, plural, multiple nature of social movements which he thinks may well be provoking a break in those totalizing, excluding, singular conceptualizations of Latin American destiny that prevailed for so long. In those small, everyday, cultural spaces of resistance, it may be possible to discern the emergence of collectivist values and the social forms of self-government and solidarity – the seeds perhaps of continuing oppositions and the reconstruction of 'historical subjects'. Our optimism of the will encourages us to agree with Calderón, whilst our 'realism of the intellect' leads us, as it does in his own analysis, to emphasize that there is great diversity and unpredictability. It is here too that the role of the 'observer', researcher, academic is rather central. There are those who have been chided for 'romantically listening to the movements', whilst conversely others have been criticized for remaining 'trapped in the well-worn grooves of class analysis'. Increasingly, more questions have been posed concerning the applicability of Western-based social theories for non-Western contexts. It is certainly the case that much of the Western discussion of social movements has proceeded as if such phenomena have never surfaced in the societies of the South. But not only is the object of knowledge confined to a Western or Northern terrain, the agents of knowledge are also predominantly of Occidental origin. The prevailing regime of representation is Euro-Americanist, whereby the driving assumption is that knowledge, and in particular theoretical knowledge, is a Western property.

The category 'Euro-Americanist' is itself symptomatic of the problem of subsuming the Americas under the heading 'America', and the modified 'Euro-North-Americanist', leaving aside its cumbersome quality, still raises the question of the difference between Canada and the United States. In relation to my above argument, what is specifically significant here is the way the inside–outside thematic connects with questions of knowledge, culture and representation. If, as Alvarez, Dagnino and Escobar (1996) have proposed, we are to think a new cultural studies of the Americas, a project that would recognize the infinite complexities of such a 'fractal structure with manifold political cultures', this invites us to consider the way we imagine different worlds, and construct new analytical meanings. Often absent in investigations of movements and mobilizations has been the continuing impact of the power over other societies, the effects of invasive discourses of control and re-ordering. The project of neo-liberal globalization represents the most recent of such discourses, and contains within it the attempted subordination of different modes of thought and interpretation. The alternative development of critical knowledge incites the crossing of borders and the connecting of inside and outside, but it does so in a frame that requires recognition and reciprocity, and in a context that transcends containment.

NOTES

1 Robertson (1992), for example, argues that much of world history can be considered as sequences of 'miniglobalization' in the sense that historic empire formation involved the unification of previously sequestered territories and social entities. He goes on to suggest a series of phases in the development of globalization and stresses the point that 'there is a general autonomy and "logic" to the globalization process, which operates in *relative* independence of strictly societal and other more conventionally studied sociocultural processes' (page 60). For other surveys, dealing with the relations between globalization and culture, and democracy and global order, see King (1991) and Held (1995) respectively.

2 See Gill, 1995; Mohan, 1996; Ould-Mey, 1994; and Trumper and Phillips, 1995.

3 Such a description first emerged in a European context, but rapidly acquired a wider geographical application. The issue of 'newness' has sparked off a variety of debates on historical continuities and discontinuities, and has provoked a number of interesting exchanges on the theoretical bases for understanding social movements in general. Themes

of North–South divergences and the problems of comparative analysis have also been introduced into the overall discussion – see Calderón (1995) for a recent text on social movements in Latin America set in an international context, and Slater (1991 and 1994) for some consideration of the question of novelty and theoretical divergences.

4 I would reserve the term 'territoriality' for contexts implying the space within nation-states, whereas 'spatiality' is used in this paper to refer to multiple contexts both within and across nation-states. Furthermore, I will argue that the term 'geopolitical' can be interpreted in a double sense, referring to external and internal instances; that it does not have to be exclusively contained within an international or transnational frame.

5 In the post-war period, Herz (1957) provided one of the first, and subsequently widely quoted papers, on the territorial state. He was particularly concerned with the peculiar unity, compactness and coherence of the modern nation-state, and related these features to what he called the substratum of statehood, where the state 'confronts us . . . in its physical, corporeal capacity: as an expanse of territory encircled for its identification and its defense by a "hard shell" of fortifications . . . in this lies what will here be referred to as the "impermeability", or simply the "territoriality" of the modern state' (page 474). Herz considered this territoriality and protection given by the modern state to its citizens to be a basic feature of the historical development of the political system in general. Interestingly, and symptomatically, generalizations are made on the basis of a certain reading of the European experience, and issues of the violation of the territorial integrity of non-European nation-states are not taken into account.

6 It is always necessary to bear in mind when referring to globalization that the processes involved are uneven, and as Mosquera (1994) has noted we live in a world of 'axial globalization' and 'zones of silence', so that, for example, in many African situations, cultural and communicative linkages tend to flow directly back to European metropolises, leaving many African countries separated from each other, or only tenuously connected.

7 Jameson (1992) usefully distinguishes Third World cultures from those of the First in the sense that the former have far more difficulty in remaining independent from the gaze and penetration of the metropolitan cultures of the North. The latter cultures can throw up their barriers and erect their fortresses; similarly the social scientists of northern lands can more easily neglect the intellectual life of the South – implicit and explicit notions of the self-contained but universally relevant nature of Occidental knowledge and especially theoretical knowledge are deeply rooted in Western culture.

8 See Munck, 1995; Jordan, 1995; Scott, 1995; Shefner, 1995 and Weyland, 1995.

9 See Fals Borda, 1992; Laserna, 1986 and Slater, 1989.

10 It also needs to be added here that discussions of the political and politics have been characterized by a wide variety of conceptual perspectives. Frequently, for instance, the political has been defined in relation to the state (Ricoeur, 1995), and there is a long history of seeing the private sphere as outside the political realm. More recently notions of the diffusion of the political in society have gained more support and reflect an expanding discussion of the 'frontiers of the political' (Morin, 1995); for an earlier and classic treatment see Schmitt (1976).

11 This was particularly evident in November 1993, when at a meeting of a majority of the country's comité's, held in Santa Cruz, differences emerged in relation to the strength of anti-centralist feeling, and also, in the context of the differential party political alignments of the committees. For some background discussion of regionalization issues in Bolivia see Slater (1995), and for a recent examination of many of the key social themes of Bolivia's current period, see Calderón and Laserna (1994).

12 For a recent analysis of the place of the Shining Path guerrillas see Starn (1995b).

13 Territory can be taken to refer to land, earth, sustenance, but the form of the word, as Connolly (1994: 24) points out can also be related to a derivation from *terrere*, to terrorize.

14 On the question of time and the colonial encounter see Fabian's (1983) pathbreaking text and Norton's recent article on 'ruling memory'. For a stimulating discussion of spatial aspects of culture and imperialism, see Said, 1993.

15 In a related argument, Derrida (1994: 82) writes that the exacerbation of the foreign debt, and connected mechanisms, are 'starving or driving to despair a large portion of humanity . . . they tend thus to exclude it simultaneously from the very market that this logic nevertheless seeks to extend . . . this type of contradiction works through many geopolitical fluctuations even when they appear to be dictated by the discourse of democratization or human rights'.

16 A copy of the Chase memorandum was obtained by the London *Independent on Sunday*;

in a connected passage, the memo stated that while the insurrection in Chiapas 'does not pose a fundamental threat to Mexican political stability, it is perceived to be so by many in the investment community . . . what Mexico needs . . . is a more authoritarian government rather than more democracy.' – quoted in Doyle, L., 'Did US bank send in battalions against Mexican rebel army?', *Independent on Sunday*, 5 March, 1995, page 14, London.

17 This is taken from a piece of writing distributed in 1992, and entitled 'The first wind' – see the Bardacke and López (1995: 32–33) translations.

18 These points are taken from an interview published in *La Jornada*, 27 August 1995, pp. 10–11, Mexico City.

19 The 'national consultation' which took place on 27 August 1995 elicited a response from just under 825, 000 people; 97.7 per cent approved of the 16 demands mentioned in the text and over 90 per cent were in favour of political reforms; in addition 56.2 per cent expressed the view that the EZLN should convert itself into an independent political force – reported in *La Jornada*, 29 August 1995, page 5, Mexico City.

20 N. Harvey (1995: 48), provides some indices of poverty for Chiapas, noting, for example, that 41.6 per cent of homes were without drinking water in 1992, whilst 33.1 per cent were without electricity, and 58.8 per cent were without drainage; the national averages were 20.6 per cent, 12.5 per cent and 36.4 per cent respectively.

21 And as Routledge and Simons (1995: 475) nicely remind us, 'social science has been a key tool for taming spirits of resistance'.

14

CONCLUSION

a changing space and a time for change

Michael Keith

A progressive politics implies something about direction. Not necessarily a Whiggish movement through history but a notion of transformation that invokes both a space of change and a time of change – with time and space themselves as contingent resources in the waging of engagements that in many different ways challenge specific social, moral and economic orders. It is this exploration of the contingent geographies of forms of political struggle that are taking place after the end of history that is shared by the contributors to this volume.

Geographies of Resistance invokes two themes, a disciplinary moment and a space of the political. By way of a conclusion what I want to do in this chapter is to explore the contradictions of the one and the occasional erasure of the other.

DISCIPLINARY MOMENTS

As Donald Moore reminds us in his chapter there is a sloppy sense in which 'resistance can become a default discourse of the left'. Deprived of the teleological certainties of revolutionary socialism, 'resistance' has that omnibus quality that can be simultaneously employed to describe both a vicarious celebration of bloodlettings of other places and a smug reassurance of the gravitas of everyday life and the political import of the transgressions of the jaywalker (see also Nigel Thrift's contribution to this volume).

In Britain and other 'western' democracies the turn away from the political mainstream – demonstrated in falling rates of electoral participation and a growing difficulty in distinguishing between the principal parties of national politics – has been accompanied by a steady upsurge in alternative forms of political involvement and mobilisation. High profile ecological campaigns around road building schemes and squatting, consumer based lobbies, neighbourhood movements, and international solidarities around civil rights have all attracted saturation media coverage and demonstrably high rates of participation. Similarly, collective action around sexuality, ethnicity and gender create alternative identity politics that likewise challenge the conventional notions of the political subject of late modernity.

In an earlier volume we attempted to explore some of the spatialities that underscored the place of identity politics (Keith and Pile, 1993a). At times the foregrounding of the spatial was a disciplinary moment of geographical credentialism and consequently an inevitable flaw was the downplaying of the simultaneous production of both spatialities and temporalities of the political. But the corollary of this was in part the contention that the processes through which political subjects were made had to be understood as constitutively linked to the production of space and time through which they were articulated. Subjectification is in part about the production of time and space as well as simultaneously about the production of political subjects, a pattern that we attempted to capture analytically in terms of a notion of 'radical contextualisation'.

Across a broader disciplinary canvas the decline of certain modernist narratives of political mobilisation finds an academic echo in what David Slater in this volume describes as a trend in which 'As the centrality of class has waned, the revival of interest in civil society and state–society relations in general has been accompanied by a search for alternative forms of "doing politics", against a backdrop of normalizing projects of global order and power.' In part such a search has moved much theoretical work towards a focus in contemporary political theory on articulation and conjuncture, leading inevitably to alternative disciplinary opportunities for the demonstration of the salience of the spatial and for the geographical tradition to lay claims to more universal relevance, at times a characteristically insecure and plaintive appeal to be taken seriously.

Yet, more constructively, the chapters in this book share a focus on the significance of the spaces through which counter-hegemonic politics are articulated. Conjunctural politics invites an exploration of the spaces of political praxis. In line with such a search, this volume highlights the necessity of considering the ensemble of institutions, mobilisations and power effects that collectively render visible the social world in terms that do not privilege one set of structures of power at the expense of others. More specifically, an interesting theme that appears common to most of the contributions to this volume – if sometimes present only through its notable absence – is the manner in which it is essential to consider simultaneously the governmentalities at the heart of the changing state with the narratives of unity that produce collective mobilisation. In short, a democratic politics of resistance, or a plausible geography of resistance, will make little sense unless considering the manner in which ostensibly sovereign political subjects are rendered visible through the negotiated sets of relations between state and civil society, between governmentality and the political subject. Through processes of subject making, institutions, individuals, nation-states and societies alike are revealed as plural rather than singular, the composite products of many subjectivities.

Sovereignty is a useful thematic through which to think through some of these missing threads. For it is the collapse of the sovereign self at several different units of analysis that always moves towards an indication that at least part of the political is elsewhere, temporally and spatially. At the heart of most theories of democratic politics is a principle of self-determination (Held, 1993), yet such principles are harder to sustain when the sovereign individual as much as the sovereign state stands unmasked as no more than theoretical shibboleths or narrative tropes through which particular governmental fictions are rendered plausible. In such circumstances the mapping of the political becomes a task in itself. But as Bhatt has recently suggested, one of the products of the influence of some trends in deconstructionist and post-Lacanian writing – after the fashion of Laclau, Zizek, Butler and others – and focusing on the nature of political identity and the multifarious forms of political antagonism, is seen through collateral damage to an identifiable political compass.

> In their more popularised form, apparent in much cultural studies work, these tendencies simulate an identity between the limitless play of the signifier and the realm of the political, economic or bodily real. The ease with which differences between signifiers can be remanufactured is aesthetically identified with the task of articulating oppositional political blocs. The harder work of the politically possible, the difficult issues of ethical accountability and moral necessity, the non-transient, often violent presence of institutional power, and the manifest and real problems of judgement and evaluation seem to disappear under the debris of multiple antagonisms, indeterminate signification or even psychosis.
>
> (Bhatt, 1997)

In this sense most of the chapters in this book are as useful in their foregrounding of empirical research as much as for bringing us back to the immediate concerns of theoretical abstraction. But there is perhaps something more substantive here also. It is precisely such empiricism that on one level has tended to mark academic labour with the stigmata of parochialism, a tension that remains at the heart of the disciplinary creation of empirically generated subjects in the social sciences such as 'geography' and 'anthropology' and distinguishes them institutionally from counterparts that have derived from rationalist traditions, evinced for example in certain conventional student texts of political science or particular genres of Marxian political economy.

Such empiricism – in no sense an unproblematic study of 'the real' – is also useful in disrupting other clumsy boundaries. Specifically, if the deconstructionist ethical moment produces particular forms of abstraction it is the case also that other categories of conventional political thinking are disrupted by such evidence. It is perhaps a salutary lesson of these essays that a politics of the possible must inevitably emerge from a sustained engagement with the empirical, not a naïve romance of the real but instead a commitment to address the specific and the particular. Drawing on Geertz's conventions of thick description (Geertz, 1973), academic labour can valorise forms of contextualisation that understand the artifactual nature of the social but do not use it as an excuse to remain in the claustrophobic – if no longer quite so privileged – territories of the ivory tower common room.

To make a particular disciplinary point, in the self-consciously geographical literature of the 1980s and early 1990s, in a move that owed much to journals such as *Society and Space*, the remorseless exploration of the empirical appears to emerge as a principal weakness of the discipline, a flawed retreat from the truly cerebral world of something more highbrow.

Such a trend might be traced to the theoretically desiccated landscape of most post-war geography and the technocratic horrors of the spatial science of the 1960s, with its legitimate evolution into increasingly sophisticated forms of spatial modelling and its bastard progeny – in and out of the academy – of the occasionally excessive claims made on behalf of the truth effects of Geographical Information Systems. This volume in part attempts to signal an end to such a retreat from the geographical tradition and reinforces a demand for a much more rigorous engagement between theoretical and empirical labour across the social sciences. It is a demand that if satisfied might ironically cut against the refrain for the privileged spatialisation of social theory and play down credentialising claims of the geographical within the academy. The beginning of geographies may in time turn out to be almost as shallow a rallying cry and as pernicious a shibboleth as the end of history.

THE STATE AND CIVIL SOCIETY

If the excesses of poststructuralist theory have at times become over-concerned with ever more refined attempts to categorise the location of antagonism there is also a territorialised version of the default politics of the left that has tended to identify civil society as a redemptive field for political activity. In his important and influential work on *Democracy and Civil Society*, John Keane suggested that civil society has 'the potential to become a non-state sphere comprising a plurality of public spheres – productive units, households, voluntary organizations and community based services – which are legally guaranteed and self organizing' (Keane, 1988: 14) and taken this on to claim that 'the separation of state and civil society must be a permanent feature of a fully democratic social and political order' (Keane, 1988: 15).

Vaclav Havel has even suggested that 'the various political shifts and upheavals within the communist world all have one thing in common: the undying urge to create a genuine civil society'. Yet if the reaction against the excesses of state power is understandable, the implication that there is somehow an authentic other, a 'genuine civil society' is one that frighteningly echoes and carelessly masks the problems of diverse forces of subject formation and 'subjectification' that are the product of any and every structure of governance.

Although it would be foolish to caricature the work of any single author here there is surely a sense in which a certain *naïveté* did appear to inform some enlightened comment in the late 1980s and early 1990s about the political capacity of civil society, particularly in the wake of the collapse of discredited forms of state socialism. Yet such political optimism surely found a consonant academic equivalent in celebrations of resistance and transgression that themselves at times steadfastly ignored the progressive evolution of state power across all points of the global compass. If too often the naïvely romantic 'we' of progressive politics falls apart around its differences then the sloppy 'they' of the opposition does likewise around the complex cartographies of the evolving state and other institutions and norms of governance.

In short it is the argument here that it is simply not plausible to address a politics of civil society without simultaneously thinking through the complicit changes in the governmentalities of particular times and places that are in part identified with the changing state. It is surely through a cartography of the ever changing territorial articulations of the form and borderlines of the division between state and civil society that we might begin to develop a calculus of resistance that might inform political practice.

There are instead of such ready distinctions a series of propositions that we want to make in the conclusion of this volume.

The first of these is that the relationship between state and civil society is more readily understood as transactional and fluid than either fixed or stable. Forces of globalisation have radically transformed the nature of state power. If sovereignty was invariably a more tendentious concept than political rhetoric might suggest (Hirst and Thompson, 1992) then there has also been a sense of the 'hollowing out' of the power of the nation-state (Jessop, 1993) through the combined forces of transnational trade, political and legal agreements above the national level and complex developments of localised patterns, techniques and forms of governance below the level of the nation-state. In one of the popular clichés of our time it is possible to assert with a degree of plausibility that the nation-state is not just too big for the little functions of government and too small for the large functions of government. It is also increasingly more usefully understood as a series of state effects (Jessop, 1990: 9) and projects that are carried through by a plurality of hegemonic projects, institutions and processes.

It makes no sense to talk of an analytical separation of state and civil society. As Jessop has suggested civil society and political society should instead be understood as two moments of the integral state, variously articulated: 'the unity, coherence and capacities of the state depend on moments and projections within its other – civil society' (Jessop, 1990: 351). Crudely, in the 'western' democracies, the former state socialist countries and the nations of the South, we are witnessing an increasing complexity of state formation that does not readily equate with many of the taxonomies of state theory of the 1970s and 1980s. This is paralleled by a decline in the legitimacy of certain key political institutions of advanced liberal democracy, notably the national systems of 'party' politics and the transcendent subject positions of both Whiggish optimism and Marxian pessimism.

We might instead talk of the histories and geographies through which state institutions and popular forms create, define and contest the borders between themselves. The territories of the one and the cartographies of the other define a landscape of the political through which the social is rendered visible. In this sense resistance itself can never be unproblematically divorced from the creation of the subjectivities through which it is defined. This point has been made well by Burchell in the suggestion that

> casual references to civil society are common today, often evoking a misplaced nostalgia. We should, I think, follow Foucault here and be a bit more nominalistic about terms like society or civil society or nation or community. Civil society was for early liberalism a kind of critical concept, an instrument of critique. It outlined the correlate or schema for a possible liberal art of government.
>
> (Burchell, 1993: 275)

Neither is it defensible to talk of a politics of transgression or of a struggle against hegemony without simultaneously thinking through the territories of antagonism against which such mobilisations are set. The 'constitutive outside' of all forms of political subjectivity may be defined by a politics of antagonism but this analytically cannot subvert the rendering visible of the architecture of such opposition in order to make sense of the ethical charge of the specific forms of oppositional politics. Too often in social analysis the heroic celebration of the transgressive moment erases the subject creation of the forces that are being transgressed. Consequently, a consideration of the geographies of resistance described in this volume needs to be considered as part of an exercise through which the texts themselves and the institutions and subjectivities with which they are concerned explore the spaces through which political subjects are rendered visible.

TRANSFORMATIVE POLITICS AND THE SPACE OF THE POLITICAL

How is the political made visible? What are the frames of representation that render open to scrutiny a transformative politics to ethical judgement? If we accept that the broad-brush understanding of the changing nature of state power needs to be supplemented by an acknowledgement of the matrix of forms of governmental practice that stretch from the globalisation of culture and markets right down to the technologies through which multiple selves are created and rendered visible to the reflexive individual then a geographical imperative is precisely about the tracing of these alternative forms and practices of governmentality.

In this sense we might accept that an ethics of the possible is in part about rendering visible what is too often taken for granted, making the familiar strange in order to create an ethics of 'a position already deposed of its kingdom of identity and substance, already in debt, "for the other" to the point of substitution for the other, altering the immanence of the subject in the depths of its identity' (Levinas, 1989: 182). Theoretically, taxonomies of the conduct of conduct can never be differentiated through humanistic appeals to agency that obscure the geographies of the creation of multiple subjectivities: 'Thus proximity is never enough; as responsible, I am never finished with emptying myself of myself' (Levinas, 1989: 182). Politically, this might be understood at least in part as an agenda through which we might bridge and transform the division between state and civil society; '"pluralizing" the former and "publicising" the latter' (Hirst, 1994: 75).

Crudely, we are rejecting both naïve celebrations of the everyday – at times characterised by certain invocations of de Certeau – and also acknowledging the flaws of the totalising certainties of some forms of state theory beyond the reach of the social. There is a discursive gaze that in rendering the political visible almost simultaneously obscures, or even inadvertently masks, the territories of institutionalised state government. Likewise, there is a mode of analysis that through foregrounding a territory of government and specific regimes of regulation ignores the micro strategies and tactics through which change is achieved, not as an implausible celebration of the everyday but through the mutations of difference which in the movement of a butterfly's wing precipitate something else.

So although there is great diversity in the circumstances explored in these chapters it is possible to identify moments of a common pattern. There may appear to be a long way between the complexities of settlement policy in Zimbabwe and the everyday realities of post-industrial Britain but perhaps it is possible to identify common threads that run through such stories and offer more than a rhetorical call to 'resist'. The challenging of state power in the global South in Moore's chapter describes the historical and geographical specificity of the negotiated and contested boundary drawing between the territories of government of the state effects – pre-colonial, colonial and post-colonial – and the proper domain of new institutional forms of governance that grow up beyond its boundaries. In such contexts, as Moore usefully emphasises, there is no 'authentic insurrectionary space "outside" of power', mobilisations and resistance occur within and are in part defined by territorialised structures of governmentality. Likewise, a precise relationship between state and 'the social', an accumulated institutional geography and history, surely shadows the 'inoperative community' that Rose evokes so powerfully in her chapter on the transgressive aesthetic of community arts projects. And as Brown quite explicitly demonstrates in thinking through the place of AIDS politics in Vancouver an unproblematic binary rendering of this transactional relation between state and civil society can itself undermine the effectiveness of political engagement. Certain versions of radical democracy cannot define a new space of the political, precisely because it is the ever-changing relationship between the institutional forms of the state and the fluid mobilisations of civil society that defines the space of the political on which such struggles are conducted. Each assumes the other and as Brown demonstrates clearly the neglect of either is a recipe for practical failure.

In reading Michael Watts' description of Matatsine it is fascinating to note the particular 'state effects' produced by the specific institutional forms that mediate the transactional negotiation of the forces of central government and forces of the local, variously expressed. It is through this contested territory that there is a mediation of oil based capitalism in which for Watts 'there was a sense in which Shell was the local government' and ethically 'civil society is the repository of justice and morality'. However, the presence in the Watts narrative of the subject category 'ethnicity' – unproblematically rendered – surely also begs certain questions about the disappearance of the universal subject, the nostalgic allure of the national subject and the complexities of post-colonial subjectivity. For the manner in which 'ethnicity' waxes and wanes as a category of political mobilisation – both in Nigeria and many other places – is surely an exemplary case of the manner in which subject formation *takes place* in part through the contested relations between state and civil society and it is only through a cartography of this terrain that we can begin to reject globally expressed forms of ethnicised politics as either *natural* or *primordial* or 'merely' *symbolic* or *ideological*. Ethnicities are surely defined neither in *essentialist* nor in *anti-essentialist* terms in this sense; they are instead *relational* subjectivities that emerge through the matrices of state–civil society negotiations of the time and the space of the political.

In this sense 'resistance' cannot be sustained analytically as a medium through which a calculus of power can be gauged. Whether in regard to the legal system in the chapter by Knopp, the tracings of urban space in the chapters by Jacobs and Hasson or the protocols of propriety in Law's piece in this volume, there is a sense almost of complicity between governing and governed that subsumes its own contradictions only for them to emerge in particular politicised forms. In Law's chapter it is perhaps in the subjectifications of identity politics that there is surely also a relationship between the rationalities through which the subjectivities of the bar workers are constructed and the broader institutional context in which bar work emerges. This is not a straightforward appeal to the notion of geographies made by subjects though not in circumstances of their own choosing, and is more than a call for a political economic context to be seen as the shadow of perspective. It is instead an appeal to consider that the subjectivities of resistance in such contexts cannot be divorced from the institutions of subjectification.

In a similar spirit we can begin to think through the practical implications of the now clichéd notion that the social is always itself a discursive construction. And in case after case the social is reinvented by left and by right alike as a territory through which the conduct of conduct will be controlled (Donzelot, 1993; Foucault, 1991; Hirst and Thompson, 1992).

There is consequently – after Foucault – an arrangement of institutions and practices through which the territorialities of forms of government are developed through their control of the 'conduct of conduct'. For example it is arguable that there is a spatial rendering of the city as a site of governmental practices through the 'promotion' of a common-sense understanding of the relationship between society and 'the inner city' (Keith and Rogers, 1991), something that Donzelot identifies in terms of the relation between 'the State', 'progress' and 'the social':

> Now the State must aim to guide progress and become positively responsible for it so as to gain the means for securing the social promotion of society and eradicating the sources of evil, poverty and oppression which prevent it from corresponding to its ideal. Society is no longer *the subject* of its evolution so much as *the object* of a promotion devised over its head and aiming to bring freedom to each and security to all.
>
> (Donzelot, 1993: 135, emphasis in original; see also Donzelot, 1991)

In the contemporary United Kingdom the problem family, the delinquent single parent, the underclass, and the economically underperforming (places and peoples) are just some of very many equivalent sites located within the social as putative territories in need of control and regulation, a control that is not necessarily exercised through conventional institutional forms of state action.

Such an analysis might allow us to think through both self-evident patterns of political mobilisation and more nuanced actions of the political subject within this matrix of patterns of governmentality. Whether considered in terms of the transformative potential of associationism (Hirst, 1994) or the expression of the inexpressible (G. Rose, this volume), the transactional form of subjectification and subject contestation mirrors a more nuanced understanding of the relation between the state and civil society that can inform a social analysis of the relationship between 'resistance' and the terrain on which it is practised:

> Foucault's comment on the concept of civil society as a 'transactional' one, an encoding of the mobile interface of the game between government and governed, has its amplest verification in the new universe of what Donzelot has called 'the social'. 'The social' designates a field of governmental action operating always within and upon the discrepancies between economy and society.
>
> (Gordon, 1991)

As new forms of collective and personal identity emerge they are paralleled with 'new governmentalities' which for Nikolas Rose is evidence of

> the imperative to fashion a revised politics, one which renders itself consonant not only with the heterogeneity of the forms in which struggles are now carried out – nationalist, ethnic, religious, ethical, environmental – but also with the new conceptions of subjectivity through which the subjects of government increasingly have come to understand and relate to themselves.
>
> (N. Rose, 1996c)

In Nikolas Rose's terms (1996c) 'The alliances forged here are always risky, provisional and revisable; nonetheless from this point onwards, projects for the government of conduct will operate on a territory marked out by the vectors of identity, choice, consumption and lifestyle.' The ever shifting relationship between techniques of governance and their institutional form create territories in which we cannot ignore the vocabulary of state formation or the increasing institutional complexity of state effects. But likewise we can neither isolate such institutional forms from their creations of subjectivities of resistance nor locate in civil society, the public sphere, communities of resistance or any other romanticised analytical territory a domain of the self-evidently good. It is not that the distinction between state and civil society is rendered useless but instead that as Slater stresses in his chapter there is an imbrication of the political and the social that never privileges particular locations of resistance; what we might consider as a politics of the Moebius strip that denies the separation of inside and outside in sovereign political subjects.

RESISTING GEOGRAPHIES OF RESISTANCE: TOWARDS A SPATIALISATION OF THE POLITICAL SUBJECT

In particular, it is important to avoid the near ethical mercantilism that characterises some Foucauldian analyses of regimes of governmentality, the perspective that almost implies a zero-sum game – governmental practices in one place are simply displaced to another through particular innovations. Only in this sense can Foucault's own search for a new '*logique de gauche*' be shaped by a genealogy of speaking subjects and a comparison of alternative regimes of governmental practice as opposed to naïve celebrations of 'democracy'.

Whether or not we regard this as – after Baudrillard – a 'post-social' territory of government (N. Rose, 1996c), we will inevitably be faced with the further collapse of some of the basic categories of analysis that underscored the rationality and rhetoric of mainstream social thinking. With the complex pluralisation of the state, and the disenchantment of the social as an undifferentiated field of government, the value of a notion of geographies of resistance makes sense principally as a cartography of the spaces through which the conduct of conduct is exercised. The contributions in this book are in part about such mapping exercises.

BIBLIOGRAPHY

Abu-Lughod, J. L. (ed.) (1994) *From Urban Village to East Village: the Battle for New York's Lower East Side*, Oxford: Basil Blackwell.

Abu-Lughod, L. (1990) 'The romance of resistance: tracing transformations of power through Bedouin women', *American Ethnologist*, 17(1), 41–55.

Achebe, C. (1983) *The Trouble with Nigeria*, Ibadan: Heinemann.

Achebe, C. (1988) *Anthills of the Savannas*, New York: Vintage.

Adorno, T. W. (1973) *The Jargon of Authenticity*, Evanston: Northwestern University Press.

Agarwal, B. (1994) 'Gender, resistance and land: interlinked struggles over resources and meanings in South Asia', *Journal of Peasant Studies*, 22(1), 81–125.

Agnew, J. A. (1987) *Place and Politics: the Geographical Mediation of State and Society*. Boston: Allen and Unwin.

Agrawal, A. (1995) 'Dismantling the divide between indigenous and scientific knowledge', *Development and Change*, 26(3), July, 413–439.

Ahearne, J. (1995) *Michel de Certeau: Interpretation and Its Other*, Cambridge: Polity Press.

Albó, X. (1996) 'Bolivia: Making the Leap from Local Mobilization to National Politics', *NACLA Report on the Americas*, 29(5), March–April, 15–20.

Alexander, C. (1995) *Fin-de-Siècle Social Theory: Relativism, Reduction and the Problems of Reason*, London: Verso.

Ali, S. (1989) 'The big squeeze', *Far Eastern Economic Review*, 144(24), 26.

Allen, J. (forthcoming) 'Economies of power and space', in R. Lee and J. Wills (eds) *Geographies of Economies*, London: Edward Arnold.

Alvarez, S., Dagnino, E. and Escobar, A. (1996) 'Introduction: the cultural and the political in Latin American social movements', mimeo, paper presented at the conference on 'Cultures of Politics/Politics of Cultures: Revisioning Latin American Social Movements', State University of Campinas, Brazil, March.

Appadurai, A. (1990) 'Disjuncture and difference in the global cultural economy', *Public Culture*, 2(2), 1–24.

Appadurai, A. (1995) 'The Production of Locality', in R. Fardon (ed.) *Counterworks: Managing the Diversity of Knowledge*, London: Routledge, 204–225.

Appiah, K. A. (1992) *In My Fathers House: Africa in the Philosophy of Culture*, Oxford: Oxford University Press.

Appiah, K. A. and Gates, H. L. (eds) (1992) 'Identities', special issue of *Critical Inquiry*, 18(4), 625–890.

Appignanesi, L. (ed.) (1987) *The Real Me: Post-Modernism and the Question of Identity*, London: ICA Documents Number 6.

Arditi, B. (1994) 'Tracing the political', *Angelaki – a New Journal in Philosophy, Literature and the Social Sciences*, 1(3), 15–28.

Arnold, D. (1993) *Colonizing the Body: State Medicine and Epidemic Disease in Nineteenth-century India*, Berkeley: University of California Press.

Asad, T. (1993) *Genealogies of Religion*, Baltimore, Johns Hopkins University Press.

Ashwe, C. (1986) *Fiscal Federalism in Nigeria*, Monograph Number 46, Canberra: Australian National University.

Atlas, J. (1988) 'The case of Paul de Man', *The New York Times Magazine*, 28 August.

Ball, E. (1987) 'The great sideshow of the Situationist International', in A. Kaplan and K. Ross (eds) *Yale French Studies*, New Haven: Yale University Press, 21–27.

Ballantyne, A. (1995) 'Doing the devil's work', *Times Literary Supplement*, 7 April.

Bammiker, A. (ed.) (1994) *Displacements: Cultural Identities in Question*, Bloomington: Indiana University Press.
Banzhat, M. (1990) *Women, AIDS, and Activism*, Toronto: Between the Lines Press.
Baral, L. R. (1977) *Oppositional Politics in Nepal*, New Delhi: Abhinav Publications.
Bardacke, F. and López, L. (eds and trs) (1995) *Shadows of Tender Fury, the Letters and Communiqués of Subcomandante Marcos and the Zapatista Army of National Liberation*, New York: Monthly Review Press.
Bardhan, P. (1988) 'Dominant proprietory classes and India's democracy', in A. Kuli (ed.) *India's Democracy*, Princeton: Princeton University Press, 76–83.
Bargery, G. (1934) *Hausa–English Dictionary*, London: Oxford University Press.
Barkindo, B. (1993) 'Growing Islamism in Kano since 1970', in L. Brenner (ed.) *Muslim Identity and Social Change in Sub-Saharan Africa*, Bloomington: University of Indiana Press, 91–105.
Bateson, G. (1977) *Steps to an Ecology of Mind*, London: Picador.
Bauer, R. and Bauer, A. (1942) 'Day to day resistance to slavery', *Journal of Negro History*, 27(4), 388–419.
Bauman, Z. (1993) *Postmodern Ethics*, Oxford: Blackwell.
Bell, D. and Valentine, G. (eds) (1995) *Mapping Desire: Geographies of Sexuality*, London: Routledge.
Bell, D., Binnie, J., Cream, J. and Valentine, G. (1994) 'All hyped up and no place to go', *Gender, Place and Culture*, 1, 31–47.
Benjamin, W. (1969) *Illuminations*, New York: Schochen Books.
Benjamin, W. (1973) *Charles Baudelaire: Lyric Poet of High Capitalism*, London: Verso.
Benjamin, W. (1979) *One Way Street*, London: New Left Books.
Benka-Cocker, M. and Ekundayo, J. (1995) 'Effects of an oil spill on soil physico-chemical properties of a spill site in the Niger Delta', *Environmental Monitoring and Assessment*, 30, 93–104.
Berlant, L. (1994) '68, or something', *Critical Inquiry*, 21(1), 124–55.
Berlant, L. and Freeman, E. (1993) 'Queer nationality', in M. Warner (ed.) *Fear of a Queer Planet*, Minneapolis: University of Minnesota Press, 193–229.
Berlin, I. (1969) *Four Essays on Liberty*, Oxford: Oxford University Press.
Bhabha, H. (1985) 'Signs taken for wonders: questions of ambivalence and authority under a tree outside Delhi, May 1817', *Critical Inquiry*, 12, autumn, 144–165.
Bhabha, H. (1990a) 'The third space: interview with Homi Bhabha', in J. Rutherford (ed.) *Identity: Community, Culture, Difference*, London: Lawrence and Wishart, 207–221.
Bhabha, H. (1990b) 'Introduction: narrating the nation', in H. Bhabha (ed.) *Nation and Narration*, London and New York: Routledge, 1–7.
Bhabha, H. (1992) 'Postcolonial authority and postmodern guilt', in L. Grossberg, C. Nelson and P. Treichler (eds) *Cultural Studies*, New York: Routledge, 56–68.
Bhabha, H. (1994) *The Location of Culture*, New York and London: Routledge.
Bhabha, H. (ed.) (1990c) *Nation and Narration*, London: Routledge.
Bhatt, C. (1997) *Liberation and Purity: Race, New Religious Movements and the Ethics of Postmodernity*, London: UCL Press.
Blackman, E. L. (1952) *Religious Dances in the Christian Church and in Popular Medicine*, London: Collins.
Blomley, N. (1994) 'Activism and the academy', *Environment and Planning D: Society and Space*, 12(4), 383–385.
Bondi, L. (1993) 'Locating identity politics', in M. Keith and S. Pile (eds) *Place and the Politics of Resistance*, London: Routledge, 84–101.
Boston, J. (ed.) (1995) *The State Under Contract*, Wellington, New Zealand: Bridget Williams Books.
Bourdieu, P. (1991) *Language and Symbolic Power*, Cambridge: Polity Press.
Boyarin, J. (ed.) (1994) *Remapping Memory*, Minneapolis: University of Minnesota Press.
Brennan, F. (1995) *One Land, One Nation: Mabo – towards 2001*, St Lucia, Queensland: University of Queensland Press.
Brown, M. (1994) 'The work of city politics: citizenship through employment in the local response to AIDS', *Environment and Planning A*, 26, 873–894.
Brown, M. (1997) 'Travelling through the closet', in J. Duncan and D. Gregory (eds) *Writes of Passage*, London: Routledge.
Brunton, R. (1993) *Black Suffering, White Guilt? Aboriginal Disadvantage and the Royal Commission into Deaths in Custody*, Institute of Public Affairs: West Perth.

Bukatman, S. (1995) 'The artificial infinite', in L. Cooke and P. Wollen (eds) *Visual Display: Culture Beyond Appearances*, Seattle: Bay Press, 254–289.

Burawoy, M. (1985) *The Politics of Production*, London: Verso.

Burchell, G. (1993) 'Liberal government and techniques of the self', *Economy and Society*, 22(3), 267–282.

Burrell, I. and Forsyth, D. (1991a) 'Scandal kept at bay', *Edinburgh Evening News*, 14 February.

Burrell, I. and Forsyth, D. (1991b) 'The gay sex scandal', *Edinburgh Evening News*, 13 February.

Butler, J. (1990a) *Gender Trouble: Feminism and the Subversion of Identity*, New York: Routledge.

Butler, J. (1990b) 'Gender trouble, feminist theory and psychoanalytic discourse', in L. J. Nicholson (ed.) *Feminism/Postmodernism*, New York and London: Routledge, 324–340.

Butler, J. (1992a) 'Discussion on Aronwitz, S. "Reflections on identity"', *October*, 61, 108–120.

Butler, J. (1992b) 'Contingent foundations: feminism and the question of "Postmodernism"', in J. Butler and J. W. Scott (eds) *Feminists Theorize the Political*, London: Routledge, 3–21.

Butler, J. (1993) *Bodies that Matter: on the Discursive Limits of 'Sex'*, New York and London: Routledge.

Buttle, J. (1990) 'AIDS protesters won't stop confrontations, protests', *Vancouver Sun*, 13 September, A-17.

Calderón, F. (1995) *Movimientos Sociales y Política*, Mexico: Siglo XXI.

Calderón, F. and Laserna, R. (1994) *Paradojas de la Modernidad: Sociedad y Cambios en Bolivia*, Fundación Milenio, Producciones "CIMA"', La Paz, Bolivia.

Calderón, F., Piscitelli, A. and Reyna, J. L. (1992) 'Social movements: actors, theories, expectations', in A. Escobar and S. E. Alvarez (eds) *The Making of Social Movements in Latin America*, Boulder: Westview Press, 19–36.

Callen, M., Grover, J. Z. and Maggenti, M. (1991) 'Roundtable', in B. Wallis (ed.) *Democracy: a Project by Group Material*, Seattle: Bay Press, 241–258.

Campbell, D. (1993a) 'Gay myths, criminal realities', *Gay Scotland*, 71, August, 9–10, 20.

Campbell, D. (1993b) 'Gay myths, criminal realities', *Gay Scotland*, 72, September, 9–10, 12, 25.

Campbell, D. (1996) 'The politics of radical interdependence: a rejoinder to Daniel Warner', *Millennium – Journal of International Studies*, 25, 1, spring, 129–141.

Canel, E. (1992) 'New social movement theory and resource mobilization: the need for integration', in W. Carroll (ed.) *Organizing Dissent: Social Movements in Theory and Practice*, Toronto: Garamond, 22–51.

Canetti, E. (1962) *Crowds and Power*, Harmondsworth: Penguin.

Carter, P. (1987) *The Road to Botany Bay*, London: Faber and Faber.

Castells, M. (1977) *The Urban Question*, New York: Edward Arnold and MIT Press.

Castells, M. (1983) *The City and the Grassroots: a Cross-cultural Theory of Urban Social Movements*, London: Edward Arnold.

Centenary of Federation Advisory Committee (1996) *2001: a Report from Australia*, Canberra: Australian Government Publishing Service.

Chakrabarty, D. (1992) 'Trafficking in history and theory: Subaltern Studies', in K. K. Ruthven (ed.) *Beyond the Disciplines: the New Humanities*, Canberra: Australian Academy of the Humanities, 101–108.

Chakrabarty, D. (1994) 'The difference-deferral of a colonial modernity: public debates on domesticity in British India', in D. Arnold and D. Hardiman (eds) *Subaltern Studies VIII*, Delhi: Oxford University Press, 50–88.

Chew, S. (1993) 'What's going down with ACT UP?', *Out*, November, 72–137.

City of Melbourne (1995a) *Another View Walking Trail: Pathway of the Rainbow Serpent*, (text by Robert Mate Mate), Melbourne: City of Melbourne.

City of Melbourne (1995b) *Melbourne: Our City Our Culture*, Melbourne: City of Melbourne.

Clifford, J. (1986) 'Introduction: partial truths', in J. Clifford and G. E. Marcus (eds) *Writing Culture: the Poetics and Politics of Ethnography*, Berkeley and Los Angeles: University of California Press, 1–26.

Clifford, J. (1988) *The Predicament of Culture*, Cambridge, MA: Harvard University Press.

Clifford, J. (1994) 'Diasporas', *Cultural Anthropology*, 9(3), 302–338.

Clifford, J. and Dhareshwar, V. (eds) (1989) 'Traveling theories: traveling theorists', special issue of *Inscriptions*, 5, 1–188.

Cocks, J. (1989) *The Oppositional Imagination: Feminism, Critique and Political Theory*, London: Routledge.

Cohen, E. (1972) 'The Black Panthers in Israeli society', *Jewish Journal of Sociology*, 14, 93–109.
Cohen, J. L. and Arato, A. (1992) *Civil Society and Political Theory*, Cambridge, MA: MIT Press.
Colburn, F. (ed.) (1989) *Everyday Forms of Peasant Resistance*, London: M. E. Sharpe.
Collier, J. and Yanagisako, S. (eds) (1987) *Gender and Kinship: Essays toward a Unified Analysis*, Stanford: Stanford University Press.
Collinson, H. (ed.) (1996) *Green Guerrillas: Environmental Conflicts and Initiatives in Latin America and the Caribbean*, London: Latin America Bureau.
Comaroff, J. (1985) *Body of Power, Spirit of Resistance: the Culture and History of a South African People*, Chicago: University of Chicago Press.
Commonwealth of Australia (1993) *Mabo: the High Court Decision on Native Title. Discussion Paper*, Canberra: Australian Government Publishing Service.
Connolly, W. (1991a) *Identity/Difference: Democratic Negotiations of Political Paradox*, Ithaca: Cornell University Press.
Connolly, W. (1991b) 'Democracy and territoriality', *Millennium*, 20(3), 463–484.
Connolly, W. (1994) 'Tocqueville, territory and violence', *Theory, Culture and Society*, 11(1), 19–40.
Cooper, F. (1994) 'Conflict and connection: rethinking colonial African history', *American Historical Review*, 99(5), 1516–1545.
Cooper, F. and Stoler A. (1989) 'Tensions of empire: colonial control and visions of rule', *American Ethnologist*, 16(4), 609–621.
Copjec, J. (ed.) (1994) *Supposing the Subject*, London: Verso.
Corngold, S. (1988) 'Letter to the editor', *Times Literary Supplement*, 26 August–1 September.
Coronil, F. (1987) 'The black El Dorado: money, fetishism, democracy and capitalism in Venezuela', Ph.D. dissertation, University of Chicago.
Council for Aboriginal Reconciliation (1993) *Addressing the Key Issues for Reconciliation. Overview of Key Issue Papers 1– 8*, Canberra: Australian Government Publishing Service.
Council for Aboriginal Reconciliation (1994a) *Walking Together: The First Steps. Report of the Council for Aboriginal Reconciliation to Federal Parliament 1991–94*, Canberra: Australian Government Publishing Service.
Council for Aboriginal Reconciliation (1994b) *Footprints Catalogue*, Canberra: Australian Government Publishing Service.
Council for Aboriginal Reconciliation (1994c) *Improving Relationships. Key Issue Paper 2*, Canberra: Australian Government Publishing Service.
Council for Aboriginal Reconciliation (1995) *Going Forward: Social Justice for the First Australians. A Submission to the Commonwealth Government*, Canberra: Australian Government Publishing Service.
Cowan, J. K. (1990) *Dance and the Body Politic in Northern Greece*, Princeton, Princeton University Press.
Craig, G. A. (1995) 'In love with Hitler', *New York Review of Books*, 2 November.
Craig, G. A. (1996) 'How hell worked', *New York Review of Books*, 18 April.
Cresswell, T. (1996a) *In Place/Out of Place: Geography, Ideology, and Transgression*, Minneapolis: University of Minnesota Press.
Cresswell, T. (1996b) 'Domination/resistance and the politics of transgression', paper presented at the 'Geographies of Domination/Resistance' conference, Glasgow, 19–21 September.
Crimp, D. (1993) 'Right on, girlfriend!', in M. Warner (ed.) *Fear of a Queer Planet*, Minneapolis: University of Minnesota Press, 300–320.
Crimp, D. and Rolston, A. (1990) *AIDS DemoGraphics*, Seattle: Bay Press.
Crow, A. and McIlwraith, G. (1992) 'Pay up or you die!', *Daily Record*, 23 September.
Crush, J. (1994a) 'Post-colonialism, de-colonization, and geography', in A. Godlewska and N. Smith (eds) *Geography and Empire*, Oxford: Blackwell, 333–350.
Crush, J. (1994b) 'Scripting the compound: power and space in the South African mining industry', *Environment and Planning D: Society and Space*, 12(3), 301–324.
Daily Record (1990) 'Two judges went to gay disco ... but one stormed out in disgust, it was claimed', 18 January.
de Certeau, M. (1984a) *The Practice of Everyday Life*, London and Berkeley: University of California Press.
de Certeau, M. (1984b) *Heterologies*, Manchester: Manchester University Press.
de Lauretis, T. (1990) 'Eccentric subjects: feminist theory and historical consciousness', *Feminist Studies*, 16, 115–150.
de Man, P. (1989) *Wartime Journalism 1940–1942*, edited by W. Hamacher, N. Hertz, and T.

Keenan, Lincoln, University of Nebraska Press.
Deleuze, G. (1988) *Foucault*, Minneapolis: University of Minnesota Press.
Deleuze, G. and Guattari, F. (1987) *A Thousand Plateaus: Capitalism and Schizophrenia, Volume 2*, Minneapolis: University of Minnesota Press.
Deleuze, G. and Guattari, F. (1994) *What is Philosophy?*, London: Verso.
Dempster, E. (1988) 'Women writing the body: let's watch a little how she dances', in S. Sheridan, (ed.) *Grafts: Essays in Feminist Criticism*, London: Verso, 35–54.
Derrida, J. (1984) *Otobiographies: L'enseignement de Nietzsche et la politique du nom propre*, Paris, Galilee.
Derrida, J. (1988) 'Like the sound of the sea deep within a shell: Paul de Man's war', *Critical Inquiry*, 14, 590–652.
Derrida, J. (1994) *Specters of Marx*, Routledge, New York and London.
Deutsche, R. (1995) 'Surprising geography', *Annals of the Association of American Geographers*, 85(1), 168–175.
Dietz, G. (1995) 'Zapatismo y Movimientos Étnicos-Regionales en México', *Nueva Sociedad*, 140, November–December, 33–50.
Donzelot, J. (1991) 'The mobilization of society', in G. Burchell and C. Gordon (eds) *The Foucault Effect: Studies in Governmentality*, Brighton: Harvester Wheatsheaf, 169–180.
Donzelot, J. (1993) 'The promotion of the social', in M. Gane and T. Johnson (eds) *Foucault's New Domains*, London and New York: Routledge.
Douglas, D. (1993) '"Magic circle" allegations thrown out', *Glasgow Herald*, 27 January.
Douglas, D. and Dinwoodie, R. (1992) 'Gays in turmoil over Fettesgate', *Glasgow Herald*, 11 August.
Douglas, S. (1992a) 'Files theft twist', *Edinburgh Evening News*, 4 August.
Douglas, S. (1992b) 'Officers probing Fettes affair swoop on hotel', *Edinburgh Evening News*, 5 August.
Dreyfus, H. L. (1991) *Being – in-the – World*, Cambridge, MA: MIT Press.
Dreyfus, H. L. (1996) 'The current relevance of Merleau Ponty's phenomenology of embodiment', in H. Haber and G. Weiss (eds) *Perspectives on Embodiment*.
Duncan, N. (ed.) (1996) *BodySpace: Destabilizing Geographies of Gender and Sexuality*, London: Routledge.
Easton, S. (1992) 'Longest survivor declined hero status', *The Vancouver Province*, 16 November, A-5.
Edinburgh Evening News (1992) 'Storm over gay article', 20 November.
Edinburgh Evening News (1993) 'Gay claim was police fantasy', 14 January.
Eisenstadt, S. N. (1967) *Israeli Society*, Jerusalem: Magnes University Press (in Hebrew).
EIU (Economist Intelligence Unit) (1990–1996, various issues) *Nigeria: Quarterly Economic Report*, London: Economist Intelligence Unit.
Elliott, A. (1996) *Subject to Ourselves: Social Theory, Psychoanalysis and Postmodernity*, Cambridge: Polity Press.
Enloe, C. (1989) *Bananas, Beaches and Bases: Making Feminist Sense of International Politics*, London: Pandora Press.
Enriquez, V. (1990) 'Indigenous personality theory', in V. Enriquez (ed.) *Indigenous Psychology*, Quezon City: Akademya ng Sikolohiyang Pilipi.
Escobar, A. (1995) *Encountering Development*, Princeton: Princeton University Press.
Escobar, A. and Alvarez, S. E. (eds) (1992) *The Making of Social Movements in Latin America*, Boulder: Westview Press.
Evans, R. J. (1995) 'The deceptions of Albert Speer', *The Times Literary Supplement*, 29 September.
Fabian, J. (1983) *Time and the Other*, New York: Columbia University Press.
Falk, R. (1993) 'The making of global citizenship', in J. Brecher, J. Brown Childs and J. Cutler (eds) *Global Visions: beyond the New World Order*, Montreal: Black Rose Books, 39–50.
Fals Borda, O. (1992) 'Social movements and political power in Latin America', in A. Escobar and S. Alvarez (eds) *The Making of Social Movements in Latin America*, Boulder: Westview Press, 303–316.
Faludi, S. (1992) *Backlash: the Undeclared War against American Women*, New York: Doubleday.
Fanon, F. (1952) *Black Skin, White Masks*, repr. 1986, London: Pluto Press.
Fanon, F. (1959) *Studies in a Dying Colonialism*, repr. 1989, London: Earthscan.
Farias, V. (1989) *Heidegger and Nazism*, Philadelphia, Temple University Press.
Farquharson, K. (1992) 'Fettes target: secret dossier on top gays', *Scotland on Sunday*, 9 August.

Federal Government, Nigeria (1981) *Report of Tribunal of Inquiry on Kano Disturbances*, Lagos: Federal Government Press.
Feierman, S. (1990) *Peasant Intellectuals: Anthropology and History in Tanzania*, Madison: University of Wisconsin Press.
Feig, K. G. (1981) *Hitler's Death Camps: the Sanity of Madness*, New York: Holmes and Meier.
Ferguson, R., Gever, M., Trinh, M.-h. T. and West, C. (eds) (1990) *Out There: Marginalization and Contemporary Culture*, Cambridge: MIT Press.
Filler, M. (1994) 'Prince of the City', *New York Review of Books*, 22 December.
Fischer, M. and Abedi, A. (1990) *Debating Muslims*, Madison: University of Wisconsin Press.
Fleming, G. (1984) *Hitler and the Final Solution*, Berkeley: University of California Press.
FOPHUR (Forum for Protection of Human Rights) (1990a) *Dawn of Democracy*, Kathmandu: Forum for Protection of Human Rights.
FOPHUR (Forum for Protection of Human Rights) (1990b) *FOPHUR and Pro-Democracy Movement*, Kathmandu: Forum for Protection of Human Rights.
Forest, M. (1995) 'West Hollywood as symbol: the significance of place in the construction of a gay identity', *Environment and Planning D: Society and Space*, 13(2), 133–156.
Forrest, T. (1995) *Politics and Economic Development in Nigeria*, Boulder: Westview.
Forsyth, D. (1992) 'Gay threat to justice', *Edinburgh Evening News*, 11 September.
Forsyth, D. and Douglas, S. (1992) 'Amnesty for Fettes raider', *Edinburgh Evening News*, 25 September.
Foster, S. L. (1986) *Reading Dancing: Bodies and Subjects in Contemporary American Dance*, Berkeley: University of California Press.
Foster, S. L. (ed.) (1996) *Corporealities: Dancing Knowledge, Culture and Power*, New York: Routledge.
Foucault, M. (1978) *The History of Sexuality: an Introduction. Volume 1*, London: Penguin.
Foucault, M. (1980) 'The question of geography', in C. Gordon, (ed.) *Power/Knowledge*, New York: Pantheon Books, 63–77.
Foucault, M. (1983) 'The subject and power', in H. Dreyfus and P. Rabinow (eds) *Michel Foucault: beyond Structuralism and Hermeneutics*, Chicago: University of Chicago Press, 208–226.
Foucault, M. (1986) 'Of other spaces', *Diacritics*, 16(1), 22–27.
Foucault, M. (1991) 'Governmentality', in G. Burchell, C. Gordon and P. Miller (eds) *The Foucault Effect: Studies in Governmentality*, London: Harvester Wheatsheaf.
Foweraker, J. (1995) *Theorizing Social Movements*, London: Pluto Press.
Fox, R. (1985) *Lions of the Punjab: Culture in the Making*, Berkeley: University of California Press.
Frankenberg, R. and Mani, L. (1993) 'Crosscurrents, crosstalk: race, "postcoloniality" and the politics of location', *Cultural Studies*, 7(2), 292–310.
Fraser, K. (1992) 'Dr. Peter dies in his sleep: AIDS activist put human face to deadly disease', *Vancouver Province*, 16 November, A-1.
Freadman, A. (1988) 'Of cats, and companions, and the name of George Sand', in S. Sheridan (ed.) *Grafts: Feminist Cultural Criticism*, London: Verso, 125–156.
Freud, S. (1915) 'Repression', in *On Metapsychology: the Theory of Psychoanalysis*, repr. 1984, Volume 11, Harmondsworth: Penguin Freud Library, 145–158.
Freud, S. (1923) 'The ego and the id', in *On Metapsychology: the Theory of Psychoanalysis*, repr. 1984, Volume 11, Harmondsworth: Penguin Freud Library, 350–407.
Freud, S. (1926) 'Inhibitions, symptoms and anxiety', in *On Psychopathology: Inhibitions, Symptoms and Anxiety and Other Works*, repr. 1979, Volume 10, Harmondsworth: Penguin Freud Library, 237–333.
Fuller, J. and Blackley, S. (1995) *Restricted Entry: Censorship on Trial*, Vancouver: Press Gang Publishers.
Furro, T. (1992) 'Federalism and the politics of revenue allocation in Nigeria', Ph.D. dissertation, Clark Atlanta University.
Fuss, D. (1989) *Essentially Speaking: Feminism, Nature and Difference*, London: Routledge.
Fuss, D. (1994) 'Interior colonies: Frantz Fanon and the politics of identification', *Diacritics*, 24(2/3), 20–42.
Gal, S. (1995) 'Language and the "arts of resistance"', *Cultural Anthropology*, 10(3), 407–424.
Game, A. and Metcalfe, A. (1996) *Passionate Sociology*, London: Sage.
Gamson, J. (1991) 'Silence, death, and the invisible enemy: AIDS activism and social movement "newness"', in M. Burawoy (ed.) *Ethnography Unbound: Power and Resistance in the Modern Metropolis*, Berkeley: University of California Press, 35–57.

Gawthrop, D. (1994) *Affirmation: The AIDS Odyssey of Dr. Peter*, Vancouver: New Star Books.

Gay Scotland (1992) 'An old Fettesian speaks out', 62 (November), 5.

Geertz, C. (1972) 'Deep play: notes on the Balinese cock fight', *Daedalus*, 101, 1–37.

Geertz, C. (1973) *The Interpretation of Cultures*, New York: Basic Books.

Gibson-Graham, J. K. (1996) *The End of Capitalism (as We Know It): a Feminist Critique of Political Economy*, Oxford: Basil Blackwell.

Giddens, A. (1980) *The Class Structure of the Advanced Societies*, second edition, London: Hutchinson.

Gill, S. (1995) 'The global panopticon? The neoliberal state, economic life and democratic surveillance', *Alternatives*, 20, 1, Jan.–March, 1–49.

Gilroy, P. (1993) *Small Acts*, London: Serpent's Tail.

Goldberg, M. and Mercer, J. (1986) *The Myth of the North American City*, Vancouver: UBC Press.

Goldberger, P. (1994) 'The man in the glass house', *New York Times Book Review*, 27 November, 14.

Goldhagen, D. J. (1996) *Hitler's Willing Executioners: Ordinary Genius and the Holocaust*, New York: Knopf.

Gonzalez, J. and Habell-Pallan, M. (1994) 'Heterotopias and shared methods of resistance: navigating social spaces and spaces of identity', *Inscriptions*, 7, 80–104.

Gooding-Williams, R. (ed.) (1993) *Reading Rodney King/Reading Urban Uprising*, London: Routledge.

Gordon, C. (1991) 'Governmental rationality: an introduction', in G. Burchell, C. Gordon and P. Miller (eds) *The Foucault Effect: Studies in Governmentality*, London: Harvester Wheatsheaf.

Gordon, C. (1993) 'A maverick insider', *The Times Literary Supplement*, 27 August.

Gordon, L. R., Sharpley-Whiting, T. D. and White, R. T. (eds) (1996) *Fanon: a Critical Reader*, Oxford: Basil Blackwell.

Gramsci, A. (1971) *Selections from the Prison Notebook*, London: Lawrence and Wishart.

Grasby, A. J. and Hill, M. (1988) *Six Australian Battlefields: the Black Resistance to Invasion and the White Struggle against Colonial Oppression*, North Ryde, NSW: Angus and Robertson.

Greenpeace (1994) *Shell Shocked*, Amsterdam: Greenpeace International.

Gregoire, E. (1993) 'Islam and identity of merchants in Maradi (Niger)', in L. Brenner (ed.) *Muslim Identity and Social Change in Sub Saharan Africa*, Bloomington, University of Indiana Press, 106–115.

Gregory, D. (1978) *Ideology, Science and Human Geography*, London: Hutchinson.

Gregory, D. (1994) 'Social theory and human geography', in D. Gregory, R. Martin and G. Smith (eds) *Human Geography*, Minneapolis: University of Minnesota Press, 78–109.

Grewal, I. and Kaplan, C. (eds) (1994) *Scattered Hegemonies: Postmodernity and Transnational Practice*, Minneapolis: University of Minnesota Press.

Grossberg, L. (1992) *We Gotta Get Out Of This Place: Popular Conservatism and Post-modern Culture*, New York: Routledge.

Grossberg, L. (1996) 'Identity and cultural studies: is that all there is?', in S. Hall and P. du Gay (eds) *Questions of Cultural Identity*, London: Sage, 87–107.

Grosz, E. (1994) *Volatile Bodies: toward a Corporeal Feminism*, St Leonards, NSW: Allen and Unwin.

Guattari, F. (1996) 'Microphysics of power/micropolitics of desire', in G. Genosko (ed.) *The Guattari Reader*, Oxford, Blackwell, 172–184.

Guha, R. (1982) 'On some aspects of the historiography of colonial India', in R. Guha (ed.) *Subaltern Studies I*, Delhi: Oxford University Press, 1–8.

Guha, R. (1989) 'Dominance without hegemony and its historiography', in R. Guha (ed.) *Subaltern Studies VI*, Delhi: Oxford University Press, 210–309.

Gumbrecht, H. U. and Pfeiffer, K. L. (eds) (1994) *Materialities of Communication*, Stanford, Stanford University Press.

Gunew, S. and Yeatman, A. (eds) (1993) *Feminism and the Politics of Difference*, Boulder: Westview Press.

Gupta, A. and Ferguson J. (1992) 'Beyond "culture": space, identity, and the politics of difference', *Cultural Anthropology*, 7(1), 6–23.

Gupta, A. and Ferguson J. (eds) (1997) *Culture, Power, Place: Explorations in Critical Anthropology*, Durham: Duke University Press.

Gurr, T. R. (1970) *Why Men Rebel*, Princeton: Princeton University Press.

Guyer, J. (1994) 'The spatial dimensions of civil society in Africa: an anthropologist looks at Nigeria', in J. Harbeson, D. Rothchild, and N. Chazan (eds) *Civil Society and the State in Africa*, Boulder: Lynne Reinner, 215–229.

Hall, S. (1981) 'Notes on deconstructing "the popular"', in R. Samuel (ed.) *People's History and Socialist Theory*, London: Routledge and Kegan Paul, 227–240.

Hall, S. (1982) 'The rediscovery of ideology: return of the repressed in media studies', in M. Gurevitch, T. Bennett, J. Curran and J. Woolacott (eds) *Culture, Society and the Media*. New York: Methuen, 56–90.

Hall, S. (1990) 'Cultural identity and diaspora', in J. Rutherford (ed.) *Identity: Community, Culture, Difference*, London: Lawrence and Wishart, 222–237.

Hall, S. (1992a) 'The question of cultural identity', in S. Hall, D. Held and T. McGrew (eds) *Modernity and Its Futures*, Oxford: Polity Press, 273–325.

Hall, S. (1992b) 'Cultural studies and its theoretical legacies', in L. Grossberg, C. Nelson and P. Treichler (eds) *Cultural Studies*, New York: Routledge, 277–294.

Hall, S. (1995a) 'Fantasy, identity, politics', in E. Carter, J. Donald and J. Squires (eds) *Cultural Remix: Theories of Politics and the Popular*, London: Lawrence and Winehart, 63–69.

Hall, S. (1995b) 'New cultures for old', in D. Massey and P. Jess (eds) *A Place in the World? Places, Culture and Globalization*, Oxford: Oxford University Press and The Open University, 175–215.

Hall, S. and Jefferson, T. (eds) (1976) *Resistance through Rituals: Youth Sub-cultures in Post-war Britain*, London: Routledge and Kegan Paul.

Hammer, J. (1996) 'Nigerian crude', *Harpers Magazine*, June, 58–68.

Haraway, D. (1991) 'Overhauling the meaning machines (an interview with Donna Haraway)', *Socialist Review*, 21(2), 65–84.

Haraway, D. (1992) 'The promises of monsters: a regenerative politics for inappropriate/d others', in L. Grossberg, C. Nelson and P. Treichler (eds) *Cultural Studies*, New York: Routledge, 295–337.

Hardimon, M. O. (1994) *Hegel's Social Philosophy: the Project of Reconciliation*, Cambridge University Press: Cambridge.

Harley, J. B. (1988) 'Maps, knowledge, and power', in D. Cosgrove and S. Daniels (eds) *The Iconography of Landscape*, Cambridge: Cambridge University Press, 277–312.

Harré, R. and Gillett, G. (1994) *The Discursive Mind*, London: Sage.

Hart, G. (1991) 'Engendering everyday resistance: gender, patronage and production politics in rural Malaysia', *Journal of Peasant Studies*, 19(1), 93–121.

Harvey, D. (1989) *The Condition of Postmodernity*, Oxford: Blackwell.

Harvey, D. (1990) 'Between space and time: reflections on the geographical imagination', *Annals of the Association of American Geographers*, 80(3), 418–434.

Harvey, D. (1993a) 'Class relations, social justice and the politics of difference', in M. Keith and S. Pile (eds) *Place and the Politics of Identity*, London: Routledge, 41–66.

Harvey, D. (1993b) 'From space to place and back again: reflections on the condition of modernity', in J. Bird, B. Curtis, T. Putnam, G. Robertson and L. Tickner (eds) *Mapping the Futures: Local Cultures: Global Change*, London: Routledge, 3–29.

Harvey, D. (1995) 'Geographical knowledge in the eye of power: reflections on Derek Gregory's "Geographical Imaginations"', *Annals of the Association of American Geographers*, 85(1), 160–164.

Harvey, N. (1995) 'Rebellion in Chiapas: rural reforms and popular struggle', *Third World Quarterly*, 16(1), 39–73.

Hasson, S. (1977) 'Immigrant housing projects in the old towns of Israel: a study of social differentiation', Ph.D. dissertation, Hebrew University of Jerusalem.

Hasson, S. (1993) *Urban Social Movements in Jerusalem: the Protest of the Second Generation*, Albany: State University of New York Press.

Hasson, S. and Ley, D. (1994) *Neighbourhood Organizations and the Welfare State*, Toronto: Toronto University Press.

Haverman, C. (1995) 'In a museum of hell, qualms about decorum', *New York Times*, 7 March, A-6.

Hayles, N. K. (1996) 'Embodied virtuality: or how to put bodies back into the picture', in C. A. Moser (ed.) *Immersed in Technology: Art and Virtual Experts*, Cambridge, MA: MIT Press.

Haynes, D. and Prakash, G. (1991) 'Introduction: the entanglement of power and resistance', in D. Haynes and G. Prakash (eds) *Contesting Power: Resistance and Everyday Social Relations in South Asia*, Delhi: Oxford University Press, 1–22.

Hebdige, D. (1979) *Subculture: the Meaning of Style*, London: Methuen.
Hebdige, D. (1993) 'Redeeming witness: in the tracks of the Homeless Vehicle Project', *Cultural Studies*, 7(2), 173–223.
Hecht, D. and Simone, M. (1994) *Invisible Governance*, New York: Autonomedia.
Hecker, J. F. (1970) *The Dancing Mania in the Middle Ages*, New York: Franklin Burt.
Hegel, G. (1991) *Elements of a Philosophy of Right*, New York: Cambridge University Press.
Heidegger, M. (1976) 'Only a God can save us', *Der Speigel*, 31 May (in German); repr. in *Philosophy Today*, 20(4/4), 267–285 and in R. Wolin (ed.) *The Heidegger Controversy: a Critical Reader*, Cambridge: MIT Press.
Held, D. (1993) *From City States to a Cosmopolitan Order: In Prospects for Democracy*, Oxford: Polity.
Held, D. (1995) *Democracy and the Global Order*, Polity Press, Cambridge.
Hershatter, G. (1993) 'The subaltern talks back: reflections on subaltern theory and Chinese history', *Positions*, 1(1), 103–130.
Herz, J. H. (1957) 'Rise and demise of the territorial state', *World Politics*, 9(4), 473–493.
Hill, C. (1975) *The World Turned Upside Down: Radical Ideas During the English Revolution*, Harmondsworth, Pelican.
Hirst, P. (1994) *Associative Democracy*, Oxford: Polity.
Hirst, P. and Thompson, G. (1992) 'The problem of "globalization": international economic relations, national economic management and the formation of trading blocs', *Economy and Society*, 21(4), 357–396.
Hobsbawm, E. (1973) *Revolutionaries*, London: Quartet Books.
Hochschild, A. R. (1983) *The Managed Heart: Commercialisation of Human Feeling*, Berkeley: University of California Press.
Hoffie, P. (1991) 'Centres and peripheries', in V. Binns (ed.) *Community and the Arts: History, Theory, Practice: Australian Perspectives*, Leichhardt, New South Wales: Pluto Press, 31–44.
Honig, B. (1993) *Political Theory and the Displacement of Politics*, Ithaca: Cornell University Press.
hooks, b. (1990a) 'Homeplace: a site of resistance', in b. hooks (1991) *Yearning: Race, Gender, and Cultural Politics*, Boston: South End Press, 41–50.
hooks, b. (1990b) 'Liberation scenes: speak this yearning', in b. hooks (1991) *Yearning: Race, Gender, and Cultural Politics*, Boston: South End Press, 1–14.
hooks, b. (1991) *Yearning: Race, Gender, and Cultural Politics*, London: Turnaround.
Hoorn, J. (1992) 'The lie of the land: Joseph Lycett and Terra Nullius', *Modern Times*, August.
Huizinga, J. (1949) *Homo Ludens: a Study of the Play Element in Culture*, London: Routledge and Kegan Paul.
Human Rights Watch, (1995) *The Ogoni Crisis*, report, number 7/5, New York: Human Rights Watch.
Hutchison, A. (1992a) 'New leads on police HQ break-in', *Scotsman*, 4 August.
Hutchison, A. (1992b) 'Fettesgate: the full story of the raid and the secret deal', *Scotsman*, 15 October.
Hutchison, A. (1992c) 'Police ask Fettesgate raider if cash was offered', *Scotsman*, 31 October.
Hutchison, A. (1992d) 'Report on Fettes "deal" sent to fiscal', *Scotsman*, 18 November.
Hutchison, A. (1993a) 'Fettes suspects linked to attack on reporter', *Scotsman*, 5 January.
Hutchison, A. (1993b) 'Where the "magic circle" scandal had its origins', *Scotsman*, 28 January.
Hutton, R. (1996) *Stations of the Sun: a History of the Ritual Year in Britain*, Oxford, Oxford University Press.
Ibrahim, J. (1991) 'Religion and political turbulence in Nigeria', *Journal of Modern African Studies*, 29(1), 115–136.
Ikein, A. (1990) *The Impact of Oil on a Developing Country*, New York: Praeger.
Ikporukpo, C. (1993) 'Oil companies and village development in Nigeria', *OPEC Review*, 83–97.
Ikporukpo, C. (1996) 'Federalism, political power and the economic power game: control over access to petroleum resources in Nigeria', *Environment and Planning C*, 14, 159–177.
INSEC (Informal Sector Research Center) (1991) *Nepal and its Electoral System*, Kathmandu: Informal Sector Service Center.
INSEC (Informal Sector Research Center) (1992) *Identification of Bonded Labour in Nepal*, Kathmandu: Informal Sector Service Center.
International Defence and Aid Fund (1972) *Rhodesia: the Ousting of the Tangwena*, London: Christian Action Publications.
Irigaray, L. (1985) *This Sex Which Is Not One*, Ithaca: Cornell University Press.

Jackson, P. (1989) *Maps of Meaning: an Introduction to Cultural Geography*, London: Unwin Hyman.

Jackson, R. H. (1993) *Quasi-States: Sovereignty, International Relations and the Third World*, Cambridge University Press, Cambridge.

Jacobs, J. M. (1994) 'Earth honoring: western desires and indigenous knowledges', in A. Blunt and G. Rose (eds) *Writing Women and Space: Colonial and Postcolonial Geographies*, New York: Guilford Press, 169–196.

Jacobs, J. M. (1996) *Edge of Empire: Postcolonialism and the City*, London and New York: Routledge.

James, C. (1996) 'Blaming the Germans', *The New Yorker*, 22 April, 44–50.

Jameson, F. (1992) *The Geopolitical Aesthetic: Cinema and Space in the World System*, Bloomington: Indiana University Press.

Janik, A. and Toulmin, S. (1973) *Wittgenstein's Vienna*, New York: Touchstone.

Jaquette, J. S. (1994) 'Introduction: from transition to participation: women's movements and democratic politics', in J. S. Jaquette (ed.) *The Women's Movement in Latin America: Participation and Democracy*, Colorado: Westview Press, 1–11.

Jessop, B. (1990) *State Theory: Putting Capitalist States in Their Place*, Oxford: Polity.

Jessop, B. (1993) 'Towards a Schumpeterian workfare state? Preliminary remarks on post-Fordist political economy', *Studies in Political Economy*, 40, 7–39.

Jones, J. P. III and Moss, P. (1995) 'Democracy, identity, space', *Environment and Planning D: Society and Space*, 13(3), 153–257.

Jordan, T. (1995) 'The unity of social movements', *The Sociological Review*, 43(4), 675–692.

Joseph, G. and Nugent, D. (1994) 'Popular culture and state formation in revolutionary Mexico', in G. Joseph and D. Nugent (eds) *Everyday Forms of State Formation*, Durham: Duke University Press, 3–23.

Jowitt, K. (1992) *New World Disorders: the Leninist Extinction*, Berkeley, University of California Press.

Kala, P. (1990) 'Letter from Nepal', *Z Magazine*, 21–24 June.

Kaplan, A. (1993) *French Lessons: a Memoir*, Chicago: University of Chicago Press.

Kaplan, A. and Ross, K. (1987) 'Introduction', in A. Kaplan and K. Ross (eds) *Yale French Studies*, New Haven: Yale University Press, 1–20.

Kaplan, C. (1994) 'The politics of location as transnational feminist practice', in I. Grewal and C. Kaplan (eds) *Scattered Hegemonies*, Minneapolis: University of Minnesota Press, 137–152.

Kaplan, C. (1995) '"A world without boundaries": The Body Shop's trans/national geographics', *Social Text*, 43, 45–66.

Kaplan, C. (1996) *Questions of Travel: Postmodern Discourses of Displacement*, Durham, NC: Duke University Press.

Katz, R. (1982) *Boiling Energy: Community Healing among the Kalahari Kung*, Cambridge, MA: MIT Press.

Keane, J. (1988) *Democracy and Civil Society*, London: Verso.

Keating, J. P. (1992) *One Nation*, statement by the Prime Minister, the Honorable P. J. Keating, Canberra: Australian Government Publishing Service.

Keating, J. P. (1993) 'Australian launch of the international year of the world's indigenous people', repr. in J. P. Keating, L. O'Donoghue and S. Bellear, *Speeches to Mark the National and International Launch of the 1993 International Year of the World's Indigenous People*, Woden, ACT: Australian Institute of Aboriginal and Torres Strait Islanders.

Keith, M. and Pile, S. (eds) (1993a) *Place and the Politics of Identity*, London: Routledge.

Keith, M. and Pile, S. (1993b) 'Introduction: Part 1. The politics of place ...', in M. Keith and S. Pile (eds) *Place and the Politics of Identity*, London: Routledge, 1–21.

Keith, M. and Rogers, A. (eds) (1991) 'Introduction', in M. Keith and A. Rogers (eds) *Hollow Promises: Rhetoric and Reality in the Inner City*, London: Mansell.

Kellner, D. (1995) *Media Culture: Cultural Studies, Identity and Politics between the Modern and the Postmodern*, New York: Routledge.

Kelly, O. (1984) *Storming the Citadels: Community, Art and the State*, London: Comedia.

Kerkvliet, B. (1990) *Everyday Politics in the Philippines*, Berkeley: University of California Press.

Khan, S. A. (1994) *Nigeria: the Political Economy of Oil*, London: Oxford University Press.

Kharkhordin, D. (1995) 'The soviet individual: genealogy of a dissimulating animal', in M. Featherstone, S. Lash and A. Robertson (eds) *Global Modernities*, London: Sage, 209–226.

Kines, L. (1990) 'AIDS activist to take his life this week', *Vancouver Sun*, 20 August, B-6.

King, A. D. (ed.) (1991) *Culture, Globalization and the World System*, London: Macmillan.
Kinsey, R. (1992) 'Fettesgate: the real line of inquiry', *Scotsman*, 3 November.
Kirby, A. (1995) 'Straight talk on the PomoHomo question', *Gender, Place and Culture*, 2, 89–95.
Kirby, K. (1996) *Indifferent Boudaries: Exploring the Space of the Subject*, New York: Guilford Press.
Knopp, L. (1995) 'If you're going to get all hyped up you'd better go *somewhere*', *Gender, Place and Culture*, 2, 85–88.
Kogon, E., Langbein, H. and Rucherl, A. (eds.) (1993) *Nazi Mass Murder: a Documentary History of the Use of Poison Gas*, New Haven: Yale University Press.
Kondo, D. (1990) *Crafting Selves: Power, Gender and Discourses of Identity in a Japanese Workplace*, Chicago: University of Chicago Press.
Koonz, C. (1995) 'Blind by choice', *New York Times Book Review*, 8 October.
Kramer, J. (1995) 'The politics of memory', *The New Yorker*, 14 August.
Kramer, L. (1989) *Notes from the Holocaust: the Makings of an AIDS Activist*, New York: St Martins Press.
Kymlicka, W. (1990) *Contemporary Political Philosophy*, Toronto: Oxford University Press.
LaCapra, D. (1983) *Rethinking Intellectual History: Texts, Contexts, Language*, Ithaca: Cornell University Press.
Laclau, E. (1990) *New Reflections on the Revolution of Our Time*, Verso, London.
Laclau, E. (ed.) (1994a) *The Making of Political Identities*, London: Verso.
Laclau, E. (1994b) 'Negotiating the paradoxes of contemporary politics: an interview', *Angelaki*, 1(3), 43–50.
Laclau, E. and Mouffe, C. (1985) *Hegemony and Socialist Strategy*, London: Verso.
Lange, R. (1975) *The Nature of Dance: an Anthropological Perspective*, London.
Laserna, R. (1986) 'Movimientos sociales regionales (apuntes para la construcción de un campo empírico)', *Pensamiento Iberoamericano*, 10, 83–105.
Last, M. (1991) 'Adolescents in a Muslim city', *Kano Studies*, Special Issue, 1–22.
Latour, B. (1992) 'Where are the missing masses? A sociology of a few mundane artifacts', in M. Lynch and S. Woolgar (eds) *Representations of Scientific Practice*, Cambridge, MA: MIT Press.
Latour, B. (1993) *We Have Never Been Modern*, Hemel Hempstead, Harvester-Wheatsheaf.
Lavie, S. and Swedenburg, T. (1996) 'Between and Among the Boundaries of Culture: Bridging Text and Lived Experience in the Third Timespace', *Cultural Studies*, 10(1), 154–179.
Law, J. (1994) *Organising Modernity*, Oxford: Blackwell.
Law, L. (1997) 'A matter of "choice": discourses on prostitution in the Philippines', in L. Manderson and M. Jolly (eds), *Sites of Desire/Economies of Pleasure: Sexualities in Asia and the Pacific*, Chicago: University of Chicago Press, 233–261.
Lefebvre, H. (1974) *The Production of Space*, repr. 1991, Oxford: Blackwell.
Lefebvre, H. (1995) *Writings on Cities*, Oxford, Blackwell.
Lefort, C. (1988) *Democracy and Political Theory*, Polity Press, Cambridge.
Lehman, D. (1991) *Signs of the Times: Deconstruction and the Fall of Paul de Man*, New York: Poseidon Press.
Levinas, E. (1989) 'God and philosophy', in S. Hand (ed.) *The Levinas Reader*, Oxford: Blackwell, 166–189.
Levy, R. I. (1990) *Mesocosm*, Berkeley: University of California Press.
Lewis, C. and Pile, S. (1996) 'Woman, body, space: Rio Carnival and the politics of performance', *Gender, Place and Culture*, 3(1), 33–41.
Lewis, P. (1996) 'From prebendalism to predation: the political economy of decline in Nigeria', *Journal of Modern African Studies*, 24(1), 79–104.
Linde-Laursen, A. (1995) 'Small dilemmas – large issues: the making and rendering of a national border', *South Atlantic Quarterly*, 94, 1123–1144.
Lingis, A. (1994) *Foreign Bodies*, New York: Routledge.
Lippmann, L. (1991) *Generations of Resistance: Aborigines Demand Justice*, Melbourne: Longman Cheshire.
Loolo, G. (1981) *A History of the Ogoni*, Port Harcourt, Saros Publishers.
Loomba, A. and Kaul, S. (1994) 'Introduction: location, culture, post-coloniality', *Oxford Literary Review*, 16(1–2), 3–30.
Low, G. C.-L. (1996) *White Skins, Black Masks: Representation and Colonialism*, London: Routledge.
Lubeck, P. (1985) 'Islamic protest under semi-industrial capitalism', *Africa*, 55(4), 369–389.

Lubeck, P. (1986) *Islam and Labor*, Cambridge: Cambridge University Press.
Lubeck, P. (1995) 'Globalization and the Islamist moment', unpublished paper, Department of Sociology, University of California, Santa Cruz.
Magnusson, W. (1992) 'Decentering the state, or looking for politics', in W. Carroll (ed.) *Organizing Dissent: Contemporary Social Movements in Theory and Practice*, Toronto: Garamond, 69–80.
Malcomson, S. (1995) 'Disco dancing in Bulgaria', in M. Henderson (ed.), *Borders, Boundaries, and Frames: Essays in Cultural Criticism and Cultural Studies*, New York and London: Routledge.
Malkki, L. (1992) 'National geographic: the rooting of peoples and the territorialization of national identity among scholars and refugees', *Cultural Anthropology*, 7(1), 24–44.
Malkki, L. (1995) *Purity and Exile: Violence, Memory, and National Cosmology among Hutu Refugees in Tanzania*, Chicago: University of Chicago Press.
Mallon, F. (1994) 'The promise and dilemma of subaltern studies: perspectives from Latin American History', *American Historical Review*, 99(5), 1491–1515.
Mamdani, M. (1996) *Citizen and Subject*, Princeton, Princeton University Press.
Manderson, L. (1995) 'The pursuit of pleasure and the sale of sex', in P. Abramson (ed.) *Sexual Nature/Sexual Culture: Theorising Sexuality from the Perspective of Pleasure*, Chicago: University of Chicago Press.
Mani, L. (1990) 'Multiple mediations: feminist scholarship in the age of multinational reception', *Feminist Review*, 35, 24–41.
Mankekar, P. (1993) 'National texts and gendered lives: an ethnography of television viewers in a North Indian city', *American Ethnologist*, 20(3), 543–563.
Mannheim, K. (1936) *Ideology and Utopia*, New York: Harcourt Press.
Mannheim, K. (ed.) (1965) *Essays on the Sociology of Knowledge*, London: Routledge and Kegan Paul.
Marcus, G. (1989) *Lipstick Traces. A Secret History of the Twentieth Century*, London: Secker and Warburg.
Marris, P. (1980) *Meaning and Action*, Los Angeles: School of Architecture and Urban Planning.
Marx, K. (1867) *Capital, Volume 1*, repr. 1976, Harmondsworth: Penguin.
Marx, K. (1869) *The Eighteenth Brumaire of Louis Bonaparte*, repr. 1963, New York: International Publishers.
Massey, D. (1993a) 'Power-geometry and a progressive sense of place', in J. Bird, B. Curtis, T. Putnam, G. Robertson and L. Tickner (eds) *Mapping the Futures: Local Cultures, Global Change*, London: Routledge, 59–69.
Massey, D. (1993b) 'Politics and Space/Time', in M. Keith and S. Pile (eds) *Place and the Politics of Identity*, London: Routledge, 141–161.
Massey, D. (1994a) 'Double articulation: a place in the world', in A. Bammer (ed.) *Displacements*, Bloomington: Indiana University Press, 110–121.
Massey, D. (1994b) *Space, Place, and Gender*, Minneapolis: University of Minnesota Press.
Massey, D. (1995) 'Thinking radical democracy spatially', *Environment and Planning D: Society and Space*, 13, 283–288.
Massey, D. (forthcoming) 'unbounded spaces', mimeo.
McCarthy, J. D. and Zald, N. N. (1977) 'Resource mobilization and social movements: a partial theory', *American Journal of Sociology*, 86(1).
McCartney, B. (1992) 'Tie me up!', *Daily Record*, 22 September.
McClintock, A. (1995) *Imperial Leather: Race, Gender and Sexuality in the Colonial Context*, London: Routledge.
McCrone, D. (1992) *Understanding Scotland: the Sociology of a Stateless Nation*, London: Routledge.
McKay, G. (1996) *Senseless Acts of Beauty: Cultures of Resistance since the Sixties*, London: Verso.
McNeill, W. H. (1995) *Keeping Together in Time: Dance and Drill in Human History*, Cambridge, MA: Harvard University Press.
McRobbie, A. (1984) 'Dance and social fantasy', in A. McRobbie and M. Nava (eds) *Gender and Generation*, London: Macmillan.
McRobbie, A. (1991) 'Dance narratives and fantasies of achievement', in A. McRobbie, *Feminism and Youth Culture*, London: Macmillan.
McRobbie, A. (1994) *Postmodernism and Popular Culture*, New York: Routledge.
Melucci, A. (1989) *Nomads of the Present: Social Movements and Individual Needs in*

Contemporary Society, London: Hutchinson/Radius.
Merleau-Ponty, M. (1962) *Phenomenology of Perception*, London: Routledge and Kegan Paul.
Merleau-Ponty, M. (1968) *The Visible and the Invisible*, Evanston: Northwestern University Press.
Michael, M. (1996) *Constructing Identities*, London: Sage.
Miles, R. and Dunlop, A. (1987) 'Racism in Britain: the Scottish dimension', in P. Jackson (ed.) *Race and Racism: Essays in Social Geography*, London: Allen and Unwin.
Miller, J. (1993) *The Passion of Michel Foucault*, New York: Simon and Schuster.
Miller, J. H. (1988) 'NB', *Times Literary Supplement*, 17–23 June, 685.
Miralao, V. A., Carlos, C. O. and Santos A. F. (1990) *Women Entertainers in Angeles and Olongapo: A Survey Report*, Manila: Women's Education, Development, Productivity and Research Organisation (WEDPRO) and Katipunan ng Kababaihan Para Sa Kalayaan (KALAYAAN).
Mitchell, T. (1990) 'Everyday metaphors of power', *Theory and Society*, 19, 545–577.
Mohan, G. (1996) 'Adjustment and decentralization in Ghana: a case of diminished sovereignty', *Political Geography*, 15(1), January, 75–94.
Mohanty, C. T. (1987) 'Feminist encounters: locating the politics of experience', in M. Barrett and A. Phillips (eds) *Destabilizing Theory: Contemporary Feminist Debates*, repr. 1992, Cambridge: Polity Press, 74–92.
Mohanty, C. T. (1991) 'Cartographies of struggle: Third World women and the politics of feminism', in C. T. Mohanty, A. Russo and L. Torres (eds) *Third World Women and the Politics of Feminism*, Bloomington: Indiana University Press, 1–47.
Mohr, R. (1993) 'On some words from ACT UP: doing and being done', in *Gay Ideas: Outing and Other Controversies*, Boston: Beacon, 49–53.
Monga, C. (1995) 'Civil society and democratisation in francophone Africa', *The Journal of Modern African Studies*, 33(3), 359–379.
Monk, J. (1992) 'Gender in the landscape: expressions of power and meaning', in K. Anderson and F. Gale (eds) *Inventing Places: Studies in Cultural Geography*, London: Longman, 123–138.
Mooney, J. (1965) *The Ghost Dance Religion and the Sioux Outbreak of 1890*, Chicago: University of Chicago Press.
Moore, B. (1966) *Social Origins of Dictatorship and Democracy*, Boston: Beacon Press.
Morin, E. (1995) 'Fronteras de lo Político', *Revista de Occidente*, Madrid, 167, April, 5–18.
Morris, A. D. and McClurg Mueller, C. (1992) *Frontiers in Social Movement Theory*, New Haven: Yale University Press.
Morris, M. (1988) 'Things to do with shopping centres', in S. Sheridan (ed.) *Grafts: Feminist Cultural Criticism*, London: Verso, 193–226.
Mosquera, G. (1994) 'Some problems in transcultural curating', in J. Fisher (ed.) *Global Visions: towards a New Internationalism in the Visual Arts*, London: Kala Press, 133–139.
Mouffe, C. (1993) *The Return of the Political*, London: Verso.
Mouffe, C. (1995) 'Post-Marxism: democracy and identity', *Environment and Planning D: Society and Space*, 13, 259–265.
Moyana, H. (1987) *The Victory of Chief Rekayi Tangwena*, Harare: Longmann.
Muir, A. (1992) 'Gay judges link to palace', *The Sun*, 18 September.
Munck, G. L. (1995) 'Actor formation, social co-ordination and political strategy: some conceptual problems in the study of social movements', *Sociology*, 29(4), 667–685.
Murray, A. (1995) 'Femme on the streets, butch in the sheets', in D. Bell and G. Valentine (eds) *Mapping Desire*, New York and London: Routledge.
Naanen, B. (1995) 'Oil Producing minorities and the restructuring of Nigerian federalism', *Journal of Commonwealth and Comparative Politics*, 33(1), 46–58.
Nancy, J.-L. (1990a) 'Sharing voices', in G. L. Ormiston and A. D. Schrift (eds) *Transforming the Hermeneutic Context: from Nietzsche to Nancy*, Albany: State University of New York Press, 211–259.
Nancy, J.-L. (1990b) 'Finite history', in D. Caroll (ed.) *The States of 'Theory': History, Art, and Critical Discourse*, New York: Columbia University Press, 149–172.
Nancy, J.-L. (1991a) *The Inoperative Community*, Minneapolis: University of Minnesota Press.
Nancy, J.-L. (1991b) 'Introduction', in E. Cadava, P. Connor and J.-L. Nancy (eds) *Who Comes after the Subject*, London: Routledge, 1–8.
Nash, C. (1994) 'Remapping the body/land: new cartograpies of identity, gender, and landscape in Ireland', in A. Blunt and G. Rose (eds) *Writing Women and Space: Colonial and Postcolonial Geographies*, New York: Guilford Press, 227–250.

Nast, H. J. and Kobayashi, A. (1996) 'Re-corporealizing vision', in N. Duncan (ed.) *BodySpace: Destabilizing Geographies of Gender and Sexuality*, London: Routledge, 75–93.
Natter, W. (1995) 'Radical democracy: hegemony, reason, time and space', *Environment and Planning D: Society and Space*, 13, 267–274.
Nepal Human Rights Committee–USA (1993) *Nepal Today*, 4(1), Washington DC: Nepal Human Rights Committee–USA.
NEST (Nigerian Environmental Study Action Team) (1991) *Nigeria's Threatened Environment*, Ibadan, Nigerian Environmental Study Action Team.
Nicoll, F. (1993) 'The art of reconciliation: art Aboriginality and the state', *Meanjin*, 52(4), 705–718.
Nietzsche, F. W. (1968) *The Will to Power*, New York: Vintage.
Nietzsche, F. W. (1973) *Beyond Good and Evil*, section 6, in W. Kaufman (ed.) *Basic Writings of Nietzsche*, New York: Modern Library.
Nordstrom, C. (1995) 'War on the front lines', in C. Nordstrom and A. C. G. M. Robben (eds) *Fieldwork Under Fire: Contemporary Studies of Violence and Survival*, Berkeley: University of California Press, 129–154.
Norris, C. (1986) *Derrida*, London: Fontana.
Northrop, A. (1992) 'The radical debutante', in E. Marcus, (ed.) *Making History: the Struggle for Gay and Lesbian Equal Rights, 1945–1990: an Oral History*, New York: Harper, 474–490.
Nugent, D. and Alonso, A. M. (1994) 'Multiple selective traditions in agrarian reform and agrarian struggle: popular culture and state formation in the Ejido of Namiquipa, Chihuahua', in G. Joseph and D. Nugent (eds) *Everyday Forms of State Formation*, Durham: Duke University Press, 209–246.
O'Donnell, G. (1993) 'On the state, democratization and some conceptual problems', *World Development*, 21(8), 1355–1369.
O'Hanlon, R. (1988) 'Recovering the subject: subaltern studies and histories of resistance in colonial South Asia', *Modern Asian Studies*, 22(1), 189–224.
O'Tuathail, G. (1995) 'Political geography I: theorizing history, gender and the world order amidst crises of global governance', *Progress in Human Geography*, 19(2), 260–273.
Offe, C. (1985) 'New social movements: challenging the boundaries of institutional politics', *Social Research*, 52, 817–868.
Ogbonna, D. (1979) 'The geographic consequences of petroleum in Nigeria with special reference to Rivers State', Ph.D. dissertation, University of California, Berkeley.
Okilo, M. (1980) *Derivation: a Criterion of Revenue Allocation*, Port Harcourt: Rivers State Newspaper Corporation.
Okpu, U. (1977) *Ethnic Minority Problems in Nigerian Politics*, Stockholm: Wiksell.
Olander, W. (1991) 'The window on Broadway by ACT UP', in B. Wallis (ed.) *Democracy: a Project by Group Material*, Seattle: Bay Press, 277–279.
Olsson, G. (1991) *Lines of Power, Limits of Language*, Minneapolis: University of Minnesota Press.
Ong, A. (1987) *Spirits of Resistance and Capitalist Discipline: Factory Women in Malaysia*, Albany: State University of New York Press.
Ong, A. (1990) 'State versus Islam: Malay families, women's bodies, and the body politic in Malaysia', *American Ethnologist*, 17(2), 258–276.
Ong, A. (1991) 'The gender and labour politics of postmodernity', *Annual Review of Anthropology*, 20, 279–309.
Ortner, S. (1995) 'Resistance and the problem of ethnographic refusal', *Comparative Studies in Society and History*, 37(1), 173–193.
Ortner, S. and Whitehead, H. (eds) (1981) *Sexual Meanings: the Cultural Construction of Gender and Sexuality*, Cambridge: Cambridge University Press.
Osaghae, E. (1991) 'Ethnic minorities and federalism in Nigeria', *African Affairs*, 90, 237–258.
Osaghae, E. (1995) 'The Ogoni uprising', *African Affairs*, 94, 325–344.
Ott, H. (1993) *Martin Heidegger: a Political Life*, New York: Harper Collins.
Ould-Mey, M. (1994) 'Global adjustment: implications for peripheral states', *Third World Quarterly*, 15(2), 319–336.
Ozouf, M. (1988) *Festivals and the French Revolution*, Cambridge, MA: Harvard University Press.
Paden, J. (1973) *Religion and Political Culture in Ka*, Berkeley: University of California Press.
Paige, J. (1975) *Agrarian Revolution: Social Movements and Export Agriculture in the Undeveloped World*, New York: Free Press.

Painter, J. (1995) *Politics, Geography and 'Political Geography': a Critical Perspective*, London: Edward Arnold.

Palmer, R. (1977) 'The agricultural history of Rhodesia', in R. Palmer and N. Parsons (eds) *The Roots of Rural Poverty in Central and Southern Africa*, Berkeley: University of California Press, 221–254.

Parajuli, P. (1992) 'Beyond parliaments and elections: democracy and political culture in Nepal', *South Asia Forum Quarterly*, 5(1), 1–4.

Parekh, B. (1993) 'The cultural particularity of liberal democracy', in D. Held (ed.) *Prospects for Democracy*, Cambridge: Polity Press, 156–175.

Parry, B. (1991) 'The contradictions of cultural studies', *Transition*, 53, 37–45.

Parry, B. (1994) 'Resistance theory/theorising resistance, or two cheers for nativism', in F. Barker, P. Hulme and M. Iversen (eds) *Colonial Discourse/Postcolonial Theory*, Manchester: Manchester University Press, 172–196.

Peet, R. (ed.) (1977) *Radical Geography: Alternative Viewpoints on Contemporary Social Issues*, London: Methuen.

Peet, R. and Watts, M. (eds) (1996) *Liberation Ecologies: Environment, Development, Social Movements*, London: Routledge.

Phelan, P. (1993) *Unmasked*, London: Routledge.

Pickering, A. (1995) *The Mangle of Practice: Time, Agency and Science*, Chicago, University of Chicago Press.

Pickvance, C. G. (1976) 'On the study of urban social movements', in C. G. Pickvance (ed.) *Urban Sociology: Critical Essays*, London: Tavistock.

Pickvance, C. G. (1977) 'From "social base" to "social force": some analytical issues in the study of urban protest', in M. Harloe (ed.) *Captive Cities*, Chicago: John Wiley.

Pieterse, J. N. (1992) 'Emancipations, modern and postmodern', *Development and Change*, 23(3), 5–43.

Pile, S. (1994) 'Masculinism, the use of dualistic epistemologies and third spaces', *Antipode*, 26(3), 255–77.

Pile, S. (1996) *The Body and the City: Psychoanalysis, Space and Subjectivity*, London: Routledge.

Pile, S. and Thrift, N. (1995a) 'Conclusions: spacing the subject', in S. Pile and N. Thrift (eds) *Mapping the Subject: Geographies of Cultural Transformation*, London: Routledge, 371–380.

Pile, S. and Thrift, N. (eds) (1995b) *Mapping the Subject: Geographies of Cultural Transformation*, London: Routledge.

Pilger, J. (1984) *A Secret Country*, Vintage: London.

Pini, M. (1996) 'Dance classes – dancing between classifications', *Feminism and Psychology*, 6, 411–426.

Platzky, L. and Walker, C. (1985) *The Surplus People: Forced Removals in South Africa*. Johannesburg: Ravan Press.

Polanyi, K. (1967) *The Tacit Dimension*, London: Routledge and Kegan Paul.

Prakash, G. (1990) 'Writing post-orientalist histories of the third world: perspectives from Indian historiography', *Comparative Studies in Society and History*, 32(2), 383–408.

Prakash, G. (1992) 'Can the subaltern ride? A reply to O'Hanlon and Washbrook', *Comparative Studies in Society and History*, 34(1), 168–184.

Prakash, G. (1994) 'Subaltern studies as postcolonial criticism', *American Historical Review*, 99(5), 1475–1490.

Pratt, M. L. (1992) *Imperial Eyes: Travel Writing and Transculturation*, New York: Routledge.

Pred, A. and Watts, M. (1992) *Reworking Modernity: Capitalisms and Symbolic Discontent*. New Brunswick, New Jersey: Rutgers University Press.

Presland, G. (1994) *Aboriginal Melbourne: the Lost Land of the Kulin People*, Melbourne: Penguin Books, McPhee Gribble.

Probyn, E. (1995) 'Lesbians in space: gender, sex and the structure of missing', *Gender, Place and Culture*, 2, 77–84.

Probyn, E. (1996) *Outside Belongings*, London: Routledge.

Radcliffe, S. (1993) 'Women's place/el lugar de mujeres: Latin America and the politics of gender identity', in M. Keith and S. Pile (eds) *Place and the Politics of Identity*, London: Routledge, 102–116.

Radcliffe, S. (1996) 'Entangled geographies of identities: "popular" discourses around nation, place and difference in Ecuador', paper presented at the 'Geographies of Domination/Resistance' conference, Glasgow, 19–21 September.

Radhakrishnan, R. (1993) 'Postcoloniality and the boundaries of identity', *Callaloo*, 16(4), 750–771.
Radley, A. (1995) 'The elusory body and social constructionist theory', *Body and Society*, 1, 3–24.
Rajchman, J. (ed.) (1995) *The Identity in Question*, New York: Routledge.
Rancière, J. (1995) *On the Shores of Politics*, London: Verso.
Ranger, T. O. (1975) *Dance and Society in East Africa, 1890–1970*, London: Oxford University Press.
Raven, A. (1989) 'Introduction', in A. Raven (ed.) *Art in the Public Interest*, Ann Arbor: UMI Research Press, 1–28.
Rayside, D. M. and Lindquist, E. A. (1992) 'AIDS activism and the state in Canada', *Studies in Political Economy*, 39, 37–76.
Read, A. (ed.) (1996) *The Fact of Blackness: Frantz Fanon and Visual Representation*, London: Institute of Contemporary Arts and Institute of International Visual Arts.
Rebalski, M. (1989) 'AIDS activists stage protest at Fantasy Gardens', *Vancouver Sun*, 5 September, D-12.
Rex, J. A. (1968) 'The sociology of a zone in transition', in R. Pahl (ed.) *Readings in Urban Sociology*, London: Hutchinson.
Rex, J. and Moore, R. (1967) *Race, Community and Conflict*, London: Oxford University Press.
Rich, A. (1987) 'Notes toward a politics of location', in A. Rich, *Blood, Bread, and Poetry: Selected Prose 1979–1985*, London: Virago Press, 210–231.
Ricoeur, P. (1995) 'La persona: desarrollo moral y político', *Revista de Occidente*, 167, April, 129–142.
Rimmer, A. (1992a) 'Mr Evil's gay hush-money', *Sunday Mirror*, 27 September.
Rimmer, A. (1992b) 'Sex-slave boy's palace pervert', *Sunday Mirror*, 27 September.
Robertson, R. (1992) *Globalization: Social Theory and Global Culture*, Sage: London.
Robinson, J. (1996) 'Femininity, friendship and the "noisy look": Octavia Hill women housing managers in South Africa', paper presented at the 'Geographies of Domination/Resistance' conference, Glasgow, 19–21 September.
Rosaldo, M. (1980) 'The use and abuse of anthropology: reflections on feminism and cross-cultural understanding', *Signs*, 5(3), 389–417.
Rose, G. (1993a) *Feminism and Geography: the Limits of Geographical Knowledge*, Cambridge: Polity Press.
Rose, G. (1993b) 'The perils of selectively situating knowledges', *Women in Geography Study Group Newsletter*, 4, 9–10.
Rose, G. (1994) 'The cultural politics of place: local representation and oppositional discourse in two films', *Transactions of the Institute of British Geographers*, 19(1), 46–60.
Rose, G. (1995) 'The interstitial perspective: a review essay on Homi Bhabha's *The Location of Culture*', *Environment and Planning D: Society and Space*, 13, 365–73.
Rose, G. (1996) 'Community arts and the remaking of Edinburgh's geographies', *Scotlands*, 3 (forthcoming).
Rose, G. (1997) 'Spatialities of "community", power and change: the imagined geographies of community arts projects', *Cultural Studies*, 17(1) (forthcoming).
Rose, J. (1986) *Sexuality in the Field of Vision*, London: Verso.
Rose, N. (1996a) 'Identity, genealogy, history', in S. Hall and P. du Gay (eds) *Questions of Cultural Identity*, London: Sage, 128–150.
Rose, N. (1996b) *Inventing Ourselves: Psychology, Power and Personhood*, Cambridge: Cambridge University Press.
Rose, N. (1996c) 'The death of the social: re-figuring the territory of government', *Economy and Society* (forthcoming).
Ross, D. (undated) *Contacts: Individual and Protest Poems*, Jerusalem: private publication (in Hebrew).
Ross, K. (1988) *The Emergence of Social Space: Rimbaud and the Paris Commune*, Basingstoke: Macmillan.
Routledge, P. (1993) *Terrains of Resistance: Non-violent Social Movements and the Contestation of Place in India*, Westport: Praeger.
Routledge, P. (1994) 'Backstreets, barricades, and blackouts: urban terrains of resistance in Nepal', *Society and Space*, 12(5), 559–578.
Routledge, P. (1996) 'Critical geopolitics and terrains of resistance', *Political Geography*, 15(6/7), 509–531.

Routledge, P. and Simons, J. (1995) 'Embodying spirits of resistance', *Environment and Planning D: Society and Space*, 13, 471–498.

Royal Commission into Aboriginal Deaths in Custody, Australia (1991) *National Report: the Process of Reconciliation*, Volume 5, Chapter 38, Canberra: Australian Government Publishing Service.

Rupley, L. (1981) 'Revenue sharing in the Nigerian federation', *Journal of Modern African Studies*, 19(2), 252–277.

Russo, V. (1992) 'The film historian', in E. Marcus (ed.) *Making History: the Struggle for Gay and Lesbian Equal Rights, 1945–1990: an Oral History*, New York: Harper, 407–419.

Rutherford, J. (ed.) (1990) *Identity: Community, Culture, Diffference*, London: Lawrence and Wishart.

Ryan, A. (1993) 'Foucault's life and hard times', *New York Review of Books*, 8 April.

Saad, H. (1988) 'Urban blight and religious uprising in Northern Nigeria' *Habitat International*, 12(2), 111–128.

Saalfield, C. and Navarro, R. (1991) 'Shocking pink praxis: race and gender on the ACT UP frontlines', in D. Fuss (ed.) *Inside/Out: Lesbian Theories, Gay Theories*, London: Routledge, 291–311.

Said, E. (1983) 'Traveling theory', in E. Said, *The World, the Text, and the Critic*, Cambridge, MA: Harvard University Press, 226–247.

Said, E. (1993) *Culture and Imperialism*, London: Vintage.

Sandoval, C. (1990) 'U.S. Third World Feminism: the Theory and Method of Oppositional Consciousness in the Postmodern World', *Genders*, 10, spring, 1–24.

Saro-Wiwa, K. (1989) *On A Darkling Plain*, Port Harcourt: Saros International Publishers.

Saro-Wiwa, K. (1992) *Genocide in Nigeria*, Port Harcourt: Saros International Publishers.

Saro-Wiwa, K. (1995) *A Month and A Day*, London: Penguin.

Saunders, P. (1980) *Urban Politics*, Harmondsworth: Penguin.

Schama, S. (1989) *Citizens: a Chronicle of the French Revolution*, New York: Vintage Books.

Schechner, R. (1993) *The Future of Ritual: Writings on Culture and Performance*, London: Routledge.

Schmitt, C. (1976) *The Concept of the Political*, New Jersey: Rutgers University Press.

Schorske, C. E. (1980) *Fin-de-Siècle Vienna: Politics and Culture*, New York: Knopf.

Schulze, F. (1994) *Philip Johnson: Life and Work*, New York: Knopf.

Scotland on Sunday (1992) 'The informer file', 4 August.

Scotsman (1992) 'I knew nothing about immunity deal, says chief constable', 23 October.

Scott, A. (1995) 'Culture or politics? Recent literature on social movements, class and politics', *Theory, Culture and Society*, 12, 169–178.

Scott, J. (1985) *Weapons of the Weak: Everyday Forms of Peasant Resistance*, New Haven: Yale University Press.

Scott, J. (1986) 'Everyday forms of peasant resistance', *Journal of Peasant Studies*, 13(2), 5–35.

Scott, J. (1990) *Domination and the Arts of Resistance: Hidden Transcripts*, New Haven: Yale University Press.

Scott, J. (1992) 'Domination, acting, and fantasy', in C. Nordstrom and J.-A. Martin (eds) *The Paths to Domination, Resistance, and Terror*, Berkeley: University of California Press, 55–84.

Scott, J. and Kerkvliet, B. (eds) (1986) 'Everyday forms of peasant resistance in South-East Asia', special issue of *Journal of Peasant Studies*, 13(2).

Scottish Arts Council (1995) *Changing Places: the Arts in Scotland's Urban Areas*, Edinburgh: Scottish Arts Council.

Scruggs, J. C. and Swerdlow, J. L. (1985) *To Heal a Nation: the Vietnam Veterans Memorial*, New York: Harper and Row.

Searle, J. (1983) *Intentionality: an Essay in the Philosophy of Mind*, Cambridge: Cambridge University Press.

Sedgwick, E. K. (1990) *Epistemology of the Closet*, Berkeley: University of California Press.

Seidman, S. (1992) 'Postmodern social theory as narrative with a moral intent', in S. Seidman and D. G. Wagner (eds) *Postmodernism and Social Theory*, Cambridge, MA: Blackwell, 47–81.

Seidman, S. and Wagner, D. G. (eds) (1992) *Postmodernism and Social Theory*, Cambridge, MA: Blackwell.

Sereny, G. (1995) *Albert Speer: His Battle with Truth*, New York: Knopf.

Serres, M. and Latour, B. (1995) *Conversations on Science, Culture and Time*, Ann Arbor: University of Michigan Press.

Shaha, R. (1975) *An Introduction to Nepal*, Kathmandu: Ratna Pustak Bhandar.
Shaha, R. (1990) *Politics in Nepal 1980-1990*, New Delhi: Manohar.
Shariff, S. (1990) 'Anger alone isn't enough', *Vancouver Sun*, 11 July, A-11.
Sharp, G. (1973) *The Politics of Non-violent Action (3 volumes)*, Boston: Porter Sargent.
Sheehan, T. (1988) 'Heidegger and the Nazis', *New York Review of Books*, 16 June.
Shefner, J. (1995) 'Moving in the wrong direction in social movement theory', *Theory and Society*, 24, 595–612.
Shotter, J. (1993) *Cultural Politics of Everyday Life: Social Constructionism, Rhetoric and Knowing of the Third Kind*, Milton Keynes: Open University Press.
Shotter, J. (1996) 'Wittgenstein's world: beyond the way of theory toward a social poetics', (unpublished manuscript).
Sibley, D. (1995) *Geographies of Exclusion*, London: Routledge.
Slater, D. (1989) *Territory and State Power in Latin America: the Peruvian Case*, London and New York: Macmillan.
Slater, D. (1991) 'New social movements and old political questions: re-thinking state–society relations in Latin American development', *International Journal of Political Economy*, 21(1), 32–65.
Slater, D. (1994) 'Power and social movements in the other occident: Latin America in an international context', *Latin American Perspectives*, 21(2), spring, 11–37.
Slater, D. (1995) 'Democracy, decentralization and state power: on the politics of the regional in Chile and Bolivia', in *C.L.A.G. Yearbook 1995*, 21, Conference of Latin Americanist Geographers, edited by D. J. Robinson, Austin: University of Texas Press, 49–65.
Sluga, H. (1993) *Heidegger's Crisis: Philosophy and Politics in Nazi Germany*, Cambridge: Harvard University Press.
Smelser, N. (1963) *The Theory of Collective Behavior*, New York: Free Press.
Smith, B. F. (1981) *The Road to Nuremberg*, New York: Basic Books.
Smith, J. (1993) 'Above average: bright young artist devotes his work to charity – despite HIV', *The West Ender (Vancouver)*, 4 February.
Smith, N. (1984) *Uneven Development: Nature, Capital and the Production of Space*, Oxford: Blackwell.
Smith, N. (1992) 'Contours of a spatialized politics: homeless vehicles and the production of geographical scale', *Social Text*, 33, 54–81.
Smith, N. (1993) 'Homeless/global: scaling places', in J. Bird, B. Curtis, T. Putnam, G. Robertson and L. Tickner (eds) *Mapping the Futures: Local Cultures, Global Change*, London: Routledge, 87–119.
Smith, N. (1996) 'After Tompkins Square park: degentrification and the revanchist city', in A. D. King, (ed.) *Re-Presenting the City: Ethnicity, Capital and Culture in the 21st-Century Metropolis*, London: Macmillan, 93–107.
Smith, N. and Katz, C. (1993) 'Grounding metaphor: towards a spatialized politics', in M. Keith and S. Pile (eds) *Place and the Politics of Identity*, New York: Routledge, 67–83.
Smith, S. (1993) 'Bounding the borders: claiming space and making place in rural Scotland', *Transactions, Institute of British Geographers*, 18.
Smothers, R. (1955) 'Issue behind King memorial: who owns history?', *New York Times*, 16 January.
Snow, D. A. and Benford, R. D. (1992) 'Master frames and cycles of protest', in A. D. Morris and C. M. Muller (eds.) *Frontiers in Social Movement Theory*, New Haven and London: Yale University Press, 133–155.
Socialist Review Collective (1995) 'Arranging identities: constructions of race, ethnicity, and nation', special issue of *Socialist Review*, 91(1–2).
Soja, E. (1989) *Postmodern Geographies: the Reassertion of Space in Social Critical Theory*, London: Verso.
Soja, E. (1993) 'Postmodern geographies and the critique of historicism', in J. P. Jones III, W. Natter and T. Schatzki (eds) *Postmodern Contentions*, New York: Guilford Press, 113–136.
Soja, E. (1995) 'Heterotopologies: a remembrance of other spaces in the citadel-LA', in K. Gibson and S. Watson (eds) *Postmodern Cities and Spaces*, Oxford: Blackwell, 13–34.
Soja, E. (1996) *Thirdspace: Journeys to Los Angeles and Other Real-and-Imagined Places*, Oxford: Blackwell.
Soja, E. and Hooper, B. (1993) 'The spaces that difference makes: some notes on the geographical margins of the new cultural politics', in M. Keith and S. Pile (eds) *Place and the Politics of Identity*, New York: Routledge, 183–205.
Soyinka, W. (1996) *The Open Sore of a Continent*, London: Oxford University Press.

Sparke, M. (1994) 'White mythologies and anemic geographies: a review', *Environment and Planning D: Society and Space*, 12(1), 105–123.
Speer, A. (1970) *Inside the Third Reich: Memoirs*, New York: Macmillan.
Spencer, P. (ed.) (1985) *Society and the Dance*, Cambridge: Cambridge University Press.
Spivak, G. C. (1985) 'Subaltern studies: deconstructing historiography', in R. Guha (ed.) *Subaltern Studies IV*, Delhi: Oxford University Press, 330–363.
Spivak, G. C. (1990) 'Explanation and culture: marginalia', in R. Ferguson, M. Gever, M.-h. T. Trinh and C. West (eds) *Out There: Marginalization and Contemporary Cultures*, Cambridge, MA: MIT Press, 377–393.
Star, S. L. (1991) 'Power, technology and the phenomenology of conventions: on being allergic to onions', in J. Law (ed.) *A Sociology of Monsters: Essays on Power, Technology and Domination*, London: Routledge, 25–56.
Star, S. L. (1995) 'Introduction', in S. L. Star (ed.) *The Cultures of Computing*, London: Routledge, 1–28.
Starn, O. (1992) '"I dreamed of foxes and hawks": reflections on peasant protest, new social movements and the Rondas Campesinas of Northern Peru', in A. Escobar and S. Alvarez (eds) *The Making of Social Movements in Latin America*, Boulder: Westview Press.
Starn, O. (1995a) 'To revolt against the revolution: war and resistance in Peru's Andes', *Cultural Anthropology*, 10(4), 547–580.
Starn, O. (1995b) 'Maoism in the Andes: the Communist Party of Peru – Shining Path and the refusal of history', *Journal of Latin American Studies*, 27, 399–421.
Stein, S. J. (1992) *The Shaker Experience in America: a History of the United Society of Believers*, New Haven, CT: Yale University Press.
Steinem G, (1994) *Moving Beyond Words*, New York: Simon and Schuster.
Stevens, J. (1996) 'Another view of city's history', *The Age*, 6 January, A15.
Sturdevant, S. P. and Stoltzfus, B. (1992) *Let the Good Times Roll: Prostitution and the US Military in Asia*, New York: New Press.
Sun, The (1990a) 'Scandal', 18 January.
Sun, The (1990b) 'Scandal judge hides on isle', 22 January.
Sun, The (1990c) 'Judge not named', 29 January.
Sun, The (1992) 'Police: gay judges do fix trials', 12 September.
Sword, P. (1991) 'ACT UP pushes for anonymous AIDS tests', *Halifax Chronicle Herald*, 25 January, A-5.
Taussig, M. (1980) *The Devil and Commodity Fetishism in Latin America*, Chapel Hill: University of North Carolina Press.
Taussig, M. (1992) *The Nervous System*, New York: Routledge.
Taylor, P. J. (1995) 'Beyond containers: inter-nationality, inter-stateness, inter-territoriality', *Progress in Human Geography*, 19, 1–15.
Temko A. (1994) 'Bad boy builder', *Los Angeles Times Book Review*, 25 December.
Thomas, H. (ed.) (1993) *Dance, Gender and Culture*, London: Macmillan.
Thomas, H. (1995) *Dance, Modernity and Culture: Explorations in the Sociology of Dance*, London: Routledge.
Thomas, H. (1996) *Dance in the City*, London: Macmillan.
Thompson, E. P. (1967) *The Making of the English Working Class*, Harmondsworth: Penguin.
Thornton, S. (1995) *Club Cultures: Music, Media and Subcultural Capital*, Cambridge: Polity Press.
Thrift, N. J. (1996a) *Spatial Formations*, London: Sage.
Thrift, N. J. (1996b) 'From a straight line to a curve or, cities are not mirrors of modernity', unpublished paper.
Thrift, N. J. and Forbes, D. K. (1983) 'A landscape with figures: political geography with human contact', *Political Geography Quarterly*, 2, 247–263.
Tickner, R. (1991) *Aboriginal Reconciliation: a Discussion Paper*, Australian Government Publishing Service: Canberra.
Tiwari, S. R. (1992) 'No future for an urban past', *Himal*, 5(1), 5–7.
Tomko, L. J. (1996) 'Fete accompli: gender, "folk-dance" and progressive–era political ideas in New York City', in S. L. Foster (ed.) *Corporealities: Dancing Knowledge, Culture and Power*, New York: Routledge, 155–176.
Tonkiss, F. (forthcoming) 'Civil/political', in C. Jenks (ed.) *Core Sociological Dichotomies*, London: Routledge.
Trevor-Roper, H. R. (1962) *The Last Days of Hitler*, New York: Collier.
Trinh, M.-h. T. (1992) 'Cotton and iron', in R. Ferguson, M. Gever, M.-h. T. Trinh and C. West

(eds) *Out There: Marginalization and Contemporary Cultures*, Cambridge, MA: MIT Press, 327–336.
Trumper, R. and Phillips, L. (1995) 'Cholera in the time of neoliberalism: the cases of Chile and Ecuador', *Alternatives*, 20(2) April–June, 165–193.
Tsing, A. (1993) *In the Realm of the Diamond Queen*, Princeton: Princeton University Press.
Turner, R. and Killian, L. (1972) *Collective Behavior*, second edition, Englewood Cliffs, New Jersey: Prentice-Hall.
Turner, V. (1983) 'Body, brain and culture', *Zygon*, 18, 221–245.
Umar, M. (1993) 'Changing Islamic identity in Nigeria from the 1960's to 1980's', in L. Brenner (ed.) *Muslim Identity and Social Change in Sub Saharan Africa*, Bloomington: University of Indiana Press, 154–178.
UN (United Nations) (1996) *Report of the Fact-Finding Mission of the Secretary-General to Nigeria: Summary of Information and Views Received*, New York: United Nations.
UNPO (Unrepresented Nations and Peoples Organization) (1995) *Ogoni: Report of the UNPO Mission to Investigate the Situation of the Ogoni*, The Hague: Unrepresented Nations and Peoples Organization.
Valentine, G. (1989) 'The geography of women's fear', *Area*, 21(4), 385–390.
Valentine, G. (1993) '(Hetero)sexing space: lesbian perceptions and experiences of everyday spaces', *Environment and Planning D: Society and Space*, 11, 395–413.
Vancouver Province (1992) 'Acting stupid: AIDS group only hurts its own cause', 28 April.
Vancouver Sun (1992) 'Seven AIDS protesters arrested: occupied John Jansen's office', 1 December, A-2.
Walker, L. (1995) 'More than just skin-deep: fem(me)ininity and the subversion of identity', *Gender, Place and Culture*, 2, 71–76.
Walker, R. B. J. (1994) 'Social movements/world politics', *Millennium*, 23(3), winter, 669–700.
Walker, R. B. J. (1995) 'International relations and the concept of the political', in K. Booth and S. Smith (eds.) *International Relations Theory Today*, Cambridge: Polity Press, 306–327.
Wapner, P. (1995) 'Politics beyond the state – environmental activism and world civic politics', *World Politics*, 47 April, 311–340.
Ward, A. (1993) 'Dancing in the dark: rationalism and the neglect of social dance', in H. Thomas (ed.) *Dance, Gender and Culture*, London: Macmillan.
Watts, M. (1983) *Silent Violence*, Berkeley, University of California Press.
Watts, M. (1991) 'Mapping meaning, denoting difference, imagining identity: dialectical images and postmodern geographies', *Geografiska Annaler*, 73B(1), 7–16.
Watts, M. (1994) 'The devil's excrement', in S. Corbridge, R. Martin and N. Thrift (eds) *Money, Power and Space*, Oxford, Blackwell, 406–445.
Watts, M. (1996) 'Islamic modernities?: citizenship, civil society and Islamism in a Nigerian city', *Public Culture*, 8(2), 251–289.
Welch, C. (1995) 'The Ogoni and self-determination', *Journal of Modern African Studies*, 33(4), 635–650.
Weyland, K. (1995) 'Social movements and the state: the politics of health reform in Brazil', *World Development*, 23(10), 1699–1712.
Wigwood, P. (1992) 'Coping with AIDS' step-by-step death', *Vancouver Sun*, 11 September, A-1.
Willems-Braun, B. (1994) 'Situating cultural politics: fringe festivals and the production of spaces of intersubjectivity', *Environment and Planning D: Society and Space*, 12(1), 75–104.
Williams, R. (1976) *Keywords: a Vocabulary of Culture and Society*, London: Fontana.
Williams, R. (1977) 'Dominant, residual, and emergent', in *Marxism and Literature*, Oxford: Oxford University Press, 121–127.
Willis, P. (1981) *Learning to Labour: How Working Class Kids get Working Class Jobs*, New York: Columbia University Press.
Wilson, D. (1992) 'Militant tactics by AIDS groups dramatize frustrations with the crisis: massive die-in staged to mark loss of life to the disease', *Toronto Globe and Mail*, 24 August, A-1, A-8.
Winnicott, D. (1971) *Playing and Reality*, London: Routledge and Kegan Paul.
Winnipeg Free Press, (1992) 'AIDS protesters jostle, spit at Vander Zalm, knock wife to the ground', 26 August.
Wittgenstein, L. (1980) *Culture and Value*, Chicago: University of Chicago Press.
Wolf, E. (1969) *Peasant Wars of the Twentieth Century*, New York: Harper Torchbooks.
Wolin, R. (1993a) 'French Heidegger wars', in R. Wolin (ed.) *The Heidegger Controversy: a Critical Reader*, Cambridge, MA: MIT Press.

Wolin, R. (ed.) (1993b) *The Heidegger Controversy: a Critical Reader*, Cambridge, MA: MIT Press.

Wollen, P. (1995) 'Tales of total art and dreams of the total museum', in L. Cooke and P. Wollen (eds) *Visual Display: Culture beyond Appearances*, Seattle: Bay Press, 154–177.

Yadav, S. R. (1984) *Nepal: Feudalism and Rural Formation*, New Delhi: Cosmo Publication.

Yanagisako, S. and Delaney, C. (eds) (1995) *Naturalizing Power: Essays in Feminist Cultural Analysis*, New York: Routledge.

Young, I. M. (1990) *Justice and the Politics of Difference*, Princeton: Princeton University Press.

Young, R. J. C. (1995) *Colonial Desire: Hybridity in Theory, Culture and Race*, London: Routledge.

Yúdice, G. (1996) 'Culture and the organization of civil society in an age of global restructuring', mimeo, paper presented at the Conference on 'Cultures of Politics/Politics of Cultures: Revisioning Latin American Social Movements', State University of Campinas, Brazil, March.

Yusuf, A. (1988) *Maitatsine: Peddler of Epidemics*, Syneco: Kano.

Zabaida, S. (1995) 'Islam: is there a Muslim society?', *Economy and Society*, 24(2), 155–188.

Zermeño, S. (1995) 'Zapatismo: región y nación', *Nueva Sociedad*, 140 November–December, 51–57.

Zimmerman M, (1988) 'L'Affaire Heidegger', *Times Literary Supplement*, October 7–13.

INDEX

Abacha, General S. 34, 36–37, 61–62
Abu-Lughod, L. 89, 112
Achebe, C. 36, 45
activism 260; AIDS 154–155, 164; labour 5, 258
actor-network theory 130–132
Adorno, T. W. 220
age 177–178, 180
agency: human 111, 114, 130–131, 135–136, 148, 150, 171, 176, 239, 264, 273, 283; non-human 131, 132, 136
Agrawal, B. 267
Ahearne, J. 125
AIDS 115–116, 152–155, 158–161, 165, 189–191; education 115; medical authority 160–161; organisation 153, 160; politics of 152, 161–164, 283; *see also* Persons With AIDS Society
AIDS Coalition to Unleash Power (ACT UP) 152–166, 166n.; New York 153–154, 162; Vancouver 153, 155–166, 167n.
Ake, C. 53
Algeria: nationalist struggle 17–26; women in 17–26
Algerianness 19, 26
Allen, J. 31n.
Angela 99–100, 102–103
anger 75, 158–160, 161, 165, 208
Arditi, B. 266, 267
assemblages 132–133, 137
Australia: Aboriginal Reconciliation Act (1991) 207; Aboriginal resistance 203–204, 213, 216; Aboriginal 'tent embassy' 203–204; Council for Aboriginal Reconciliation 205–206, 209–210; history 210, 212, 213, 215; land rights struggle 203; the 'Mabo decision' 207, 215; narratives of nation 203, 206, 213, 216–217; reconciliation 204–210, 213, 215, 216; Royal Commission into Aboriginal Deaths in Custody 206–207, 212; postcolonial 213; surrender 206; *see also* Melbourne *and* rights
authenticity 220, 232
autonomy 91, 265, 273

Ballantyne, A. 230
Bammiker, A. 101
Barak, W. 212
bars 111, 116, 120, 122; Brunswick, The 112–113, 116, 118–119, 122; drinking 118; Richmond, The 117–118
Batman, J. 215
Bauman, Z. 145–147
Bell, D. 88
Benford, R. 238
Benjamin, W. 125
Berlant, L. 186–187, 191, 192
Bhabha, H. 101–102, 103, 109–111, 187, 209, 217
Bhatt, C. 279
Birendra, King 68, 73
body, the 127–128, 131, 132, 134, 136–137, 139, 141, 144, 147–149; discomfort 116; embodiment 125, 126, 128, 130, 137, 139–142, 145, 148, 150; female 18–19, 23–24, 114; performance 148; senses 147; subject 141–142, 147, 148; *see also* practices
Bolivia 265, 267–268, 272
Bondi, L. 87
borders 16, 261, 272–273, 274; border zones 264
boundaries 28–29, 71, 79, 91, 97, 112, 268–269; inside/outside 269–274; national 258, 259, 267; *see also* borders
Bourdieu, P. 226
Brown, M. 11, 283
Burchell, G. 282
Butler, J. 28, 121, 170, 263–264

Calderón, F. 273
Campbell, D. 173, 177
Canel, E. 165–166
capital 6
Capital (Knorr) xi
capital 6
capitalism 6, 8; petrolic 38, 284; space of 11
Castells, M. 9–13, 237
Cebu City (The Philippines) 108, 112, 116, 120

Chiapas (Mexico) 52, 269–272; Zapatistas (Ejército Zapatista de Liberación Nacional / Zapatista Front of National Liberation) 269–272
Churchill, W. 219
city, the *see* urban space
City and the Grassroots, The (Castells) 237
civil society 12, 152–153, 155, 156, 158, 160, 163, 166, 259, 263, 282
civil society–state relations *see* state–civil society (relations)
class politics *see* politics
class (relations) 7, 18, 90, 220, 240–241, 245–246, 264, 274; *Talakawa* (Nigeria) 48–49
colonialism 21–22, 64, 94–95, 114, 204, 206, 207–209, 215, 216–217, 269; anti-colonial struggle 17–26, 96–97, 203–204; British, in Australia 203–204, 206, 209, 211–213, 214, 215; British, in Nigeria 50; British, in Zimbabwe 93–95; French, in Algeria 16–26; internal 23–24
communication 233, 271
community 4, 9–10, 29, 36, 49, 78, 79, 84, 144–145, 185–202, 209, 210; arts 184, 189–202, 213–214, 217, 251, 254, 255, 260, 261, 263, 267, 271; gay 153–154, 160, 173, 286; inoperative 187–202, 283; myth of 187–188, 190, 195, 197, 200, 202; spatiality of inoperative 199–201; of struggle 13, 96; violences of 188; workers (Israel) 248–249, 251, 254
Connolly, W. 260, 261
Conroy, S. 172
Copjec, J. 187
Cora 112–114
Corngold, S. 223
Cresswell, T. 31n.
Crystal Cathedral (Los Angeles, USA) 229–230
Cuba 265
cultural sabotage 155
culture of criticism 225
cyberspace 141; *see also* virtual reality

Dalyell, T. 172
dance 124, 142–150, 151n.; and community 143–144; dancing 119–120, 121; Ghost Dance 144; practices 143–144
de Certeau, M. 15, 71, 84, 90, 105n., 125, 137–138, 283
de Man, P. 221–226
Dear, M. 24
Deleuze, G. 69, 132–133, 137, 150n.–151n.
democracy 260, 269, 271–273, 279, 286; democratization 265, 266; liberal 281; practices of democratization 260, 261; spatialities of democratization 260; territorializing 268; *see also* radical
Democracy and Civil Society (Keane) 280
Dempster, E. 145
Department of Health and Social Security (DHSS) (UK) 171, 172, 174–176, 180; fraud 176, 178; gay scandals 172
Derrida, J. 220, 224
Dervaid, Lord (J. Murray) 171, 172, 175, 179
desire 119, 120–121, 122, 128, 134
Devay, R. 212
development 269
difference 185, 202, 220, 265
discourse 187–188, 191–192, 194, 195, 197, 200, 201, 220, 225, 264, 269; *see also* practices
disguise 220, 224
Dolzura Movie, The (Philippines) 115
domination 10, 12, 70, 75, 82, 83, 84, 85n., 95, 203, 209, 220, 233; practices of 15–16, 70, 80–82, 180; spaces of 15–16, 17, 23, 70, 80, 95–96
domination/resistance 2–3, 11; spatialities of 12, 15, 26, 169
Domination and the Arts of Resistance (Scott) 91
Donzelot, J, 286
Dracula 8
dress 21, 31n.; *see also* veil
Duncan, I. 139, 144
Duncan, N. 173

Edinburgh (Scotland): AIDS 189–191; community arts 189–202; Craigmillar 189, 190, 194, 196, 198, 199–200; Fettesgate (Fettes Avenue break-in) 168, 169, 172, 174, 177, 178, 180; land and housing markets 177–178, 180; Lothian and Borders Police (Scotland) 169, 172, 174, 179; magic circle, alleged gay 171, 175–182; Muirhouse 189, 190; Operation Uranus/Planet 172, 173, 175, 183n.; Pilton 190, 196, 198; video project 196–198; Wester Hailes 190
Edna 112–114, 116–117, 118
embodiment *see* body
empiricism 279–280
empowerment 181, 189, 193
End of History, The (Knorr) xi, xiii
Escobar, A. 35
everyday life 69, 90, 149, 257, 283
excess 187–188, 194, 195, 200
exclusion 16, 30; racial 95–96
exteriority *see* interiority/exteriority

Falk, R. 259–260
Fanon, F. 4, 16–26, 31n.
Farias, V. 222
Fely 113, 114–115, 117–118
Filler, M. 229
Flinders, Matthew, statue of (Australia) 211–213
Forrest, T. 40
Foucault, M. 71, 101, 112, 130, 131, 133–134, 137–138, 151n., 226, 234n., 282, 284, 285, 286

Four Quartets, The (Eliot) 124
Freud, S. 24
friendship 122
Full Metal Jacket (Kubrick) xi

Galeano, E. 85, 270
Gamson, J. 154
gay criminality, alleged 172–173, 174–175, 181
gay political organisation: Edinburgh (Scotland) 170–171, 174, 181–182; San Francisco (USA) 11, 16; Vancouver (Canada) 152–167
Gay Scotland 173, 174, 179
gaze, the *see* looks
gender (relations) 18, 20, 90, 100, 109, 111–112, 179
Generations of Resistance (Lippmann) 203
geography 87, 88, 130, 133; radical 4
geopolitical, the 260, 262, 267–269, 270
Giddens, A. 241
globalisation 258, 261, 267, 274, 274n., 278, 281, 282; and the local 270
Goldberger, P. 228
Goldhagen, D. 231–232
Goriely, G. 224
governance 265, 266, 280; governmentality 278, 283, 285, 286
Gregory, D. 88
grief 161, 165
Grosz, E. 128, 132, 139–140
Guattari, F. 69, 132–133, 137, 151n.
guerrilla war 22, 96, 268, 272–273
Guha, R. 91–92
guilt 23, 207, 208, 214

haïk, the *see* veil
Hall, S. 88, 92, 185
Hans 118–119
Haraway, D. 103
Hardimon, M. 206
Harley, J. 94
Harvey, D. 5–8, 10, 11–12, 88, 185
Hasson, S. 9, 284
Havel, V. 280
Hayles, N. 141
Hecht, D. 68
Hegel, G. 206
Heidegger, M. 221–226, 232
Heidegger et le Nazisme (Farias) 222
Held, D. 260–261
Herz, J. 275n.
Het Vlaasmsche Land (in Flemish) 222
HIV *see* AIDS
homophobia 11, 173, 178, 180–181
homosexuality 169, 181; *see also* gay political organisation *and* sexuality
hooks, b. 71, 88, 96
Hooper, B. 102
Hopkins, T. 177
human rights *see* rights
hunting (girls) 117–118

Husserl, E. 222
hybridity 110–111, 114, 130, 209, 239

identity 18, 87–88, 101, 108, 109–112, 113–114, 116, 121, 139, 150, 169–171, 181–182, 186, 188, 190–191, 201, 211, 232, 236, 252, 260, 263, 272, 286; collective 185–186, 201; cultural 9, 17, 48; dancer 117; ethnic 40, 95; gay 176; identification 110, 113, 185, 188, 254, 261; local 192, 197; Muslim 35, 37; national 30, 37–38, 42, 64, 220, 231; political 4, 7, 8, 10, 11–12, 30, 181, 278–279; politics of 29, 35, 37, 103, 201, 284; prostitute 107; spaces of 108, 185
Imperial Foods chicken processing plant (Hamlet, USA), fire at 5–6
injustice 7, 11, 30, 178, 207, 214
interiority/exteriority 144
intimacy 115, 119
Islam, in Nigeria 34–35, 54–59, 64
Israel 239–257; Black Panthers, 239, 241–247, 252–256; Dai (Enough) movement 240, 243–247, 249–250, 252; Haifa 239; Holocaust memorial (Jerusalem) 230–232; Jerusalem 236, 240–241; Ohalim (Tents) movement 240–250, 252, 254–256; Ohel (Tent) Shmuel 247, 256; Ohel (Tent) Yosef 240, 242–248, 252–253, 255; Shahak (Improvement of Community Life) 240, 243–246, 253, 255; social inequality 241–243, 253; Tsalash (Young For the Neighborhood) 240, 242, 246–247, 249–252, 255–256; urban housing segregation 240–241, 253–254; urban protest 9, 236, 239–257

Jackson, P. 31n.
Jacobs, J. M. 24, 28–29, 284
Jessop, B. 281
Johnson, P. 221, 228–230
joint action 128–129, 149
Joseph, G. 92
Jowitt, K. 138
justice 169, 171, 175, 176, 177, 179, 203, 205, 211, 214–217, 260, 271, 272; gay 'threat' to (Edinburgh) 169, 172, 176, 179

Kano (Nigeria) 34, 36, 47–49, 55, 57–58
Kaplan, C. 87, 92, 104
Kathmandu (Nepal) 77, 78, 80, 81, 83–85, 86n.; massacre 81, 85
Katz, C. 88
Keane, J. 280
Keating, Prime Minister P. 207
Kellner, D. 89
Khan, S. 44, 45
Kharkhordin, D. 138
King, R. 6
Kinsey, C. 219
Knopp, L. 11, 284

knowledge 126, 133, 232, 255, 257, 259; and power 17, 187
Kondo, D. 90, 111
Koonz, C. 227
Kramer, J. 231
Kramer, L. 153
Kulin people, the (Australia) 210, 211

labour: conditions 5–6, 12; exploitation of 12; interests 7; vulnerability 6
LaCapra, D. 253
Laclau, E. 164–166, 262, 266
Last, M. 48
Last Days of Hitler, The (Trevor-Roper) 227
Latour, B. 131, 134
Law, J. 131–132, 134–135, 151n.
Law, L. 3, 17–18, 284
Le Soir (Belgium) 222
leaders 248–252; leadership 247, 252, 271
Lefebvre, H. 22, 70
Leton, G. 59–60
Lewis, C. 121
Lewis, D. 156
Lippmann, L. 203
local, the *see* globalisation *and* politics
London, the City of xi–xii
looks 17, 20, 31n., 111; scrutiny, sexual 20–21; tourist 116; visibility and invisibility 19, 23, 36, 139
Lorna 116–117
Lubeck, P. 64

McNeil, W. 143
McRobbie, A. 89
Maitatsine 33–36, 38, 49–50, 54–59, 62, 64, 65, 66n., 284
Mamdani, M. 64
Mannheim, K. 241
Marian 219
Marris, P. 238
Marwa 58
Marx, K. 6, 104
Maryann 113, 114–115
Massey, D. 87, 185
Mate Mate, R. 210, 216
Maulboyhenner, J. 212
May 118–119
media, the 161, 163, 169, 172, 174, 179, 180, 190
Melbourne (Australia) 210, 215; aboriginalisation of 213, 214, 217; 'Another View Walking Trail: pathway of the rainbow serpent' 205, 210–217; new geography of 210, 213–214; old gaol 212, 214; old gaol burial poles 214, 216
memory 103, 112, 230
Memory of Fire (Galeano) 85
Merleau-Ponty, M. 127–128, 139, 141
metaphor 141; geographical 87; spatial 88, 102, 172–173, 175–176
Mickey Mouse x–xiii; March xi
Milk, H. 11
mistakes 139
Mitchell, T. 90–91
mobilisation: class 9, 237; gay 11; mass 59, 75; political 4, 8, 12, 25, 90, 114, 258, 265, 274, 278, 283, 285; social 10; territorial 241
Mohanty, C. 27–30, 90
Mohr, R. 165
Monga, C. 265
Moore, D. 26–27, 277, 283
morality 119–120, 122, 160, 180–181; spatial order of 181
Mouffe, C. 152, 153–154, 164–166, 167n., 185, 264
movement(s) 12, 29, 71, 78, 79, 237–239, 258, 261–262, 265, 269, 274; control 245; militant 256; Movement for the Restoration of Democracy (MRD) (Nepal) 72–75, 77–78, 80, 82–85, 86n.; Movement for the Survival of the Ogoni People (MOSOP) (Nigeria) 33–35, 38, 59–63, 64, 67n.; reform 245; *see also* social movements
Mugabe, R. 96
Muslim community (Nigeria) 34–35, 37, 55–59; alternative 55

Nancy, J.-L. 187–188, 195, 199–220, 202
nation, the 19, 25–26, 30, 63, 208, 210, 213, 261, 267–269, 281–282; sovereignty 272; and identity *see* identity
Nazism (Germany) 222, 226–228, 230–232, 233n.; anti-semitism 222, 224, 226, 228, 230–232, 235n.; and de Man 222–226; and Heidegger 222–226; Holocaust, the 230–232; and Johnson 228; and Speer 226–228
neighbourhood organisation 10, 241–256
Nepal 17, 68–86; army, the 80–82; inequality 72–73; Kathmandu Valley 70, 72, 75, 79–80, 83, 86n.; Nepali Congress (NC) (Nepal) 73, 74, 81, 82; non-violent resistance *see* resistance; popular movements 70, 74–80, 81; popular revolution 68, 72, 77, 83; society 72–73, state 73, 80–82; United Left Front (ULF) (Nepal) 74, 81, 82; *see also* movement(s)
networks 130–131, 134, 137, 260; location in 132; of power 149
new social movements *see* social movements
Nicaragua 265, 267
Nietzsche, F. W. 219, 233n.
Nigeria 2, 33–67; Biafra 41–43; civil government 37; civil war (1967–1970) 37, 42, 51–52, 59, 64; corruption 45, 48–49, 52, 56; Delta People's Republic 50; ethnicity 40, 284; federal 40, 41, 42, 59, 62–63, 64; foreign exchange crisis 46; Maitatsine insurrection 54–59, 64; military government 34, 36, 61–63; Niger Delta 38, 50; Ogoni genocide 54; oil 38–44;

post-colonial 35, 38–44, 63, 64; poverty 46–47, 49; regionalism 40–44, 45, 47; Rivers State 50–52, 60, 66n.; state violence 63–65; structural adjustment program 46
Nigerian National Oil Company (NNOC) / Nigerian National Petroleum Company (NNPC) 38–40, 42, 65n.
non-representational theory 125–133, 202
Norris, C. 232
North American Free Trade Association (NAFTA) 270–272; *see also* Chiapas
Northrop, A. 154
Nugent, D. 92

Ogoniland (Nigeria) 34, 49–54, 59–63, 66n.; crisis in 35, 61
oil production (Nigeria) 37, 38–44, 46–47, 52–53, 59–63, 64, 284; boom 35, 45, 46, 48–49, 52, 64; ecological consequences of 53–54, 59–60, 63; and modernisation 49, 50, 54; politics of 38–44; revenue allocation 42–43, 47, 64
Ojukwu, Colonel 41
Okilo, M. 50
Olsson, G. 220
oppression 135, 150, 180–181, 258; class 8; hierarchies of 7; spatial practices of 3
Order of Things, The (Foucault) 131
Otobiographies (Derrida) 220
Ott, H. 222

Parry, B. 89, 102
Patan (Nepal) 77, 78, 79–80, 81, 83–85, 86n.
Paton, T. 175
patriarchy: Algeria 18–19, 21, 24, 26; Zimbabwe 101
People's Security Committees (Nepal) 79
performance 122, 170, 181, 193, 197, 199, 201, 205, 210, 211, 213; bodily 148–149; spatiality of 197, 200, 201
Persons With AIDS (PWA) Society 155, 159–160, 161, 162, 163–164, 167n.
Peru 265, 272; *Sendero Luminoso* (Shining Path) 268, 269, 273; regional movements 268
Philippines, the: family relations 117; Manila 107; political discourses 109; poverty 113, 119; prostitution 113, 116–117; sex tourism 107–123
Phillips, Justice J. H. 214
Pieterse, J. 89
Pile, S. 102, 109–110, 121
Pilger, J. 207
Place and the Politics of Identity (Keith and Pile) x, 221
play 145–147, 149–150, 253; carnival 253; location of 146–147; and power 149
politics 16, 221, 259, 262–269, 274, 277, 279–282; Canadian urban culture 161–163, 165; class 7, 8, 263; global 262; Islam 56–59; Israeli urban culture 246–257; lesbian 11; local 89–90, 101, 103, 242; of locality 36–37, 256; of location 27–30, 74, 181; oppositional 208–209, 281–282; and philosophy 221; of place 87–106, 232; progressive 277; radical 4, 7, 25, 185, 202; sexualised 19; of signification 238; spatial 261; of turf 157, 181; world 262; struggle *see* resistance; *see also* identity
power (relations) 2–4, 5, 7, 10–14, 24, 26, 28–29, 31n., 41, 65, 69–70, 82, 83, 85, 90, 97, 102, 107–108, 111–112, 115, 121, 122, 133–135, 145, 149–150, 169, 178, 182, 190–191, 192, 194, 202, 208, 242, 252, 259, 262, 263–264, 269, 271, 274, 278, 281, 284; geographies of 15–16; and language 192, 194; lines of force 134, 139; social 70; spaces of 8, 30, 31n., 91; strategies of 15, 71; territoriality of 259; will to 187; *see also* state
practices 136, 266; bodily 69, 143–144; dialogic 128–129; discursive 171; everyday 125, 126–127; gay 176; of politics 221, 264; of 'process' (community arts) 195–199, 201; sexual 180; spatial 71; theory of 125; transformative 150
Prakesh, G. 92
prostitution 113, 116–117, 120, 174–175, 177
protest 9–10, 12, 74, 77, 157, 251, 265; blackouts (Nepal) 74, 78, 84–85; die-ins 156, 162, 167n.; militant 256; neighborhood 248, 251; occupations 156; poem 254; urban 9, 77–80, 236–257; violent 256

race (relations) 95–96, 109, 207, 240, 241, 245–246; fantasies of 17; racism 11, 203, 208, 216; *see also* Nazism
Radcliffe, S. 31n.
Radhakrishnan, R. 102
radical: citizenship 156, 164; democracy 152–154, 165–166, 167n., 266, 271; intellectuals 250; politics *see* politics
Radley, A. 126, 137, 139, 141, 147–149
Rancière, J. 264–265
resistance x, xii, 1–5, 9, 12–16, 18, 20, 24, 27–30, 59, 65, 68–72, 74–80, 83–85, 89–93, 101–102, 109–112, 113, 119, 120, 121–123, 124–125, 135–136, 141–142, 151n., 152–153, 157–159, 165–166, 181, 188, 201, 203–205, 209, 211, 213, 216–218, 238, 244, 258–259, 265, 267, 269–274, 277, 281–286; civil disobedience 73; counter-narrative 212, 216–217; definitional uncertainty 194; direct action 59–60; geographies of x, xii–xiii, 2, 3–4, 22, 26–27, 30, 108, 111, 278, 286; labour 7; lesbian and gay 12–13, 161, 176; mobility 71, 75–80, 83–84; non-violence 73, 74–76, 81; packs 75, 84; and place 78–80, 83–85; psychic 24–25, 29, 114; rhizomatic 69, 75; sites/spaces of 14, 15–16, 23, 28–29, 31n., 74–75, 78, 91, 93,

resistance *continued*
102, 109, 208, 273; solidarity 198; spatiality of 70–72; struggle 4, 23, 26, 27–29, 37, 104, 263; survival 200–201; swarms 75, 84, 125; symbolic 98, 158; tactics of 15, 71, 84–85, 125, 138, 158–159, 195; transgression 155, 282; *see also* gay political organisation; movement(s); protest; social movements; women's
Rich, A. 27, 30
rights 59–60, 174, 218, 236, 260; Aboriginal land 203–204, 207, 208, 214–215; human 37, 81; land 271, 273
Robb, K. 155
Robinson, G. A. 212
Robinson, J. 31n.
Rose, G. 4, 101, 283
Rose, N. 133, 285–286
Routledge, P. 17
rules of the game 138, 146
Russo, V. 154–155

Sandoval, C. 92–93
San Francisco (USA): Castro District, the 11; Latino community organisation 9–10, 16; Mission District, the 9
Saro-Wiwa, K. 33–38, 50, 52, 54, 59–60, 62–63, 64; *see also* movement(s) *and* Nigeria
scale, geographical 13–14, 19
Schechner, R. 145–146
Schulze, F. 228, 229, 230
Scotland: judicial system 169–183, 179, 180, 181; Scottish Arts Council 190; Scottish Homosexual Rights Group (SHRG) 174, 177, 180, 181; Scottish Office 190; *Scottish Screen Digest* 193; *see also* Edinburgh
Scott, J. C. 89–92, 105n., 138
Secret Country, A (Pilger) 207
self, sense of *see* subjectivity
semblances 147–148
Sereny, G. 227
sex 120, 122, 175, 177, 263–264; fantasies of 17–18; industry 120, 122n.; politics of 19; tourism 107–123
sexuality 109, 114–115; dancer 120; female 107, 179; gay 173–174, 178; Johnson's 229
shame 23, 207, 218n.
Sheehan, T. 222, 223
Shell (oil company) 33, 38, 39, 40, 49, 53, 59, 60–61, 63, 66n., 67n., 284; *see also* oil production (Nigeria)
Shotter, J. 125–130, 140
silence 184, 192–193, 201–202, 233
Simone, M. 68
Six Australian Battlefields (Grasby and Hill) 204
skill 126–127
Slater, D. 3, 13, 262, 274n., 278, 286
Sluga, H. 221, 223, 228
Smith, N. 13, 88
Snow, D. 238
social movements 70, 182, 259–265, 274, 274n.; gay liberation 11; new 7, 8, 9–10, 166, 259; Nigerian 37; urban 236–257
Soja, E. 88, 102
Soyinka, W. 36, 64
space 13–14, 21–23, 28–29, 70–72, 75, 76, 78, 84–85, 87, 91, 93, 101, 115–116, 130, 131, 133, 149, 153, 156, 171, 208, 210, 259, 266; bounded 268–269; gay 176, 188–189, 202; heterosexualised 155; impossible 120; inoperative 189; paradoxical 201; third 71, 83, 101–104, 108, 109–112, 116, 122 and time 129–130, 132, 134, 148, 269, 277, 278, 284
Speer, A. 221, 226–228, 229
Star, S. 135, 138, 140
state, the 12, 49, 56, 73, 82, 97, 100–101, 153, 156, 160, 165–166, 180, 237, 240, 252, 257, 261, 265, 267–269, 273, 281–286; heterosexist 178, 180; patronage 41, 42–43; violence 36, 58, 63–65, 80–81, 96, 100, 103, 203; welfare 161–162
state–civil society (relations) 12, 152, 157, 159, 161, 163, 165, 166n., 259, 266, 278, 280–284
Steinem, G. 230
struggle *see* resistance
Studies in a Dying Colonialism (Fanon) 17–26
subjectivity 23, 91, 108, 112, 121, 197, 208, 220, 221, 262, 278–279, 284, 286; dancer 119; female 108; political 3, 182, 278, 282; subjectification 127, 128, 135–136, 278, 280, 284, 285
Sufism, in Nigeria 56–57; anti-Sufism 57

Tangwena, Chief R. 94–95, 98, 103
Temko, A. 229
territorialisation 29–30, 71, 75, 200, 258, 267–269, 271, 284
third space *see* space
Thomas, H. 144
Thompson, E. P. 126
Thrift, N. 3, 277
Tomko, L. 149
Tompkins Square Park (New York, USA) 13–14, 31n.
Trevor-Roper, H. 227
Trouble with Nigeria, The (Achebe) 45
Tsing, A. 90
Tucker, C. 171–172, 173

unintended consequences 139
urban protest *see* protest
urban social movements *see* social movements
urban space 9, 11, 16, 18–21, 47–49, 68, 70–72, 74–75, 77–80, 82, 83–85, 137, 157–158, 175, 210, 212–213, 217, 256, 284

Valentine, G. 88
Vancouver (Canada) 152–167

veil, the (*haïk*, the) 18–22
Victoria, Queen 211, 215
virtual reality (VR) 140–141, 147
visibility *see* looks
voice 131, 193–194, 195, 201; *see also* silence

Walker, R. 263
Walking Together: the first steps (Council for Aboriginal Reconciliation, Australia) 206
Watts, M. 2, 284
Weapons of the Weak (Scott) 89–90, 138
Williams, R. 208
Woiworung people, the (Australia) 212
Wolin, R. 223
Wollen, P. 149
women's: liberation 21–22, 24–25, 260–261; participation in struggles 17–26, 100

Young, I. M. 185–186

Zalm, Prime Minister V. 156, 158, 162
Zapatistas (Mexico) *see* Chiapas
Zimbabwe: defiance 96; Eastern Highlands 93, 103–104; ethnicity 95; Gaeresi Ranch 93–94, 96, 99, 100, 103; Kaerezi 93–95, 97–98, 100, 101, 103; Kaerezi Resettlement Scheme 93, 98–100, 103, 283; nationalist struggle 96–97; post-colonial 97; resistance 96, 98–99; state violence 96, 100–101; Tangwena, the 98, 100